INTRODUCTION TO I_{DDQ} TESTING

FRONTIERS IN ELECTRONIC TESTING

Consulting Editor
Vishwani D. Agrawal

Books in the series:

Multi-Chip Module Test Strategies
Y. Zorian
ISBN: 0-7923-9920-X
Testing and Testable Design of High-Density Random-Access Memories
P. Mazumder, K. Chakraborty
ISBN: 0-7923-9782-7
From Contamination to Defects, Faults and Yield Loss
J.B. Khare, W. Maly
ISBN: 0-7923-9714-2
Efficient Branch and Bound Search with Applications to Computer-Aided Design
X.Chen, M.L. Bushnell
ISBN: 0-7923-9673-1
Testability Concepts for Digital ICs: The Macro Test Approach
F.P.M. Beenker, R.G. Bennetts, A.P. Thijssen
ISBN: 0-7923-9658-8
Economics of Electronic Design, Manufacture and Test
M. Abadir, A.P. Ambler
ISBN: 0-7923-9471-2
I_{DDQ} Testing of VLSI Circuits
R. Gulati, C. Hawkins
ISBN: 0-7923-9315-5

INTRODUCTION TO I_{DDQ} TESTING

by

Sreejit Chakravarty
State University of New York at Buffalo

and

Paul J. Thadikaran
Intel Corporation

SPRINGER SCIENCE+BUSINESS MEDIA, LLC

Library of Congress Cataloging-in-Publication

Introduction to IDDQ Testing
by Sreejit Chakravarty and Paul J. Thadikaran

ISBN 978-1-4613-7812-9 ISBN 978-1-4615-6137-8 (eBook)
DOI 10.1007/978-1-4615-6137-8

Printed on acid-free paper.

To the loving memory of my father (Prof. Sunil Chakrabarti), my mother (Maya), Kohinoor and Ayush for their love and understanding.

SC

To my father (T. P. Joseph), mother (Mary) and Lini for support and encouragement.

PJT

CONTENTS

FOREWORD

The completeness of this book signals a maturity for that innovation in electronic testing of CMOS integrated circuits (ICs) called I_{DDQ} testing. Today we recognize the I_{DDQ} test as the single most sensitive method for detecting defects in CMOS ICs. I_{DDQ} testing has a simple concept; the quiescent or steady state portion of a CMOS circuit clock period should have a very low power current since there are no continuous circuit paths between the VDD power rail and the VSS ground. Quiescent currents of good circuits are at the sub microAmp level and typically can be tens of nanoAmps. If high μA or mA currents exist at a particular vector state, then the circuit isn't correct. Frank Wanlass, who patented the CMOS technology in 1963, described CMOS circuits as a nanoWatt logic. One could extrapolate from Wanlass that if a CMOS circuit didn't measure low quiescent currents, then it must have a defect, or a design or fabrication error.

Early in the history of CMOS IC development in the 1960's, RCA practiced I_{DDQ} testing on its CD4000 line of SSI/MSI parts. Philips Labs, Rome Labs, and Sandia National Labs actively used this technique in the 1970s, despite the fact that an electronic basis for the I_{DDQ} test practice had yet to be demonstrated. Except for George Nelson, E. King, and William Boggs of the US Naval Research Labs, publications in the 1970's were rare and little exchange among people occurred. That changed around 1980 when CMOS began to replace nMOS as the dominant IC technology. The International Test Conference (ITC) attracted an initially small, but growing number of test engineers interested in the topic. Mark Levi in 1981, and Yashwant Malaiya and Stephen Su in 1982, presented aggressive papers at ITC outlining the benefits of I_{DDQ} testing. Around 1981, Toshio Maruyama of Advantest Corp. designed the first I_{DDQ} monitor for ATE that was fast and sensitive. His I_{DDQ} Bit Current Option could read a 1 μA sensitivity at a 3-5 kHz measurement rate. Activity picked up in the mid-1980's. John Zasio's (then at AIDA Corp.) succinct COMPCON paper in 1985 was a precursor to modern thinking about defect-based testing, I_{DDQ}, and stuck-at faults. In a series of papers from 1985-91, Jerry Soden of Sandia National Labs and Chuck Hawkins of the University of New Mexico and colleagues at Sandia presented data showing the effectiveness of I_{DDQ} testing for specific CMOS defects. Mircille Jacomino, Jean Luc Rainard, and Rene David of the Lab d'Automatique de Grenoble and Centre Nat. d'Etudes des Telecom. also made comprehensive conclusions on varieties of test methods including I_{DDQ}.

In the 1990 era, major contributions to understanding the I_{DDQ} technique have come from Peter Maxwell, Rob Aitken, and Doug Josephson of Hewlett-Packard, Keith Baker, Manoj Sachdev, Eric Bruls, and Bas Verhelst of Philips Labs, Wociech Maly and Anne Gattiker of Carnegie Mellon University, Alan Righter, Chris Henderson, Ron Fritzemeier, and Dick Beagle of Sandia National Labs, Phil Nigh of IBM, Steve McEuen, Ravi Gulati, and Weiwei Mao of Ford Microelectronics, Sreejit Chakravarty of the State University of New York at Buffalo, Ken Wallquist and Grayson Garrett of Philips Semiconductors, Keiichi Sawada of Mitsubishi, Roger Perry of Storagetek, Ric Gayle of NCR, Joan Figueras, Antonio Rubio, Jaume Segura, Rosa Rodriguez-Montanes, Victor Champac, and Eugene Isern from the Universitat Politecnica de Catelonia, Paul Wiscombe of VLSI Technology, Carol Tong of Sunrise, Ben Bennetts of LogicVision, Michael Keating of GenRad, Dan Burns of Rome Lab, Ed McCluskey and Hong Hao of Stanford University, Tom Storey of Loral Federal Systems, Vic Kulkarni of Crosscheck, Rochit Rajsuman of LSI Logic, Simon Johnson of the University of Durham, Hans Manhaeve at the University of Oostende, Jacob Abraham of the University of Texas, Wayne Needham, Tony Miller, Wendy Whitman, John Acken, Tim Henry, and Thomas Soo of Intel, Joel Ferguson and Tracey Larabee of the University of California-Santa Cruz, Anura Jayasumana of Colorado State University, H. Vierhaus of the German National Research Center for CS, Sankaran Menon of the South Dakota School of Mines & Technology, Jos van Sas and colleagues from Alcatel, Adit Singh of Auburn University, Kozo Kinoshita of Osaka University, Andrew Richardson of Lancaster University, Bob Gruebel, Ken Butler and Theo Powell of Texas Instruments, M. Renovell and G. Cambon of the Universite des Sciences et Techniques du Languedoc, Scott Davidson at AT&T, Jim Frenzel of the University of Idaho, Jeff Beasley of New Mexico State University, Shobha Mallarapu, Albert Hoffman, S. Duey, and colleagues from Delco Electronics, and Rinya Kawahara, Osamu Nakayama, and Tatsuru Kurasawa from Kawasaki Steel.

These collective studies from around the world showed the following. I_{DDQ} is presently the only practical and guaranteed method of detecting the dominant defect in IC manufacturing, namely the bridging defect, and certain forms of CMOS open circuit defects are only detected by I_{DDQ} testing. Several manufacturers reported that defect levels at next stages of assembly dropped by factors of 10 to over 100 when I_{DDQ} testing was added to the normal voltage based IC testing. Advances followed in I_{DDQ} instrumentation and commercial software tools that would select vectors or generate I_{DDQ} vectors from a netlist. The targeting of specific defects by I_{DDQ} has led to a defect-based test approach that also includes those defects detected by voltage based testing. Refinements are now occurring in optimizing the roles of the various test methods such as I_{DDQ}, delay fault, at-speed functional, and stuck-at fault testing.

I_{DDQ} testing has a peculiar property of occasionally engaging people in intense debate. Two items are debated. The first is the implication of the reduced yield when ICs that fail only the I_{DDQ} test are rejected. This has stimulated an increase in failure analysis for these classes of defects. The second concern is what to do about the increase in transistor off-state leakage that accompanies deep submicron transistors. Several I_{DDQ} design for test approaches are being studied in response to this challenge.

The original, simple, I_{DDQ} idea that measuring current instead of voltage might produce better test results has extended to other base manufacturing activities. Studies of large numbers of CMOS ICs have shown that lower reliability failure rates are achieved if ICs that fail only the I_{DDQ} test are removed from the population of shipped product. Recent studies have also made progress in determining under what conditions burn-in can be eliminated when I_{DDQ} measurements are used as a screen. Millions of dollars in savings and lowered failure rates are reported.

Failure analysis is another beneficiary of I_{DDQ} measurements. An IC with a single defect related leakage path against a low quiescent background of current in the other paths is significantly easier to failure analyze. Recently, noninvasive analysis of ICs using I_{DDQ} and voltage based test vectors has been demonstrated to localize the site of a defect with high probability. This supports a trend in failure analysis called "lids-on" diagnosis that seeks to identify the defect site from noninvasive test data.

The authors of this work, Sreejit Chakravarty and Paul Thadikaran, have blended the best of academic and manufacturing styles. The book is direct in describing the process, particularly with its emphasis on defect-based testing, I_{DDQ} test pattern generation, and the instrumentation chapter by Ken Wallquist. The academic discipline is noticeable for its in-depth and painstaking work by the authors in review of all relevant literature. They have done a service for all of us and the result is a book written for any engineer or manager interested in this intriguing test technique.

Chuck Hawkins
University of New Mexico

PREFACE

Testing techniques for VLSI circuits are undergoing many exciting changes. The predomionant method for testing digital circuits consists of applying a set of input stimuli to the IC and monitoring the logic levels at primary outputs. If, for one or more inputs, there is a discrepancy between the observed output and the expected output then the IC is declared to be defective.

A new approach to testing digital circuits, which has come to be known as **I_{DDQ} testing**, has been actively researched for the last fifteen years. In I_{DDQ} testing, the steady state supply current, rather than the logic levels at the primary outputs, is monitored. Years of research suggests that I_{DDQ} testing can significantly improve the quality and reliability of fabricated circuits. This has prompted many semiconductor manufacturers to adopt this testing technique. Philips Semiconductors, Ford Microelectronics, Intel, Texas Instruments, LSI Logic, Hewlett-Packard, SUN microsystems, Alcatel, SGS Thomson, are a few of them.

This increase in the use of I_{DDQ} testing should be of interest to three groups of individuals associated with the IC business.

Product Managers and Test Engineers. Quality and reliability have important business implications. Since I_{DDQ} testing can play an important role in meeting quality and reliability goals, product managers and test engineers must understand the important issues of I_{DDQ} testing.

CAD Tool Vendors. Since more and more semiconductor manufacturers are using I_{DDQ} testing to test their products, there is an increase in the need for Computer Aided Design (CAD) tools for I_{DDQ} testing. This makes it important for CAD tool vendors to understand the fundamental issues of I_{DDQ} testing.

Circuit Designers. The advantages of I_{DDQ} testing can best be harnessed if some design techniques are followed. Another important issue to be addressed during design is the instrumentation to be used for I_{DDQ} measurement. Thus it is important that circuit designers have a good knowledge of these aspects of I_{DDQ} testing.

The purpose of this book is to educate the community of product managers, test engineers, designers and CAD tool developers. We have tried to summarize, under one title, the main findings of more than fifteen years of research in this area. We tried hard to include all the research contributions but some may have missed our attention. Our apologies to researchers whose work we may have missed. We will be grateful if such omissions are brought to our attention.

Besides I_{DDQ} testing this book has another dimension. Testing attempts to detect defective ICs. However, all testing tools use fault models. Defect detection has been ignored in the context of logic testing. Researchers in I_{DDQ} testing have paid considerable attention to it. While discussing I_{DDQ} testing we will study what is known about commonly occuring defects. We will investigate the correlation between fault models in use and the commonly occuring defect mechanism. This investigation, which will bring out the inadequacy of fault models to model defective ICs, will suggest that research in testing ought to shift focus from fault models to defects. We believe that all educators in testing owe it to the testing community to train students to think "defects" and not "fault models". We hope that this book will help you achieve that goal and help foster research in making testing more "defect-oriented".

We have to thank many of our professional colleagues. First, we thank Carl Harris of Kluwer Academic Publishers for providing us with the opportunity to write this book. Dr. Vishwani Agrawal of Bell Labs provided valuable feedback on the manuscript. Prof. Charles Hawkins of University of New Mexico deserves special thanks for writing a foreword to the book. His words of advice on many aspects of the book is greatly appreciated. Ken Wallquist, of Philips Semiconductors, contributed the chapter on I_{DDQ} monitors. His industrial experience in building current monitors adds very significantly to the quality of the book. We are very grateful to him for writing the chapter. The support from the National Science Foundation, administered by Dr. Robert Grafton, for our research on I_{DDQ} testing is gratefully acknowledged. We are grateful to Prof. Janak Patel, University of Illinois at Urbana-Champaign, for supporting our work during our visit to the Center for Reliable and High-Performance Computing.

An initial draft of this book was used in a seminar on VLSI testing at the State University of New York at Buffalo. Graduate students in the seminar (daringly) pointed out many mistakes in our examples and suggested several changes to the manuscript. We thank Vinodh Gopal, Sreenivas Mandava and Sujit Zachariah for their suggestions and many lively discussion. Our thanks to Dr. Robert Aitken (HP Design Center, Palo Alto, CA), Dr. Warren Debany (Rome Laboratories, NY), Prof. Joel Ferguson (UC Santa Cruz, CA) and Prof. Joan Figueras (UPC, Barcelona, Spain) for their feedback.

Finally, we hope that you will enjoy reading this book and benefit from it as much as we did.

Sreejit Chakravarty
State University of New York
Buffalo, NY

Paul J. Thadikaran
Intel Corp.
Santa Clara, CA

INTRODUCTION TO I_{DDQ} TESTING

1

INTRODUCTION

I_{DDQ} **Testing** is a modern, high interest testing technique for CMOS digital ICs whose roots go back to the first CMOS process. We start by understanding what is meant by I_{DDQ} *testing* and then briefly address the resurgent motivation for using I_{DDQ} *testing*. This will lead us into a discussion of why the conventional method for testing ICs, **logic testing**, is inadequate. We conclude this chapter by giving a brief overview of the book.

1.1 WHAT IS I_{DDQ} TESTING?

To answer this question we first explain what **fully complementary static CMOS** circuits are. Henceforth the term CMOS is used to refer to this class of circuits. Next the fundamental idea of I_{DDQ} *testing* is presented. This section ends with a brief history of I_{DDQ} *testing*.

1.1.1 CMOS Circuits

Figure 1.1(a) depicts a NAND gate with two inputs a, b and one output c. This logic gate is implemented in CMOS using nFETs and pFETs as shown in Figure 1.1(b).

If a and b are both assigned the logic value 1 then the two nFETs conduct and c is connected to GND. Thus, c is set to logic value 0. Note that, for this assignment of values to inputs, none of the two pFETs conduct and there is no conducting path from VDD to GND. During steady state (**quiescent**

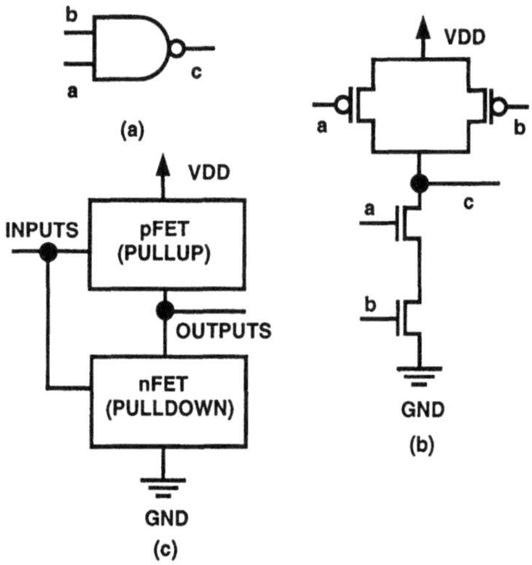

Figure 1.1 Static CMOS circuits.

state) current is drawn from the power supply. This *quiescent current*, from VDD to GND, is referred to as I_{DDQ}. Only leakage current through the FETs contributes to I_{DDQ}. Thus, in fault free circuits, I_{DDQ} is very small.

The assignment $\langle a = 0$[1]$, b = 1\rangle$ creates a conducting path from VDD to c, via the pFET driven by a, and no conducting path from c to GND. Thus, $c = 1$ and I_{DDQ} is very small.

Make a note of the following important points from this example. As shown in Figure 1.1(c), every CMOS gate consists of two networks: a pFET network (**pullup**) and an nFET network (**pulldown**). For any assignment of logic values to the inputs of the gate, in the fault free case, there are conducting paths either from VDD to the gate output or from the gate output to GND. Thus, for any input assignment, I_{DDQ} is negligible. This also implies that power consumption during the *quiescent state* of CMOS circuits is very small.

[1] From now on we drop the phrase "a is assigned logic value 0(1)" and simply use "$a = 0(1)$" to denote the same.

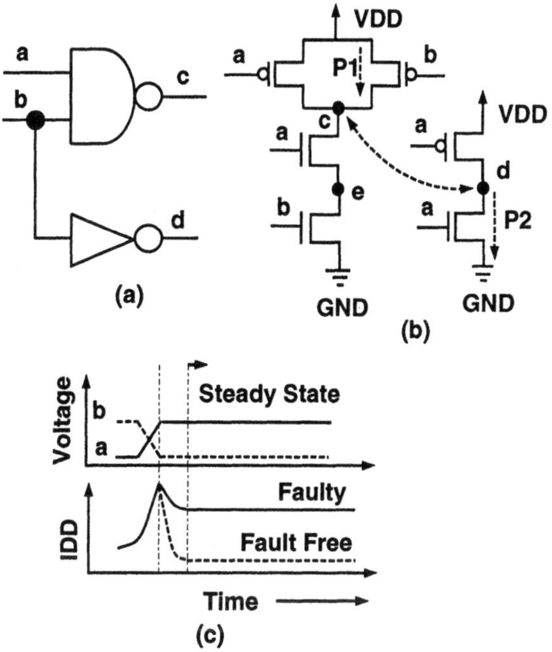

Figure 1.2 Example illustrating I_{DDQ} testing.

1.1.2 I_{DDQ} Testing

Consider the circuit in Figure 1.2(a). If $\langle a = 1, b = 0 \rangle$, as explained above, in the fault free circuit I_{DDQ} is negligible. Consider a **defect** that causes lines c and d to be "connected". The switch level circuit is shown in Figure 1.2(b). If $\langle a = 1, b = 0 \rangle$, paths $P1$ in the *pullup network* of the NAND gate and $P2$ in the *pulldown network* of the NOT gate create a conducting path from VDD to GND. Consequently, in the presence of the above fault and on application of $\langle a = 1, b = 0 \rangle$, there will be a large I_{DDQ}. This large I_{DDQ} signals the presence of a *defect* in the circuit. That is the basic idea behind I_{DDQ} *testing.*

The nature of the current waveform is shown in Figure 1.2(c). Note that there is a transient period when both good and faulty circuits will have high I_{DD}. It is important that the **current measurements be made, after the transients die down**, in the region marked as steady state.

To summarize, in I_{DDQ} testing, a set of input vectors is applied to the circuit. After some preselected vectors, or after each input vector, I_{DDQ} is measured. If any of the I_{DDQ} *measurements* register a higher than some expected reading the circuit is declared to be faulty.

What we discussed above has become known as **single threshold** I_{DDQ} testing [101, 102]. A related technique, known as **current signatures** [101, 102], also monitors I_{DDQ}. We will discuss it in Section 3.3.

1.2 WHY I_{DDQ} TESTING?

To understand the need for I_{DDQ} *testing* we have to understand: the limitations of the **conventional testing method**, usually referred to as **logic testing**; and how I_{DDQ} *testing* overcomes these limitations. We hasten to add that, our present knowledge about I_{DDQ} *testing* suggests that it will **not replace** *logic testing* but will be used in conjunction with it. Here we present, very briefly, the main points and leave the elaboration to later chapters.

Logic testing assumes that every **physical defect** can be modeled as a **fault** at the logic gate level description of the circuit. In the most widely used form of *logic testing*, it is assumed that faults are such that only one line in the circuit is either permanently stuck-at 0 or 1. This is known as the **single line stuck-at fault model.** It is often referred to as the **stuck-at fault model.**

Stuck-at faults can be detected by an input vector (or a sequence of input vectors for sequential circuits). For example, in Figure 1.3 assume: line a is stuck-at 0 (henceforth abbreviated a *s-a-0*); and input vector $\langle a = 1, b = 1, c = 0 \rangle$ is applied. In the presence (absence) of the fault $g = 1(0)$. This change in the logic value at g indicates the presence of the fault.

Logic testing consists of computing a set of tests T. Tests used with *logic testing* are henceforth referred to as **logic tests** and the tests used in I_{DDQ} *testing* are referred to as I_{DDQ} **tests.** When logic tests are applied to fabricated chips, the logic values at the output(s) are monitored. If there is a discrepancy between the expected logic value at the output(s) of the chip and the observed value, the chip is declared to be faulty.

If the *logic test set* T is such that it is computed by targeting *stuck-at faults* we refer to the *test strategy* as **stuck-at testing.** This is the predominant

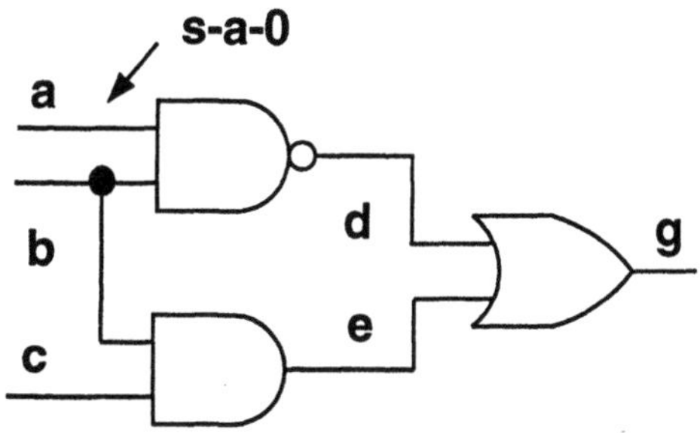

Figure 1.3 Example illustrating logic testing

mode of testing in use today. *Stuck-at testing* originated with TTL circuits. However, as we will see in Chapter 4, *stuck-at faults* do not properly model the *physical defects* that frequently occur in CMOS circuits. Consequently, *logic tests* computed using the *stuck-at fault model* are not guaranteed to detect such *defects*.

One is tempted to think that, in principle, *logic tests* for some such defects can be computed and added to the *logic test set*. However, good techniques to model important defects at the logic level have eluded us to this day. Examples are **bridging defects** which cause two or more distinct nodes in the circuit to be "connected". To properly model *bridging defects* at the gate level a good estimate of the *bridge resistance* is required. Otherwise, the value at the circuit output(s), on application of a sequence of **logic tests**, in the presence of the *defect* cannot be predicted. However, such defects can be easily detected by I_{DDQ} *testing*.

Even if defects are properly modeled at the gate level, *logic tests* are often difficult to compute. It is a well known fact that very few sequential circuit test generators can compute *stuck-at logic tests* having 100% **efficiency** (i.e., a *logic test set* guaranteed to detect all *detectable single stuck-at faults*). On the other hand, there are indications that I_{DDQ} *tests* can be computed with considerably reduced effort.

All the positive attributes of I_{DDQ} *testing* stated above suggest that it will contribute significantly to improving product **quality**. Another major benefit of I_{DDQ} *testing* is improvement of **circuit reliability**. There are physical defects, like gate oxide shorts, that do not necessarily change the functional behavior of the circuit. Therefore, *logic testing* cannot detect them. However, many of these defects can cause havoc by: draining the power supply (a serious problem with portable systems); and causing the circuit to fail very early in its life cycle. Many such defects can be detected by I_{DDQ} testing, thereby improving circuit reliability.

The extremely competitive IC market has prompted manufacturers to constantly improve the quality, reliability and fabrication yield. We claim that considerable progress can be made towards achieving these goals by including I_{DDQ} *testing*, along with *logic testing*. In Chapter 2 we take a look at empirical data gathered from engineering experiments that supports this claim. We will also look at data that show the impact this has on IC and system cost.

I_{DDQ} testing is not without its problems. Some researchers[336, 335] have raised serious questions about the feasibility of using I_{DDQ} for future generations of ICs. We will discuss this in Section 3.2.

1.3 OUTLINE OF THE BOOK

The idea of using CMOS circuits was first discussed in early 60s [332]. Many IC manufacturers, including RCA [229, 313], developed this technology. Since it consumed less power it slowly took over greater market share and by late 70s completely dominated the market.

The earliest publications that mentions the use of quiescent current monitoring for testing are [159, 221]. The two earliest published work on discussing I_{DDQ} *testing* are [174, 182]. However, as pointed out in [300], the idea of monitoring supply current started in the mid 70s at Sandia and RCA Laboratories in USA and Philips Laboratories in The Netherlands. Subsequent research has determined that many physical defects that occur in CMOS circuits could not be detected by *stuck-at testing*. This finding, which was quiet controversial at the time, and the popularity of CMOS circuits motivated further research on I_{DDQ} *testing*.

A decade and a half of intense research has led to a better understanding of I_{DDQ} *testing*. Today we have a much better idea of the impact it has on quality, reliability and yield. I_{DDQ} *testing* has matured to the point where many IC manufacturers will soon adopt it, if they have not already done so. There are two primary hindrances to the large scale adoption of I_{DDQ} *testing*: lack of good commercial grade "current measuring instruments"; and lack of good commercial grade "I_{DDQ} CAD tools". Although, much has been done to eliminate these two drawbacks more research is required.

With this background, let us explain the organization of the book. In **Chapter 2** we elaborate on the answer to the question: Why I_{DDQ} testing? *Empirical data* from a variety of system assembly and IC fabrication lines, showing the tremendous impact I_{DDQ} *testing* has on IC *quality, reliability* and cost, are presented. Studies are underway to design I_{DDQ} testing based "reliability screens" that will replace burn-in. We will discuss such "reliability screens" in Chapter 2.

In **Chapter 3** we address issues that test engineers face when using I_{DDQ} *testing.* These include issues pertaining to the speed of I_{DDQ} measurements, number of such measurements, sensitivity of the measurements, etc. In addition, we discuss issues that designers must address in order to make I_{DDQ} *testing* both feasible and effective. Critics of I_{DDQ} have raised serious questions about the usefulness of this technique for future technologies. This is closely related to the issue of setting the I_{DDQ} threshold. This has also led to a discussion of a variation of I_{DDQ} testing known as current signatures. We will discuss the concerns of the critics as well as current signatures in Chapter 3.

In **Chapter 4** we will look at commonly occurring physical defects and understand how I_{DDQ} testing can be used to detect such defects. This material forms the basis for the discussion in **Chapter 5** where we discuss fault models and test sets. This gives us an idea of the **characteristics** of I_{DDQ}, *delay* and *logic tests.* We will also discuss how tests based on these fault models have serious limitations in detecting certain defects.

Computer-Aided Design (CAD) tools to support I_{DDQ} *testing,* like all testing tools, use fault models to compute and evaluate I_{DDQ} test sets. These fault models are discussed in Chapter 5 which forms the basis for the next few chapters.

Chapters 6, 7 and **8** address the important task of **evaluation, compression,** and **generation** of I_{DDQ} *test sets,* respectively. Compression of I_{DDQ} *test sets* is closely related to an important task of **selecting I_{DDQ} measurement points** and is addressed in Chapter 7. Published algorithms, data structures and experimental results for these tasks are presented.

Another use of I_{DDQ} *testing,* that is slowly receiving increased attention, is its potential usefulness in **diagnosis (failure analysis)**. It has the potential to contribute very significantly to **yield improvement**. I_{DDQ} measurement based diagnosis is discussed in Chapter 10.

Our discussion in Chapters 6, 7, 8 and 9 presents two important perspectives of I_{DDQ} *CAD tools*. The **first perspective** is useful in evaluating I_{DDQ} *CAD tools* and is important to both test engineers and I_{DDQ} *CAD tool* developers. The **second perspective** is important to I_{DDQ} *CAD tool* developers. This perspective of I_{DDQ} *CAD tools* is important to the understanding of the algorithmic techniques that have been proposed.

An important issue in I_{DDQ} *testing* is the design of **current monitoring** devices. Without successful development of good *on-chip current monitoring* devices the full potential of I_{DDQ} *testing* cannot be harnessed. Great strides have been made in this direction. These are discussed in **Chapter 10**.

To summarize, in Chapters 2 and 3 we start by answering the very basic questions about I_{DDQ} *testing*. The ideas discussed in these chapters are put in proper perspective in Chapters 4 and 5. There we also present fault models that I_{DDQ} *CAD tools* use. The next few chapters discuss in details the algorithmic aspects of I_{DDQ} *CAD tool* development. Finally, we end by addressing important instrumentation and design problems.

2

WHY I_{DDQ} TESTING?

The 80s saw a strong emphasis on *quality* and *reliability* primarily because end users of semiconductor ICs, like system houses, auto industry, etc., demanded it [333]. The transition to a globally competitive market in the 80s was the main cause of this added emphasis, by end users, on improving quality. These market forces have led to the standardization of very high quality,levels. These quality levels cannot be achieved without I_{DDQ} *testing.*

The argument for I_{DDQ} *testing* can be summarized as follows. The "bottom line" in the semiconductor business demands high quality. Testing techniques have to be augmented to achieve and maintain quality levels as circuit density and area will continue to grow. I_{DDQ} *testing* contributes very significantly to achieve this end. Therefore, I_{DDQ} testing should be included as part of all future testing paradigms!

In this chapter we first look at data that help us understand the motivation behind insisting on high quality and reliability. Next, an analysis showing the increased difficulty in achieving quality goals if, as the current trend indicates, circuit area and density continues to grow. That is followed by empirical data, from various sources, showing that inclusion of I_{DDQ} *testing* improves quality. Data showing the potential effect I_{DDQ} *testing* can have on reliability is presented. Finally, some I_{DDQ} based "reliability,screens" under study are presented.

A study at Ford Microelectronics [207] notes that the impact of I_{DDQ} testing on quality will not be the same for all components. Assuming that I_{DDQ} will detect all "leakage failures" it was estimated that 70% of ASIC field failures could be eliminated. Data for EPROM, CPU and RAMs were respectively 44%,

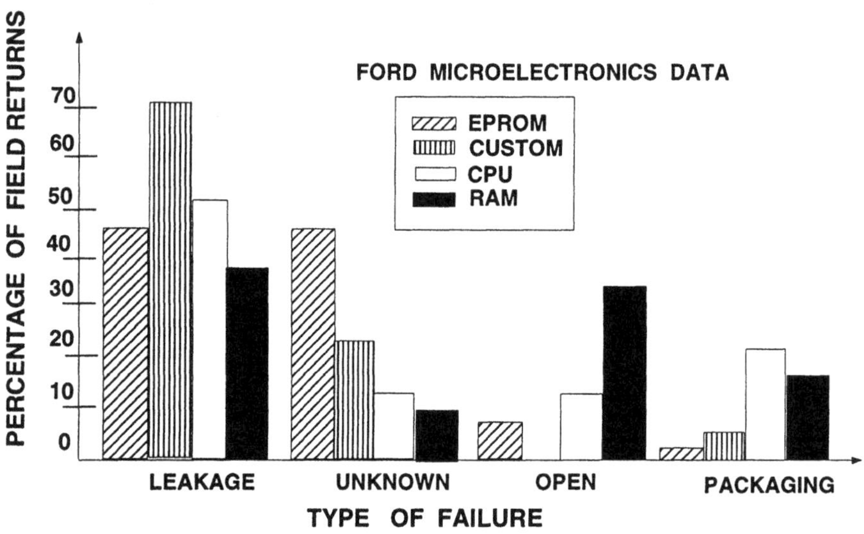

Figure 2.1 Histogram showing I_{DDQ}'s potential effectiveness in reducing field returns

50.5% and 42%. This is shown in Figure 2.1. Indications are RAM products stand to gain the least and ASIC products the most from I_{DDQ} testing.

2.1 DOLLARS AND CENTS

To understand the impact of IC quality on cost we need some background information. **defect,coverage** is the percentage of all defects detected by the test strategy in use. Reject ratio (**reject ratio (RR)**) is defined as follows.

$$rejectratio(RR) = \frac{Number\ of\ defective\ ICs\ shipped\ as\ good}{Number\ of\ ICs\ shipped} \quad (2.1)$$

We prefer reject ratio (RR) instead of the term **defect level** which is also in use. reject ratio (RR) is a measure of IC quality.

Defect cost is the total cost incurred because of defective ICs. The higher the value of reject ratio (RR) the larger the defect cost. An important task, when formulating a test strategy, is to set a target reject ratio (RR) in parts per million (**ppm**). This is done by estimating the impact of reject ratio (RR) on defect cost using a **defect cost model**. The test strategy, on the other hand, is defined by fault coverage (which is assumed to have some relationship to defect,coverage). Therefore, methods to estimate reject ratio (RR), given the fault coverage, are needed. These issues are addressed in the next two subsections. In the final subsection we present some data.

2.1.1 Estimating Defect Cost

The **defect cost model**, presented here, estimates the additional cost due to defective ICs on the total workstation cost. The three phases that ICs go through after procurement are: board assembly, system assembly and field use. The IC could fail during board assembly, system assembly or field use (**field failures**). These failures are collectively known as **line fallout**. A part of the line fallout is due to defective ICs, which is what we are interested in. In NCR's workstation products division 50% of line fallout is due to defective ICs [100].

The impact on IC cost increases as the identification of the defective IC is delayed. For example, the greatest impact is from field failures. Let C be the defect cost, C_i the cost contribution if the defective IC is detected in phase $i \in \{board,\ system,\ field\}$, f_i the probability of detecting a defective IC in phase i, R the reject ratio (RR) and V the total volume. Then,

$$\begin{aligned} C &= V \times R \times \{C_{board} \times f_{board} + C_{system} \times f_{system} \\ &\quad + C_{field} \times f_{field}\} \end{aligned} \tag{2.2}$$

In this model it is assumed that V is known. If we are computing C for an existing test paradigm then R is known from the line fallout data and failure analysis. If we are trying to evaluate a new test strategy then R has to be estimated. This is discussed in the next subsection.

The values of $f_{board}, f_{system}, f_{field}$ are estimated over time by monitoring the manufacturing process. A typical set of values for these parameters is [100]: $f_{board} = 0.85, f_{system} = 0.13, f_{field} = 0.02$. The values of C_i depend on the manufacturer and are assessed on an individual basis.

2.1.2 Estimating reject ratio (RR)

For existing test strategies or large volume ICs, fairly accurate values of reject ratio (RR) can be computed by process monitoring and collecting and analyzing all field rejects. By constantly tracking the reject ratio (RR) the testing method can be fine tuned and improved. For small volume ICs, quite prevalent today, this method cannot be used and reject ratio (RR) has to be estimated.

Reject ratio (RR) is a function of **yield**, the fraction of fabricated ICs that are good (not merely tested to be good), and the defect,coverage of the test set(s) used. Great strides have been made towards developing sound methods for estimating reject ratio (RR) [9, 290, 291, 334]. In spite of these, experimental evaluation suggests [76] that existing methods should be improved. The difficulty of accurately computing defect, coverage contributes to this. To circumvent this difficulty fault coverage, often based only on stuck-at faults, is used instead. This does not solve the accuracy problem but is the only approach discussed in the literature.

The method in [334] is simple but over estimates reject ratio (RR) [76]. We present two other methods and henceforth refer to them as the **ASA method** [9] and the **SA method** [291]. There is yet another method for solving this problem [202], but the chosen methods should suffice. The following notations are used:

- y is the yield;
- A is the chip area;
- f is the fault coverage of the test set(s);
- n_0 is the average number of faults caused by a defect (ASA method);
- a, b, c are the parameters that determine yield in the SA method;
- $P(f)$ is the fraction of chips rejected at wafer test when a test set with fault coverage f is used; and
- $R(f)$ is the reject ratio (RR) when a test set with fault coverage f is used.

The two methods differ in the assumptions made about the manifestation of a defect as a fault. In the ASA method it is assumed that yield is known using other techniques [302]. The expressions for $P(f)$ and $R(f)$ are given below:

$$P(f) = (1-y)[1-(1-f)e^{-(n_0-1)f}] \tag{2.3}$$

$$R(f) = \frac{(1-f)(1-y)e^{-(n_0-1)f}}{y+(1-f)(1-y)e^{-(n_0-1)f}} \tag{2.4}$$

The parameter n_0 is estimated experimentally and a method for it is discussed later.

In the SA method, it is assumed that yield is not known. Using the negative binomial distribution of defects the following expressions are derived:

$$y = [1 + Ab(1 - e^{-cf})]^{-a} \quad (2.5)$$

$$R(f) = 1 - \left[\frac{1 + Ab(1 - e^{-c})}{1 + Ab(1 - e^{-cf})}\right]^{-a} \quad (2.6)$$

$$P(f) = 1 - [1 + Ab(1 - e^{-cf})]^{-a} \quad (2.7)$$

Constants, a, b and c, are estimated through experiments. Note that, for both methods, if $f = 1$ then $R(f) = 0$. If f is the stuck-at fault coverage then this leads to the conclusion that: "100% stuck-at fault coverage is good enough to reduce reject ratio (RR) to 0". As we will see in subsequent sections this conclusion is grossly inaccurate and the models need to be augmented to reflect this. In fact, even if we have "logical fault coverage of 100%" this conclusion is still inaccurate! The methods in [202, 334] also suffer from this problem.

We next present a method for experimentally estimating n_0, a, Ab and c. The idea is to experimentally compute an approximation for $P(f)$ at a number of discrete values of f. Then data fitting techniques are used. The experimental data gathered is the following. Let S be a sample of ICs that is tested. The larger the value of S the more accurate are the estimates. For each IC in S, record the number of the first test pattern on which it fails. Let us assume that the failing ICs in S fails for the first time at vector numbers $N_1, N_2, \ldots, N_t$ where, $N_i > N_j$ if $i > j$. Let P_i be the cumulative fraction of chips that failed on any one of the first N_i vectors. P_i's can be easily computed from the above information. Using a fault simulator we compute f_i, the fault coverage for the first N_i vectors. Thus we have the following information $\{\langle P_1, f_1\rangle, \ldots, \langle P_t, f_t\rangle\}$ which is used as an approximation for $P(f)$.

Such experiments are reported in [9, 76]. The values of f_i and P_i reported in [9] are tabulated in Table 2.1. It was a process with a very low yield (about 7%) and the sample size was 277.

Once such an approximation for $P(f)$ is available we can use any known data fitting techniques. For example, we can use the **least mean square estimate** as discussed in any standard statistics text. In [290] **least mean square estimate,weighted** is discussed. In that data fitting technique, values of the unknown parameters are estimated by minimizing the error function E.

$$E = \sum_{i=1}^{t}[P(f_i) - P_i]^2 \quad (2.8)$$

Table 2.1 Example of experimental data for parameter estimation

Fault Coverage (f_i)	Cumulative No of Failed Chips	Cumulative Fraction of Failed Chips (P_i)
5	113	0.41
8	134	0.48
10	144	0.52
15	186	0.67
20	209	0.75
30	226	0.82
36	242	0.87
45	251	0.91
50	256	0.92
65	257	0.93

$P(f_i)$'s are computed using expressions 2.3 and 2.7 for ASA and SA, respectively. There are statistical packages that compute n_0 (or a, Ab, c) to minimize E in equation 2.8. Values of n_0 reported are 8 [9], 1.94 [76] and between 4 to 5 [100]. Values of a, Ab, c reported are: $a = 0.748, Ab = 1.77, c = 2.777$ [291]; and $a = 1.636, Ab = 0.471, c = 0.67$ [76]. Once the parameter values are known $R(f)$, the *rejectratio*(RR), is computed using expressions 2.6 and 2.4, respectively, for the SA and ASA methods. Note that for the SA method, if we know the values of a, Ab and c we can also compute the yield using 2.5. This is an added strength of the SA model.

2.1.3 Defect Cost Data

Data on the impact of quality on product cost is presented in Table 2.2 [100]. It uses the ASA method to estimate reject ratio (RR) and assumes V to be 50,000. The value of n_0 used, as derived using the ASA model, is 4.

Note that ASIC #4 is critical and its reject ratio (RR) needs to be reduced. By increasing the fault coverage from 73% to 90% its reject ratio (RR) reduced to 11,800 and the repair cost to $600,000. Stuck-at fault coverage is only an approximation of the actual defect coverage. In the next section we will show that there is a limit on the impact of increasing stuck-at fault coverage on the

Table 2.2 Defect cost summary

Main Processor Board				
ASIC #	ASIC Cost	Fault Coverage	Estimated RR (ppm)	Repair Cost
1	$5.92	0.98	309	$2,216.78
2	$28.56	0.932	4,134	$34,376.91
3	$30.50	0.80	13,502	$113,578.11
4	$166.18	0.73	50,981	$774,694.90
5	$41.21	0.91	4,409	$78,887.58
6	$4.44	0.96	428	$3,039.03
7	$53.03	0.86	10,921	$104,164.11
8	$17.10	0.935	1,683	$13,026.97
Total Defect Cost: $1,110,957.42				

defect coverage. However, defect coverage can be increased very significantly if I_{DDQ} *testing* is included.

2.2 AREA, COVERAGE AND QUALITY

We are well into the submicron era and the number of devices that are packed onto a chip continues to grow. To test engineers this means that defect density (or fault density) per unit area may increase, the area itself may increase or the product of the two may increase. This brings an added challenge for test engineers because, as we will see shortly, to maintain quality with increase in area the testing strategy has to be improved. Such improvements are possible with I_{DDQ} *testing*.

Examine eqn 2.6 for $R(f)$ in the SA model. It can be shown that, for certain combination of values of a and c, $R(f)$ is an increasing function of Ab. We use data from [76] where $a = 1.636, Ab = 0.471, c = 0.67$ to illustrate this. Some sample calculations are shown in Table 2.3 assuming fault coverage to be 0.9. There is a continuous decline in the quality level as Ab grows. The RR level grows from 12907 ppm to 49792 ppm. A plot of $R(f)$ versus Ab can also be used to arrive at a similar conclusion. These observations were also made in [202] using an alternative analysis.

Table 2.3 Data Illustrating growth in RR with area

Ab	0.25	0.75	1.0	1.25	1.5
$R(f)$ in ppm	12907	31684	38726	44685	49792

In order to maintain quality, as the area increases fault coverage must be increased. To maintain an RR of about 12907 when Ab increases to 1.5 from 0.25 then, the reader can verify that, the fault coverage should be increased beyond 0.97. One can derive the expression for f as a function of Ab from equation 2.6.

2.3 EMPIRICAL DATA ON QUALITY

Experiments have been performed by many IC manufacturers to ascertain the impact of I_{DDQ} testing on quality [202, 279, 303, 304, 337]. The experiments, described next, are typical and the conclusions are consistent. A sample of ICs, of size S, was randomly selected. Then, several test sets, which usually included either a scan test set T, high speed test (H) or functional test set (F) or both, as well as an I_{DDQ} test set (I) were computed. Every IC was subjected to I and some of H, T and F. The ICs that failed specific tests were identified. We present these data using Venn diagrams. The following notations are used to represent the respective sets. FA: failed all test; FF: failed functional test; FC: failed contact test; FT: failed scan test; FH: failed high speed test; FI: failed I_{DDQ} test. In addition to the data that follows, similar data is also available from a number of other sources like **VLSI Technology** [337], **Storage Technology** [238, 239, 240] and others [100].

Before we present the data we note that most of the earlier experiments did not consider at-speed testing as a component of the test suite. Recently, some data that do include at-speed testing has been presented. We will examine these results in Chapter 5.

HP data [202]. The IC tested was a fully static standard cell design containing 35904 transistors. The set F had 59,000 vectors applied at 2 MHz, H was a smaller test set applied at 20 MHz and 32 MHz, T had 357 vectors and I had 141 vectors. The current limit was set at $30\mu A$ and off-chip precision measuring unit (PMU) was used to measure I_{DDQ}. A total of

Table 2.4 HP reject ratio data for various combinations of test sets

	Scan (T) and Functional Tests (F)			
RR (%)	None	F	T	F,T
Without I	16.46	6.36	6.04	5.80
With I	0.80	0.09	0.11	0.00

26,415 dies from three lots were tested. All tested dies had passed the initial contact test.

The raw data is shown in Figure 2.2(a). In addition, 21 ICs failed high speed test (H) but passed both scan and functional tests. Out of these 21 ICs, 11 failed I_{DDQ} testing. From this early data it was not clear what impact the addition of high speed testing would have on reducing test escapes. In Chapter 5 we will revisit this issue and look at more recent data.

Table 2.4 is computed using Figure 2.2 as follows. To compute RR the residual reject rate is assumed to be zero. That is, we assume that if there is a defective IC then one of the three tests (I, T or F) will detect it. This is highly unlikely. So quantitative evaluations should be used.

Under the above assumption, the total number of defective ICs, which can be computed from Figure 2.2, is 19 + 134 + 25 + 122 + 2,655 + 36 + 1,358 = 4,349. The total number of ICs is 26,415. This gives us the RR value, if none of the three tests is applied, to be $\frac{4,349}{26,415} \times 100 = 16.46\%$. If we only use I then, from Figure 2.2, a total of 122 + 2,655 + 36 + 1,358 = 4,171 ICs are detected to be defective. Therefore, 26,415 - 4,171 = 22,244 ICs are shipped out as good. Out of these 22,244, 19 + 134 + 25 = 178 ICs are defective. Thus, RR is $\frac{178}{22,244} \times 100 = 0.8\%$. The rest of the entries of Table 2.4 are calculated in a similar manner.

An important point about the data is the following. Note that if we use only I_{DDQ} Testing then RR is 0.8%. Whereas, if we only used scan and functional tests then RR is 5.8%. This implies that I_{DDQ} testing is better at screening out defective parts than a combined screen of scan and functional testing.

National Semiconductor data analyzed by CMU [304]. For definition of A, B, C see Figure 2.2(b). Two chips, *Chip 1* and *Chip 2* were fabricated. For *Chip 1*, for lot 1 $A = 72.4\%, C = 3.1\%, B = 24.5\%$ and for lot 2 $A = 82.0\%, C = 0.6\%, B = 17.4\%$. Once again similar conclusions as above can be drawn.

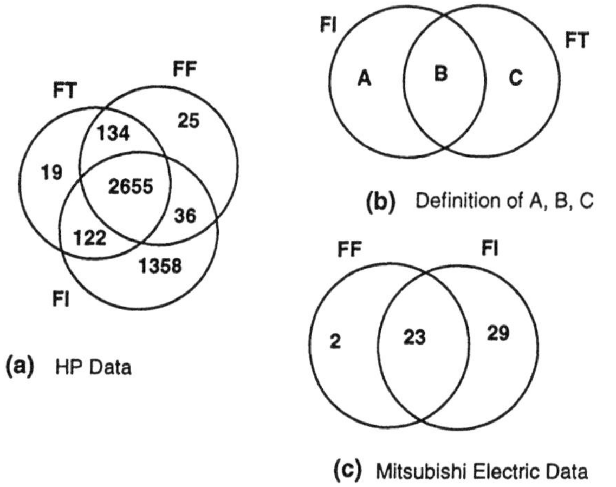

Figure 2.2 Empirical data on quality

For *Chip 2*, $A = 16.7\%, C = 20.4\%, B = 62.9\%$. The results are a little different from the other data. If we are to select either I_{DDQ} or scan testing then this data, unlike the rest, suggests that we should select scan over I_{DDQ} testing.

Mitsubishi Electric data [279]. Experiments were done on two separate chips. We summarize the results from one of those experiments. The DC test included one I_{DDQ} measurement. The core chip and the surrounding high I_{DDQ} part of the chip had two separate power sources. I_{DDQ} limits of $40\mu A$ and $2\mu A$, respectively, were set for these two power sources. The coverage of the I_{DDQ} test set and the number of measurements were not reported. The chip had an embedded RAM and the functional test set had 24,576 vectors applied at 100 MHz. The rest of the chip was tested by a functional test set of 4,096 vectors applied at 2 MHz.

Of the 400 chips tested, only 154 passed the DC test that included one I_{DDQ} measurement. Of these, the numbers failing functional, I_{DDQ} or both are shown in Figure 2.2(c). Even if we ignore the one measurement during DC test, the data strongly supports the use of I_{DDQ}.

Storage Technology data [240]. This study, like the study in [100], was done for an extended period. Storage Technology uses ICs supplied by ASIC vendors. Prior to introducing I_{DDQ} testing, all incoming ICs were tested using a test set with 99.6% stuck-at fault coverage. This was found inadequate because the failure rate reached 10,000 ppm against a target

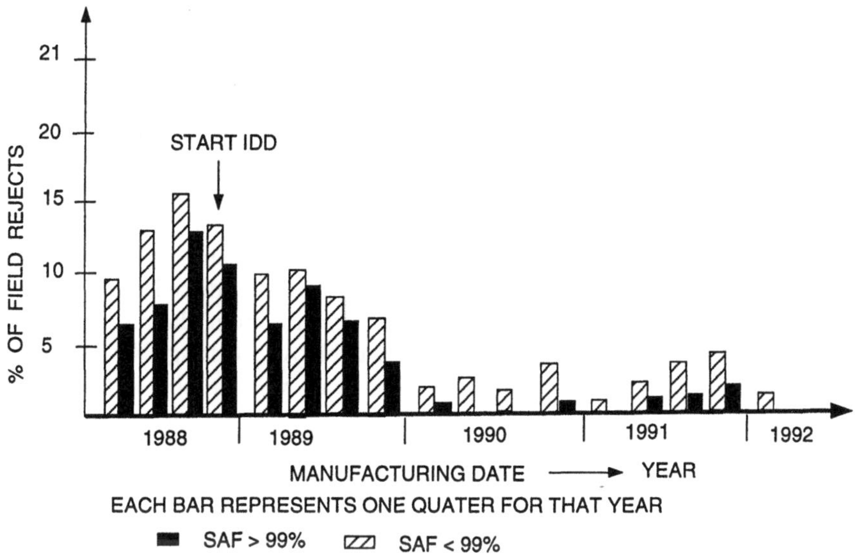

Figure 2.3 Storage Technology data on effect of I_{DDQ} testing on line fallout

of 100 ppm around the fourth quarter of 1988. I_{DDQ} was then introduced and the result is shown in Figure 2.3. Bars depict the percentage of total chips failing in the field during a quarter. Clearly, the actually measured reject ratio fell soon after the introduction of I_{DDQ}. Four months after I_{DDQ} was introduced the manufacturing process was modified. This partly contributed to a reduction in the reject ratio.

The long term effect of I_{DDQ} testing on reject ratio is best understood by plotting the percentage of total rejects against the manufacturing date. The two types of bars demonstrate the value of higher fault coverage. The reject ratio fell to an acceptable level very soon and remained stable after that.

To summarize, all but the *Chip 2* data for National Semiconductors point to the following conclusions:

- If the choice is between using only stuck-at testing or only I_{DDQ} testing then I_{DDQ} testing should be used.

- Neither stuck-at testing nor I_{DDQ} testing is guaranteed to detect all defective ICs and therefore should be used in conjunction with each other.
- Including I_{DDQ} testing significantly reduces RR.

A disadvantage of I_{DDQ} testing that is often pointed out by critics is that not all ICs showing high I_{DDQ} are actually defective. So, screening out all ICs having high I_{DDQ} may adversely affect the yield and hence the cost. This is not so much of a problem with existing fabrication processes but might become so in the future [335, 336]. Modifying I_{DDQ} testing from using single-threshold to using current signatures is a possible answer to this predicted problem. We will return to this issue in Section 3.1. This concern about the adverse effect on yield should be weighed against two impacts of I_{DDQ} testing. We have seen how it impacts IC quality. In the next section we will see how it also has a great impact on IC reliability.

2.4 EMPIRICAL DATA ON RELIABILITY

While considering quality we have concentrated on the proportion of defective chips that escaped testing. This results in field rejects, after the IC has been shipped out, with a great impact on IC cost.

There is another phenomena that impacts IC cost. To understand it refer to Figure 2.4, known as the **bathtub curve**, which plots failure probability versus time of use [60, 338]. Failure probability is very high during the early stages of an IC's life. This probability falls and increases again after the IC has been used for a long time. Early life failure is known as **infant mortality** and is due to "weak ICs" [338]. "Weak ICs" function properly long enough to pass functional testing. However, they have manufacturing defects that cause them to fail soon after they are put to use.

The standard method to weed out such "weak ICs" is to apply stress tests [60, 127], which is often referred to as **burn-in** [127]. ICs that pass burn-in are shipped out. Burn-in is an "expensive" process in that it takes a very long time, often as long as 7 days, and also needs special equipment. I_{DDQ} screens have been proposed to eliminate the cost of burn-in. The proposal is based on several experimental results which we describe next.

Sandia Labs data [118]. The fact that "weak ICs" exhibit high I_{DDQ} was first pointed out in [118] based on experiments performed at Sandia Na-

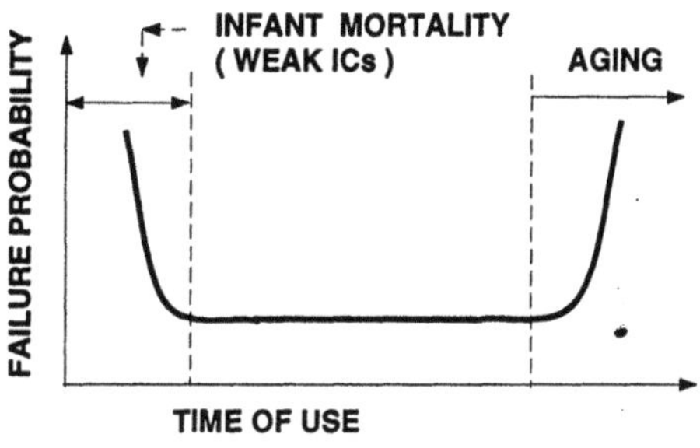

Figure 2.4 The bathtub curve

tional Labs. A set of 35 ICs, found to have I_{DDQ} above 100 μA, was used in that experiment. They were known to have defects (gate-oxide shorts as discussed in Chapter 4) that caused high I_{DDQ} but did not cause a functional failure. So, all of them passed functional testing. These ICs were then subjected to a series of six stress tests. The stress tests were applied at 150° C and at 13 V and lasted one day. After each stress test ICs were tested using functional test vectors. The experimental results are summarized below.

After the first stress test, 13 ICs failed functional test. Of the remaining, 6 failed a functional test between the 2nd and 6th stress test. A total of 11 ICs never failed the functional test. Five ICs initially passed then failed and then again passed functional test with subsequent stresses. All ICs were also subjected to high frequency tests at three different voltages (4.5V, 9.0V and 11V). The highest frequency at which an IC passed the functional test was recorded. This frequency progressively fell with the number of stress tests applied. For 4.5V, for some ICs it went down from 3.5 MHz, prior to the first stress test, to about 2.5 MHz after the sixth stress test. For all ICs, I_{DDQ} was measured at 11V after each stress test. The measured value of I_{DDQ} increased after each stress test. It increased from about 54 nA to about 6.0 mA. More detailed data are available in [118].

Philips Semiconductor data [330]. In this experiment ICs from several lots were subjected to the following process. The ICs were a high volume micro-controller having 120,000 transistors fabricated using a 1 μm CMOS process. The I_{SSQ}[1] screen used a current threshold of 15μA. For each lot

[1]similar to I_{DDQ} except that the current is measured at the V_{SS} instead of the V_{DD} node.

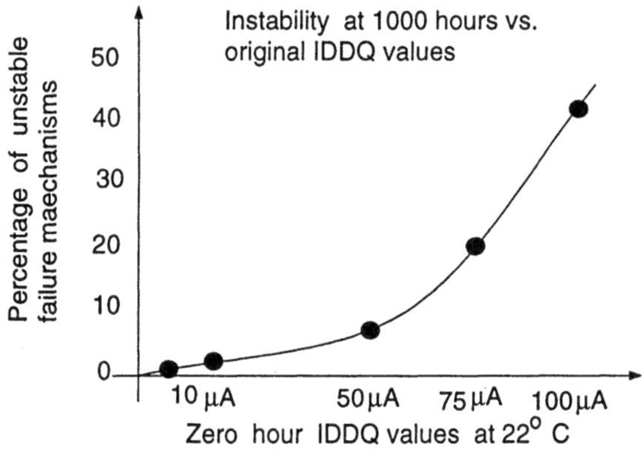

Figure 2.5 Ford Microelectronics data on relationship between high I_{DDQ} and instability after 1000 hours

2,000 ICs that passed the final production test were randomly sampled. These 2,000 ICs were subjected to the I_{SSQ} screen. Let A (B) be the set of ICs that passed (did not pass) the screen. ICs in both A and B were subjected to burn-in at 150° C and 6.5V for 24 hours. All these ICs were then tested using the production test sets. Let $C(D)$ be the subset of ICs in $A(B)$ that failed the production test after burn-in. The cardinalities of A, B, C and D are as follows: 39,389; 733; 1 and 17. That is, 1 in 39,389 ICs that passed the I_{SSQ} screen failed during burn-in and 17 of the 733 ICs that did not pass the I_{SSQ} screen failed during burn-in. Another interesting observation was that 11 of the 17 failures were from the same lot. Therefore, I_{SSQ} screens can also be used to identify "rogue lots".

Ford Microelectronics data [207]. This study tried to correlate an initial high value of I_{DDQ} with unstable behavior after 1000 hours of lifetest. All ICs having I_{DDQ} above 10 μA were tested after 1000 hours of lifetest. The results are plotted in Figure 2.5 and Figure 2.6. The group with most rejects had very high initial I_{DDQ}.

Data similar to those from Sandia Laboratories and Philips Semiconductors were also reported. Several experiments were performed. In one experiment on a 7,000 transistor IC, 59,024 dies from 26 wafer lots were used. The material screened out using a 10 μA I_{DDQ} screen had 51% fewer rejects after burn-in. In a second experiment 7,000 units of a 43,000 transistor IC were used. The post burn-in,failure of those that did not pass the I_{DDQ} screen was 10 times that of the good parts.

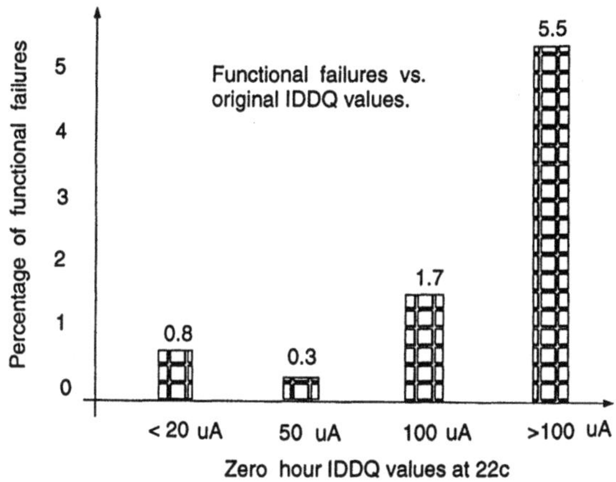

Figure 2.6 Ford Microelectronics data on functional failures after 1000 hours of lifetest vs. original I_{DDQ} value

Kawasaki-Steel data [153]. In this study, a 25k gate gate-array LSI fabricated using 1 μm technology was used. Two lots of ICs were experimented with. In lot #1, 176,459 ICs were put through an initial IC test prior to burn-in. The initial IC test did not include I_{DDQ} testing. Those that passed this test were sent for burn-in. The burn-in conditions used were 125°C, $V_{DD} = 7.0V$, 6 hours and semi-dynamic. At the end of burn-in all these ICs were subjected to a final test.

In lot #2, 31,919 ICs were put through an initial IC test prior to burn-in. The initial IC test included I_{DDQ} testing. Those that passed the test were subjected to the same burn-in test. After that they were subjected to the same final test.

The data obtained are as follows. For lot #1, 456 ICs failed the initial test and 401 failed the final test. Thus, the post burn-in failure rate was $\frac{401}{176,459-456} \times 100 = 0.2278\%$. For lot #2, 178 ICs failed the initial test and 32 failed the final test. This gives the post burn-in failure rate to be $\frac{32}{31919-178} \times 100 = 0.1008\%$. Thus, including I_{DDQ} testing in the test suite reduces the post burn-in failure.

All these data point to the fact that high I_{DDQ} is probably a sign of "weak ICs". Shipping out ICs with high I_{DDQ} is a reliability risk. burn-in is the predominant screen used to improve the reliability of shipped ICs. It adds very significantly to IC cost and IC manufacturing time. There has been proposals to replace burn-in by including I_{DDQ} testing in the test suite. Present evidence does not support replacing burn-in with I_{DDQ} testing. This is because some ICs with high I_{DDQ} are not necessarily "weak ICs". For the Sandia experiment 11 of the 35 ICs with high I_{DDQ} passed the functional test after six stress tests. In the Philips Semiconductors experiment 716 ICs that failed I_{SSQ} test did not fail during burn-in. Weeding out these ICs would decrease the yield by about 1.78% (716 out of 40,122). The Kawasaki-Steel data also shows a reduction in yield due to I_{DDQ} testing. Similar data were presented by Ford Microelectronics. However, there are some points that we have to keep in mind.

- For critical application, where it is wise to incur the added cost of reduced yield for the improvement in reliability, all ICs that do not pass the I_{DDQ} screen should be weeded out.
- For less critical applications, I_{DDQ} can reduce the cost of the burn-in process. Instead of subjecting all ICs to burn-in only the ICs having high initial I_{DDQ} are subjected to burn-in. This is followed by testing the ICs. Those ICs that pass this test are shipped out. Although a very small percentage of ICs that pass I_{DDQ} testing fail burn-in this would still enhance IC reliability. Since, only a small percentage of ICs would be subjected to burn-in the cost of burn-in test would be reduced.
- New screens to replace burn-in that uses I_{DDQ} testing are being evaluated. We discuss such screens in the next section.

2.5 I_{DDQ} AND HIGH VOLTAGE STRESS SCREENS

Recent studies have brought into focus a method, that combines high voltage stress (HVS) and either low voltage testing or I_{DDQ} testing [63, 82, 150, 153], which has the potential for replacing burn-in. This is important because it can eliminate the costly and time consuming step of burn-in from the IC fabrication process.

The method, sometimes known as SHOVE (SHOrt Voltage Elevation) test [63], has been in use in the industry for sometime [82, 150] and has been recently adopted by some [124, 153]. The method consists of applying a higher than normal voltage, for a short duration of time and applying a set of input vectors to the IC under stress. This is followed by the usual I_{DDQ} testing at the normal operational voltage. Several important points need to be discussed. They are:

Table 2.5 Lifetime of oxides

V_{OX}	E_{OX}(MV/cm)	Time to Fail	
		Flawless	Defective
3.3	3.67	9.04×10^{22} yrs	169.3 yrs
6.0	6.67	20,041 yrs	2.5 sec
7.0	7.78	11.1 yrs	60 msec
8.0	8.89	14.6 days	3.6 msec

why does this method screen out weak ICs? what should be the stress voltage? what should be the duration of the stress? and what kind of input vectors must be applied during this stress test? We do not address the problem of computing stress tests in this book. The problem of computing stress test is discussed in [74]. The rest of the questions are answered below.

Some defects show time-dependent behavior. They include: imperfections in the oxide layer that seperates the gate from the channel region of transistors; imperfection in the insulation layer that seperates adjacent conducting layers, primarily metal layers. Such imperfections, when subjected to voltage stress, deteriorate over a period of time and ultimately lead to a conducting path between layers that should be isolated. When this happens, functional failures occur. This has been identified to be one of the main causes of early life failure. They are often referred to as time dependent failures. As the feature sizes shrink the oxide thickness decreases and such defects will become more prevelant. The purpose of subjecting the IC to HVS is to accelerate this process of breakdown so weak ICs are exposed before being shipped out.

Let E_{OX} be the electric field across the oxide. The value of E_{OX} under normal operating conditions for a 3.3V process is in the neighbourhood of 3.5MV/cm.

An important point to remember is that if the IC is stressed so that E_{OX} is well above the normal value then weak ICs fail much sooner than the flawless ICs. The potential for damaging the flawless ICs increases as the voltage increases. Some quantitative estimates are presented in Table 2.5 [63].

The idea is to pick a stress level so that defective oxides fail quickly but flawless oxides are not affected. Typical values of E_{OX} selected are 6.0MV/cm [153], 7.5 MV/cm [63]. The stress voltage required to generate this level of E_{OX} is a function of the oxide thickness T_{OX}. Values of T_{OX} reported are 10 nm [153] and 9 nm [63]. These values are a function of the process being used. If T_{OX} is in nm and E_{OX} of 6 MV/cm is selected then the stress voltage V_{stress} is given by:

$$V_{stress} = T_{OX} \times 0.6V/nm. \quad (2.9)$$

To understand the process of computing the stress test time we use the following notation. Let T be a stress test set, $n = \|T\|$, D be the set of targeted defects, d be a member of D, $N(d)$ be the number of tests in T that stresses d, T_{ST} the stress time for each transistor and T_{SL} be the overall stress time for the entire IC. Furthermore, let $m = MIN_{d \in D}\{N(d)\}$. Then, the input vectors are applied at a rate of $\frac{T_{ST}}{m}$. Thus, $T_{SL} = n \times \frac{T_{ST}}{m}$.

Kawasaki-Steel has replaced their burn-in process with this approach for their $0.5\mu m$ process [153]. The input vectors are applied at a rate of 1 μs and the voltage applied is 7.5V. The normal operating voltages for these ICs is 3.3V. They have so far not been able to use it for processes with thicker gate-oxide because of the difficulty of applying higher than 7.5V from the tester.

Intel has recently reported the use of this screen for their i960JX CPU [124]. During HVS the built-in self-test is run at 7V. The DPM (defects per million) level fell to 68 after the introduction of this screen and burn-in is no longer used for the product. It is estimated that this saved the company $1.25M on that product.

3

PUTTING I_{DDQ} TESTING TO WORK

Having seen the benefits of I_{DDQ} testing we start answering the following questions which arise in the context of setting up the I_{DDQ} test environment.

- How are I_{DDQ} measurements made?
- How is the I_{DDQ} threshold value, separating the good from the faulty circuit, determined?
- How many I_{DDQ} measurements should be made? At what speed should I_{DDQ} tests be applied?
- How do we create I_{DDQ} test sets?
- What considerations for I_{DDQ} testing are required during design?

I_{DDQ} is measured by attachments to the existing automatic test equipment. They are referred to as **off-chip monitors** or **precision measuring units** (PMUs). Current measuring device can also be integrated on the chip itself. They are referred to as **built-in current sensors** (BICS). We discuss the design of I_{DDQ} measuring devices in Chapter 10.

Single threshold I_{DDQ} testing assumes that there is a single threshold that seperates the good and the faulty ICs. Methods for determining the I_{DDQ} threshold are discussed in Section 3.1. Questions have been raised about using single threshold I_{DDQ} testing for future generation of ICs. They are discussed in Section 3.2. Current signature is one variant of I_{DDQ} testing that can be used for such ICs. It is discussed in Section 3.3.

Issues pertaining to the speed and number of I_{DDQ} measurement are addressed in Section 3.4. Chapter 8 is devoted to I_{DDQ} test set generation. Finally, design considerations for I_{DDQ} testing is discussed in Section 3.5 as well as in Chapter 4.

3.1 I_{DDQ} THRESHOLD

I_{DDQ} **threshold** is the value of I_{DDQ} such that if the I_{DDQ} of an IC is above it then the IC is declared to be faulty. If the threshold is set too high then many faulty ICs will escape detection and IC quality will suffer. On the other hand, if it is set too low then effective yield will reduce because some good ICs will be declared as faulty. Therefore, care must be taken to determine an appropriate value of the threshold.

It is difficult to determine a good value of the I_{DDQ} threshold because it varies for the same design, from one IC to the next. This is a result of process variations that affect device parameters. This variation is more pronounced when ICs in different lots are compared. Two approaches to determining the threshold have been suggested: experimental and analytical.

3.1.1 Experimental Approach

In this approach [240, 300, 330] a reasonably large sample S of ICs is taken. It is assumed that the I_{DDQ} test set T is known. Next, for each IC $X \in S$ the vectors in T are applied and I_{DDQ} is measured after the application of each input vector. The maximum of these I_{DDQ} values, denoted by I_X, is recorded. The measured values of I_X are used, as described below, to determine the I_{DDQ} threshold.

The plot of the sample of values of I_X can be approximated by Gaussian or normal distribution like the one shown in Figure 3.1. Figure 3.2 is the I_{DDQ} distribution of a 32 bit microprocessor [300]. Note that the number of ICs with I_{DDQ} value less than 2.5 μA and greater than 500 μA are lumped together. Similar data for an ASIC are shown in Figure 3.3 [240]. The sample size was 1057 and instead of an I_{DDQ} test set a set of randomly generated input vectors was used.

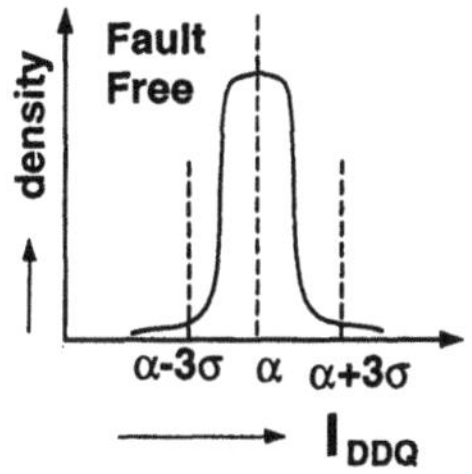

Figure 3.1 Expected I_{DDQ} distribution

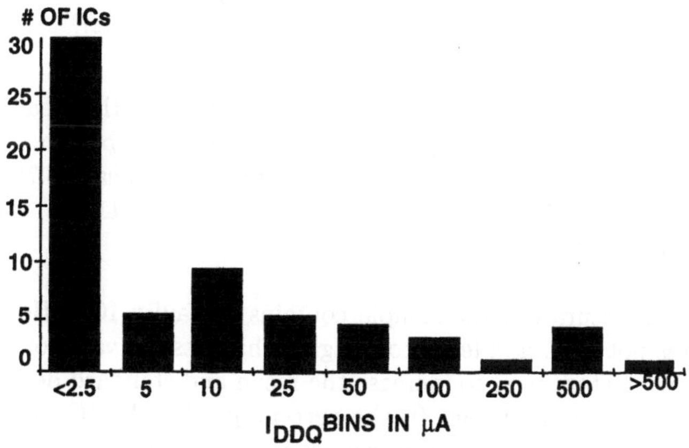

Figure 3.2 Example of I_{DDQ} distribution

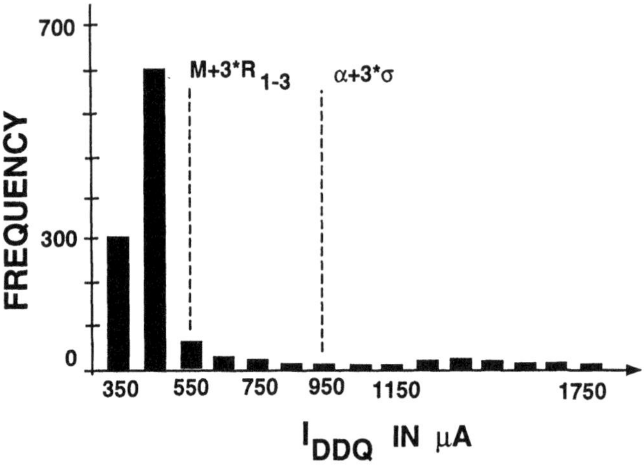

Figure 3.3 Example of I_{DDQ} distribution with a "long" tail

The I_{DDQ} threshold is calculated as follows. Let the sample size be n and $I_{X1}, I_{X2}, \ldots, I_{Xn}$ be the sample of values of I_X. Then, two parameters, the mean (α) and the standard deviation (σ), are computed using equation 3.1.

$$\alpha = \frac{1}{n}\sum_{j=1}^{n} I_{Xj} \quad ; \quad \sigma^2 = \frac{1}{n}\sum_{j=1}^{n} (I_{Xj} - \alpha)^2 \tag{3.1}$$

The "six sigma" range, evenly distributed on both sides of the mean, as shown in Figure 3.1, is taken as the significant part of the distribution. The I_{DDQ} threshold is set to $\alpha + 3 \times \sigma$. For a micro-controller, the mean (5.721 μA) and standard deviation (3.395 μA), based on a sample of size 11,405, was used to compute a threshold of 15μA [330].

The method is accurate if the sample contains no faulty ICs. However, this assumption is not valid and leads to a higher than desired value of the threshold [240]. Two factors tend to increase the mean and standard deviation from that of a fault-free distribution: (i) the percentage of faulty ICs in the sample can be significant, especially if the process is not mature; and (ii) the faulty I_{DDQ} is significantly higher than the fault-free I_{DDQ}.

Using the data in Figure 3.3, the above method gives a threshold of 950 μA. This was considered to be too high and a new threshold was computed using the "interquartile range statistics" [240]. Let A be the I_{DDQ} value such that 25% of the ICs have I_X less than A. This is known as the 25% point on the current distribution. Let B be the 75% point on the current distribution, $R_{1-3} = B - A$, M be the median (not mean) of the current distribution. Then the computed threshold is $M + 3 \times R_{1-3}$. The threshold using this method was 550 μA.

Several points should be kept in mind when using experimental methods:

- The threshold should be recomputed as the process matures. Initially, the percentage of faulty ICs in the sample will be high and the I_{DDQ} threshold computed will be high. But, as time progresses and the process yield improves, the percentage of faulty ICs in the sample decreases. The threshold value should be recalculated to get a better I_{DDQ} screen.
- Design of built-in current sensors requires the value of the threshold. This method therefore assumes that PMUs are used.

3.1.2 Analytical Approach

As discussed above, the experimental approach has its limitations especially if a BICS is used. To circumvent these difficulties, alternative approaches using computer-aided design tools have been discussed. These methods use the design data of the circuit as input and estimate the static current consumption. We next discuss some such methods.

The **first method** consists of adding up the nominal I_{DDQ}, available as part of the design data, for each cell used in the design [240]. A guard band (about 10%) is added to this value to derive the threshold. The guard band takes care of the uncertainties due to process variations. Assuming that the nominal I_{DDQ} is the average current drawn by the cell, the sum of nominal values of all cells is an upper bound on the average I_{DDQ} over all input vectors. Estimates based on the average I_{DDQ} for the given test set will be more accurate. We are unaware of any evaluation of this approach.

The **second method** uses current estimates originally proposed for reliability analysis [35, 218, 219] of VLSI circuits. A method to compute the average and the variance of the current, at each node, has been discussed. These values for the V_{DD} node can be used to estimate the threshold. The method computes the **expected waveform**. This is a waveform whose value at a time is the weighted average of all possible current values at that time. Assuming that the value will stabilize after the settling time, to get an estimate of the mean value of I_{DDQ}, a reasonably large value of time is assumed. Limitations of

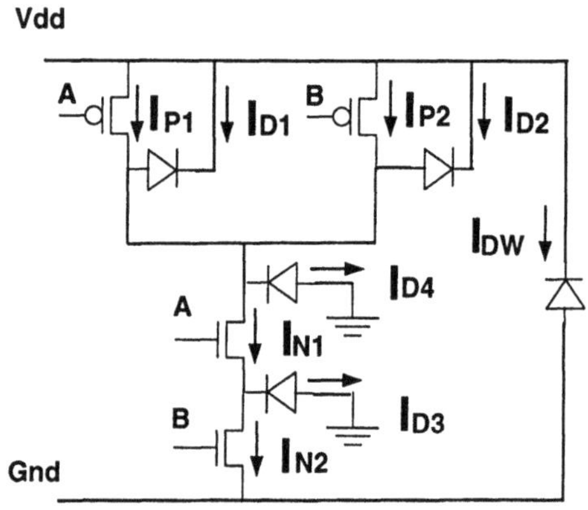

Figure 3.4 Model to estimate I_{DDQ} for a 2-NAND gate

this approach are: it is time consuming; it can only be used for combinational circuits; it computes the mean and variance over the entire set of input vectors; and it does not take process variations into account. Nonetheless, it can be a starting point for further investigation to estimate the threshold value.

The **third method** [101] models the design at three different levels: device, electrical and gate levels. At the device level, for each cell, the leakage current of all reverse-biased *pn* junctions and the threshold currents through non-conducting FETs are computed. These currents are functions of several technology dependent parameters. Expressions to compute them are given in [101].

At the electrical level, the cells are represented as an interconnection of FETs and diodes. The diodes model the corresponding reverse biased *pn* junctions. In Figure 3.4 an electrical model of a 2-NAND (two-input NAND) gate and the leakage currents are shown. In Figure 3.4, I_{DW} always contributes to the leakage current. However, the contribution of the rest of the current component is a function of the state of A and B as shown in Table 3.1. Similar tables can be derived for other gate types like NOR, XOR, etc.

A 2-NAND gate has minimum I_{DDQ} when both inputs are set to 0. I_{DDQ} values for all input combinations are normalized using this minimum value. A set of normalized I_{DDQ} values, for the technology parameter assumed in [88], is shown in Table 3.2.

Table 3.1 I_{DDQ} for input combinations to a 2-NAND gate

A	B	I_{DDQ}
0	0	$I_{DW} + I_{D3} + I_{D4} + I_{N2}$
0	1	$I_{DW} + I_{D4} + I_{N1}$
1	0	$I_{DW} + I_{D3} + I_{D4} + I_{N2}$
1	1	$I_{DW} + I_{P1} + I_{P2} + I_{D1} + I_{D2}$

Table 3.2 Normalized I_{DDQ} values for two-input gates

A	B	NAND	NOR	XOR
0	0	1	44	66
0	1	22	12	66
1	0	12	22	47
1	1	44	1	25

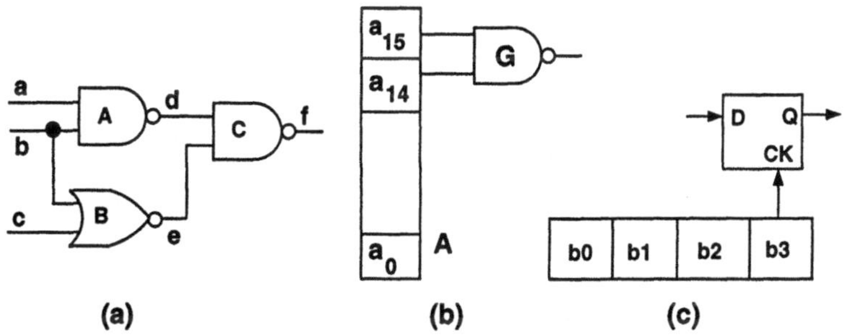

Figure 3.5 Illustrating controllability

Once the above information for each cell type is known, the maximum fault-free I_{DDQ} for a given circuit is computed. We use the example of Figure 3.5(a) to illustrate the process. For each possible input vector to the given circuit, we compute the state of the inputs of each gate. This is shown in Table 3.3 where, for each input vector to the circuit and for each gate, the logic levels at the input to the gate and the contribution to the total I_{DDQ} for that input vector are shown. Contributions of individual gates are added up to get the total I_{DDQ} of the circuit, shown in the last column, for that input vector. Next, the maximum of these, which is 100 for our example, is taken to be the estimate of the I_{DDQ} threshold.

Table 3.3 Example calculation of maximum I_{DDQ}

				A			B			C			Total
	a	b	c	a	b	I	b	c	I	d	e	I	
T1	0	0	0	0	0	1	0	0	44	1	1	44	89
T2	0	0	1	0	0	1	0	1	12	1	0	12	25
T3	0	1	0	0	1	22	1	0	22	1	0	12	56
T4	0	1	1	0	1	22	1	1	1	1	0	12	35
T5	1	0	0	1	0	12	0	0	44	1	1	44	100
T6	1	0	1	1	0	12	0	1	12	1	0	12	36
T7	1	1	0	1	1	44	1	0	22	0	0	1	67
T8	1	1	1	1	1	44	1	1	1	0	1	22	57

An exhaustive enumeration of all input vectors is not feasible for large circuits. For that, a method that computes an upper and a lower bound was proposed. The procedure for computing the lower bound is illustrated with the example of Figure 3.5(a). This circuit is first modified by inserting decoders, comparators, etc., as shown in Figure 3.6. Observe that if Y is set to 1 then there is an input vector for which I_{DDQ} is larger than I_{REFC}. The method therefore starts with a low value of I_{REFC} and uses an ATPG to determine if there is a test for the fault Y s-a-0. If it succeeds in finding a test then I_{REFC} is incremented. This process continues until some time limit is exceeded or the ATPG fails to find a test. The last value of I_{REFC} for which a test was computed is used as a lower bound on the threshold value.

To compute an upper bound, the circuit is partitioned into disjoint pieces $C_1, \ldots C_K$. The individual pieces are assumed to be independent of each other. The exhaustive method is used to compute the maximum I_{DDQ} $I_1, \ldots, I_K$ for $C_1, \ldots, C_K$ and $\sum_{j=1}^{K} I_j$ is used as upper bound on the maximum I_{DDQ} for the entire circuit. The threshold value used is the mean of the upper and lower bounds.

Readers should note several points about the above method.

- The method has severe limitations for sequential circuits which has many unreachable state. Therefore, even if the exhaustive method is used, it will overestimate the I_{DDQ} threshold unless care is taken to eliminate the unreachable states.

- Process variations change the process parameters used to calculate the components of the leakage currents at the device level. The method must

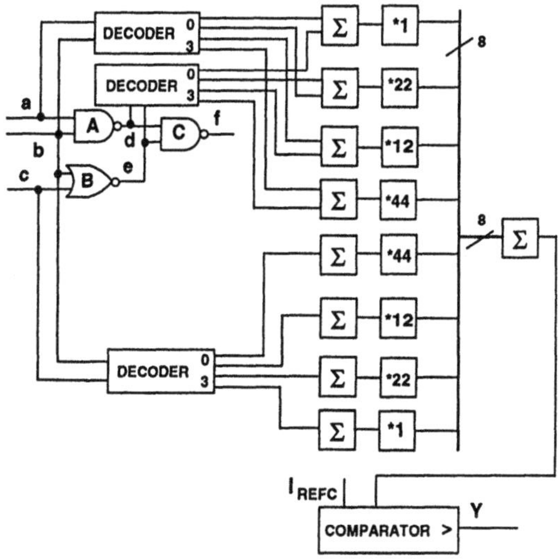

Figure 3.6 Modified circuit to be used by ATPG to estimate I_{DDQ}

be modified to get a good estimate by computing the threshold values for several values of the process parameters. Once this is done, the mean and standard deviation of these estimates can be computed and the threshold determined as explained in the previous subsection.

- Using an estimate based on the entire set of possible input vectors is an overkill. It is computationally very expensive and severely overestimates the threshold. To understand this, consider the example of Figure 3.5(a). The estimate for maximum I_{DDQ}, from Table 3.3, is 100 units. For the same circuit, as the reader can verify, we only need tests $\{T6, T7, T8\}$ to detect all possible bridging faults between lines that are either primary inputs or gate outputs. For this test set the maximum I_{DDQ} is 67 units. Thus, it seems more appropriate to put the estimate closer to 67 than to 100 units.

We conclude this section by noting that there is a need to calculate the I_{DDQ} threshold. Methods proposed so far make significant progress toward a sound methodology for computing the threshold value. However, they all have limitations and research is still needed to develop better methods.

3.2 I_{DDQ} TESTING OF DEEP SUBMICRON DEVICES

Single threshold I_{DDQ} testing is based on the assumption that the fault free I_{DDQ} of the IC under test is small. System and circuit level design practices tend to increase the fault-free I_{DDQ}. We will discuss this in Section 3.5 and Chapter 4. In this section we discuss the technological trends, which imply that future ICs may have high fault-free I_{DDQ}, making it difficult to use I_{DDQ} testing for such ICs. In the next section we will discuss a variant of I_{DDQ} testing that has been proposed to combat this problem.

Let V_t be the threshold voltage of the switching device, T the temperature, W and L the channel width and length, respectively, of the switching device, and x_c and κ, two constants. Then, I_{off} the subthreshold current, which contributes to the fault-free I_{DDQ} of the IC, is given by equation 3.2 [315].

$$I_{off} = \frac{W \times x_c}{L} \times e^{-\kappa \times V_t / T} \tag{3.2}$$

The current technological trends can be summarized as follows [77]. There is a continuous attempt to put more functionality and extract more performance from the ICs being fabricated today. Thus far, "scaling" has helped us achieve this end. It is the primary technique that will be used in the near future. Scaling, in simple terms, consists of: reducing the supply voltage (V_{DD}); reducing the width (W) and length (L) of switching devices; reducing the length and width of the interconnecting wires; and adjusting the necessary doping [77]. All these changes contribute to improved device density, higher speed, and reduced power dissipation.

However, for reasons discussed below, scaling increases I_{off}. Delay through the circuit decreases as V_t decreases. For high performance devices, $V_t \leq V_{DD}/4$ [211]. Thus, for high-performance circuits, V_t must be decreased as V_{DD} is decreased. Equation 3.2 implies that a linear decrease in V_t results in an exponential increase in I_{off}. On the basis of this, it has been predicted that the fault-free I_{DDQ} of high performance ICs of the near future will be considerably higher than the current ICs. We next see what implication this has on I_{DDQ} testing.

Refer to Figure 3.7, which shows the normal distribution of I_{DDQ} for fault-free ICs and faulty ICs. Here α and β are, respectively, the means of the I_{DDQ} values for the fault-free and faulty ICs, and σ and δ are, respectively,

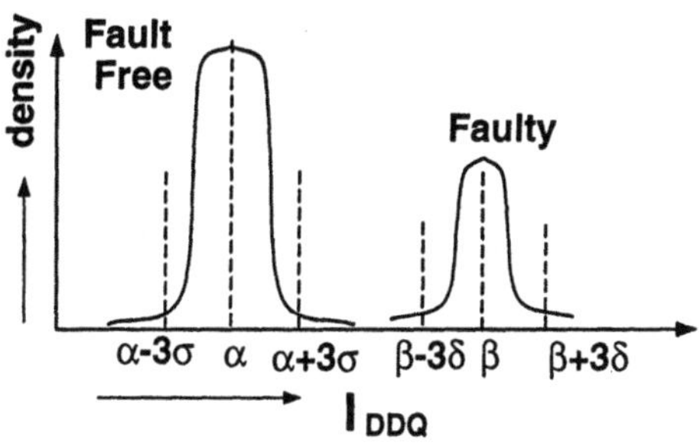

Figure 3.7 Distribution of faulty and fault-free I_{DDQ}

Table 3.4 CMOS technology trend.

Year (G)	Gates $\times 10^6$	V_{DD} V	L_{eff} μm	Ave. I_{off} $nA/\mu m$ (I_{AVE})	$S_M \times 100$ 25^oC	85^oC
1995	5	3.3	0.45	0.004	1358	25
1998	14	2.5	0.25	0.12	22.1	0.88
2001	26	1.8	0.15	2.5	0.68	0.04
2004	50	1.5	0.10	14.2	0.08	0.01
2007	210	1.2	0.07	82	0.00	0.00

the standard deviations of the I_{DDQ} values for the fault-free and faulty ICs. The faulty current, and hence β, decreases with V_{DD}. On the other hand, I_{off}, and hence α, increases as V_{DD} decreases. This implies that the two normal distribution curves of Figure 3.7 will move towards each other. When the two curves overlap, it will be difficult to differentiate the good and the defective circuits using I_{DDQ} testing. The relative separation between the two means, $S_M = \frac{\beta-\alpha}{\alpha}$, is used as a measure of the effectiveness of I_{DDQ} testing [335].

To quantify the above argument sample calculations from [335] are presented. The technology trends used are summarized in the first five columns of Table 3.4. These values were calculated assuming that V_t is scaled at the same rate as V_{DD} and temperature is 25°C. To calculate α it is assumed that the W/L ratio is 20. Equation 3.3 and values from Table 3.4 are used to compute α.

$$\alpha = G \times 20 \times L_{eff} \times I_{AVE} \tag{3.3}$$

The average faulty current, given by $\beta - \alpha$ is computed as follows. Only bridging defects are assumed and a bridging defect is assumed to have zero bridge resistance. The circuit is assumed to be made up of two input NAND gates only. The pullup network is assumed to consist of one pFET and the pulldown network is assumed to consist of two nFETs in series. The "on resistance" of a pFET (nFET) is denoted by R_{nFET} (R_{pFET}). We have

$$\begin{aligned} R_{nFET} &= \frac{6}{W/L}\ K\Omega; \quad R_{pFET} = \frac{15}{W/L}\ K\Omega; \\ \beta - \alpha &= \frac{V_{DD}}{2 \times R_{nFET} + R_{pFET}}. \end{aligned} \tag{3.4}$$

The above calculation underestimates the faulty current. Only the primary I_{DDQ} flowing through the two gates involved in the bridging defect has been used in the calculation. In Chapter 4 we will see that secondary sources also contribute to the total faulty I_{DDQ}. In addition, since only one pFET is assumed to be conducting equation 3.4 is a lower bound on the primary I_{DDQ} source in the presence of the defect.

Equations 3.3 and 3.4 are used to compute the value of S_M for 25^oC from the parameters given in the first five columns of Table 3.4. These values are shown in column 6. The value of S_M for 85^oC, taken from [335], is given in the last column. This table points out a number of trends. Firstly, the value of S_M shows that the separation between the means of the fault-free and the faulty currents decreases rapidly. Secondly, the separation is very sensitive to the temperature at which I_{DDQ} testing is performed. Additional data showing that, even for 1995 technology, if the V_t/V_{DD} ratio is changed from 1/6 to 1/20 then S_M becomes very small can be found in [335].

The above predictions are worth noting and do point out some serious limitations of single threshold I_{DDQ} testing. However, it should not be used to argue that I_{DDQ} testing will cease to exist as a viable testing method in the years to come. There are several reasons for this:

- The rise in I_{off} implies that the power dissipation of these high performance ICs is also going to be very high. If we use the parameter values of Table 3.4 then by year 2007 the average current drawn by the core of an IC (without the I/O buffers etc.) would be in excess of 24 amperes. Hence, there are reasons other than I_{DDQ} testing for which the predicted trend in the growth of I_{off} has to be changed.
- The above calculations assume that only one threshold voltage V_t will be used for all devices. To minimize power dissipation and improve noise

immunity, multiple thresholds may be used. In this scenario, the V_ts of devices along critical paths are set to a lower value than the rest. Thus the fault-free I_{DDQ} will be considerably lower than those predicted above.

- Partitioning the circuit and testing individual partitions can alleviate the problem.
- Cooling the circuit is another technique that can be used to reduce the fault-free current.

Note that, from empirical data given in the previous section, a threshold value of about 4 to 5 times the α is adequate. The statistical variation in the faulty current will be minimal. Therefore, a value of S_M in the range of 5 to 6 should be considered adequate. Given this, and moderating the extreme predictions made above, we believe that single threshold I_{DDQ} testing will be feasible for most ICs well into the first decade of the next century.

It is clear that fault-free I_{DDQ} will grow, although not as dramatically as suggested above. Keeping in mind the dramatic impact I_{DDQ} testing have on IC quality and reliability, two different modifications to single threshold I_{DDQ} testing has been proposed [101, 275]. The solution in [101] does not require any modification to the circuit under test and only minimal changes to the existing test equipment are needed. This is discussed in the next section.

3.3 CURRENT SIGNATURE

Current signature is explained with the help of Figure 3.8 [101]. The defect free circuit is shown in Figure 3.8(a). Figures 3.8(b) and (c) show two different defective circuits. Let the test set be $\{T1, T2, T3, T4\}$ as defined in Table 3.5. This table also lists the I_{DDQ} measurement for each of the three circuits of Figures 3.8(a), (b) and (c) respectively in columns 5, 6 and 7.

A plot of the I_{DDQ} measurement for the defect-free circuit is shown in Figure 3.9(a). If we sort the vectors with repect to the I_{DDQ} values and plot them in increasing order we get the plot of Figure 3.9(b). This step function is a property of the defect free circuit and is its **current signature**, with respect to the given I_{DDQ} test set.

It is characterized by: the different I_{DDQ} levels; and the number of vectors for each level. I_{DDQ} values, on application of the same test set, for the defective

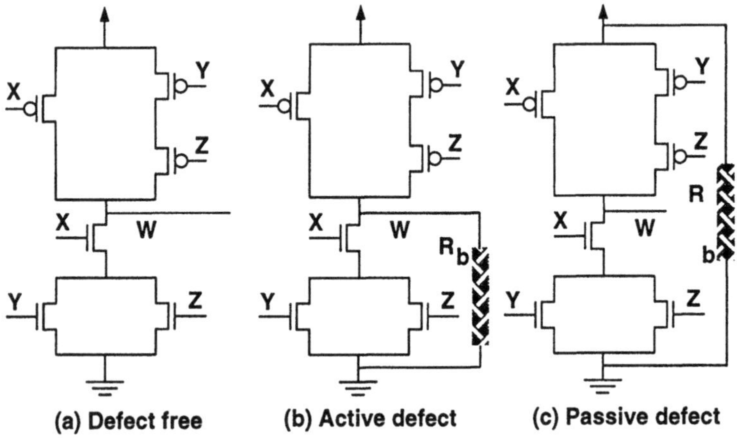

Figure 3.8 Examples to illustrate current signature

Table 3.5 Test set for example illustrating current signature.

	x	y	z	I_{DDQ} (a)	I_{DDQ} (b)	I_{DDQ} (c)
T1	0	1	1	I1	I3	I5
T2	1	1	1	I2	I2	I5
T3	1	1	0	I1	I1	I5
T4	0	0	0	I2	I4	I5

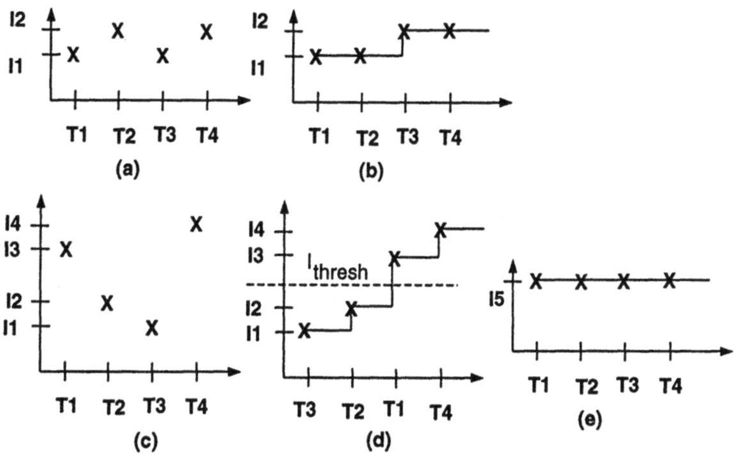

Figure 3.9 Plots of the I_{DDQ} values for the current signature examples

circuit of Figure 3.8(b) are plotted in Figure 3.9(c). From this, its current signature is easy to derive and is plotted in Figure 3.9(d). This current signature is very different from that of Figure 3.9(b) and can be used to identify the faulty circuit.

The defective circuit of Figure 3.8(b) is such that the faulty current dependents on the input vector applied. Such defects are referred to as "active defects". Other defects, like that of Figure 3.8(c), are such that every input vector leads to a high I_{DDQ}. Such defects are "passive defects". Then current signature, as shown in Figure 3.9(d), will contain no steps. Thus, ICs with such defects can be identified by their current signatures but cannot be identified by the single threshold I_{DDQ} testing. It is important to distinguish between the two kinds of defects because ICs with "passive defects" function properly. Unless power dissipation is of concern such ICs need not be discarded thereby improving the yield. This is the first advantage of using current signature.

A problem with the single threshold I_{DDQ} testing is that if the fault-free I_{DDQ} is high then it is difficult to set a threshold that seperates the good IC from the faulty IC. To put the argument against the single threshold in perspective, in Figure 3.9(d), I_{thresh} the threshold seperating the good from the faulty circuit is based on statistical estimates of $I1, I2$. Vectors $T1$ and $T4$ lead to high I_{DDQ}. The additional current due to the defect is given by $I3 - I1$ and $I4 - I2$, respectively, for $T1$ and $T4$. When, $I1$ and $I2$ are sufficiently large, $I3$ and $I4$

fall within the statistical spread of $I1$ and $I2$. Therefore, $I4$ and $I3$ will fall below I_{thresh}. Hence, with rise in the fault-free current, singlethreshold I_{DDQ} cannot seperate the good IC from the faulty IC without significantly affecting the true yield.

Even if the fault-free I_{DDQ} is high, it is conjectured that current signature can still be used. To understand the intuition behind this claim, consider the current signatures of Figures 3.9(b) and (d). The very fact that one step function has two distinct levels whereas the other has four distinct levels is sufficient to distinguish the two circuits. The actual I_{DDQ} values are not important. However, methods to compute the defect free current signature, the number of steps, the step size, etc., needs further study.

The method, alluded to in the above discussion, for computing the current signature involves the following steps: measure I_{DDQ} for each vector in the test set; sort vectors with respect to the I_{DDQ} values; and derive the value and length of the distinct levels of the step function. This is a time consuming process. In addition, this process has to be executed for each IC tested. It is therefore not appropriate for production testing. Methods to refine the computation of the signature need to be investigated and some such refinements have been discussed in [102].

3.4 SPEED AND NUMBER OF I_{DDQ} MEASUREMENTS

First, let us consider the speed at which I_{DDQ} measurements can be made. Two constraints determine the speed of I_{DDQ} measurements. The first constraint is the speed at which the I_{DDQ} measuring device can respond. This will be discussed in more details in Chapter 10. But, as a rule of thumb, using modern instrumentation 10,000 to 100,000 I_{DDQ} measurements can be made in a second. In fact, it has been proposed as a standard by QTAG (Quality Test Action Group) of IEEE Technical Committee on Test Technology. If built-in current sensors (BICS) are used the frequency of measurements is a few tens of MHz [192, 294].

The second constraint is the speed at which the circuit under test settles after an input vector is applied. Figure 3.10(a) shows how I_{DD} varies with time soon after an input vector is applied. Measurements are made after all transients in the circuit have settled. The transient part is determined by the propagation

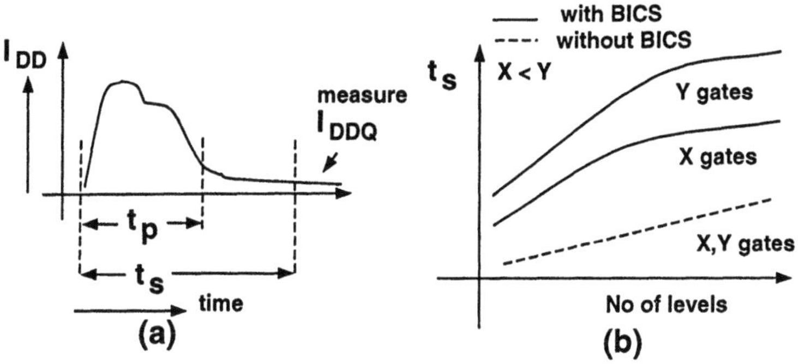

Figure 3.10 Settling time

Table 3.6 Propagation/settling time versus number of levels and gates (no BICS)

Levels	Number of Gates							
	50		100		500		1000	
	t_p(ns)	t_s(ns)	t_p(ns)	t_s(ns)	t_p(ns)	t_s(ns)	t_p(ns)	t_s(ns)
1	0.6	0.66	0.6	0.66	0.6	0.66	0.6	0.66
3	4.04	7.8	4.04	7.8	4.04	7.8	4.04	7.8
5	7.56	10.8	7.56	10.8	7.56	10.8	7.56	10.8
10	16.4	21.0	16.4	21.0	16.4	21.0	16.4	21.0

delay, denoted by t_p, of the circuit. It is a function of the order in which input vectors are applied. Therefore, the worst case propagation delay is used. After the propagation delay, additional time is required for I_{DD} to reach its quiescent state. The total time after which I_{DDQ} measurements can be made is referred to as the **settling time** and is denoted by t_s.

First consider the case when PMUs and not BICs are used. **Propagation delay**, and hence t_s, is a function of the number of levels in the circuit. Spice simulation results shown in Table 3.6 confirm this [210]. The circuit simulated consisted of inverters only. For example, the circuit with 100 gates and 10 levels had 10 copies of an inverter chain of length 10. If we plot the data its nature will be as shown by dotted lines in Figure 3.10(b). The exact nature of the plot is given in Figure 3 of [210].

The settling time increases with the number of levels but remains constant with the number of gates. In practice, because of longer wires that the individual

Table 3.7 Settling/propagation time versus number of levels and gates (with BICS)

Levels	Gates							
	50		100		500		1000	
	t_p(ns)	t_s(ns)	t_p(ns)	t_s(ns)	t_p(ns)	t_s(ns)	t_p(ns)	t_s(ns)
1	1.15	2.0	1.41	3.0	4.22	10.01	13.46	15.23
3	6.6	9.02	7.22	10.21	11.57	19.46	16.4	39.56
5	11.51	16.06	11.76	17.0	16.71	31.12	21.42	47.0
10	23.8	28.26	24.19	29.92	28.19	39.62	32.62	53.5

gates have to drive, the load capacitance on the gates will increase with the number of gates. Thus, the settling time will grow a little with the number of gates, even if we keep the number of levels the same.

An important fact of these data is that the **settling time is less than twice the maximum propagation delay**. Propagation delay estimates are made to ascertain the maximum clock frequency. From these an upperbound on the value of the settling time can be determined. Let f be the frequency of I_{DDQ}, f_{PMU} the frequency at which the PMU can measure I_{DDQ} for the circuit under test (CUT), and t_s the settling time of the CUT. Then, $f = minimum(f_{PMU}, \frac{1}{t_s})$. If we look at present day circuits, the PMU is the constraint and therefore the frequency of I_{DDQ} measurements is between 10 and 100 kHz.

For circuits with BICS the settling time increases both with the number of levels and the number of gates for the following reason. BICSs are placed between the ground of the CUT and the physical ground. Consequently, the source and drain capacitances of all the transistors in the CUT appear as one lumped capacitance at the input of the BICS. This capacitance increases as the number of gates increases. This conjecture has been confirmed using SPICE simulation [210]. The results are presented in Table 3.7. The nature of the plot of this data is shown in Figure 3.10(b). **Note that even in this case, the settling time is about twice the propagation time. However, the propagation time itself has increased considerably.**

The question of how frequently I_{DDQ} measurements can be made using BICS is more difficult to answer. It is a function of the BICS design, the CUT under test as well as the size of the CUT being tested by each BICS. Nonetheless, experimental implementations of BICS can measure I_{DDQ} at a considerably higher rate than PMUs.

Next, let us consider the issue of how many I_{DDQ} measurements must be made. During the early years of I_{DDQ} testing many IC manufacturers used only one or two I_{DDQ} measurements. This number has increased since then and many ATEs allow 20 or more I_{DDQ} measurements. In addition, the vectors for which the measurements are to be made can also be programmed. Currently many production facilities allow I_{DDQ} test sets of small size like 20 or so.

However, as ATEs and BICS improve such restrictions will be eliminated. When this happens test application time and fault coverage are two important factors that will determine the number of I_{DDQ} measurements. If the I_{DDQ} test set detecting all targeted faults is such that it can be applied within an acceptable time then all vectors in it should be applied. However, in many case we have to reduce the number of I_{DDQ} measurements. An example will help put this in proper perspective.

Assume that: we have a functional test set of size 400,000; a tester that can apply these vectors at 40MHz; and an I_{DDQ} measurement is made for each vector. The total test time will be 0.01 second. If the frequency of I_{DDQ} measurements is 10 kHz, the total test application time is 40 seconds which is not acceptable. However, if we select 1% of these vectors (4,000) and measure I_{DDQ} only for those, then the total test application time will be 0.4099 second which is certainly acceptable. This example leads us to the following question: how do we select 1% of the possible measurement points? We will get back to this issue in Chapter 7.

3.5 I_{DDQ} TESTING RELATED DESIGN ISSUES

I_{DDQ} testing assumes that fault-free ICs have negligible I_{DDQ}. However, many standard design practices lead to ICs with high fault-free I_{DDQ}. Thus, an important design consideration for I_{DDQ} testing is to avoid design practices that lead to high fault-free I_{DDQ}. In addition, like all other testing techniques, if care is not taken during design then the effectiveness of testing is compromised. In the next two subsections we address these I_{DDQ} testing related design issues. I_{DDQ} detection of defects in some components like flip-flops, shift registers, scan chains, RAMs, etc., pose additional challenges. We will study them in Chapter 4.

Another design related issue is, how large a circuit can we test with one current sensing device? We do not discuss this issue here. Some discussion of this issue can be found in [185].

3.5.1 High Current States

If 100% of a CMOS circuit is made up of static, fully complementary MOS gates then the fault-free I_{DDQ} is very small. However, many designs do not conform to this design style. Some approaches, discussed below, have been used to circumvent this difficulty.

- **Mask High Current States:**

  Mask out vectors resulting in high I_{DDQ} during I_{DDQ} testing. I_{DDQ} measurements are made only for the other vectors. For example, for the 80C51 8 bit microprocessor out of the 20,000 functional vectors about 300 resulted in design related high current states [129]. The high I_{DDQ}, on application of these 300 vectors, was not used as an indicator of a faulty circuit.

  This approach has its limitations. Firstly, this solution is useful if there are only a few vectors for which the circuit has high fault-free I_{DDQ}. It is not guaranteed that for a given design there will be a small percentage of such vectors. Secondly, the vectors left after masking out vectors with high fault-free current may not have good defect coverage. These issues have to be properly considered prior to using this technique.

- **Static Versus Non-Static Implementations:**

  If the design is not static and fully complementary then redesign the circuit using static and fully complementary logic [300]. The usual argument against this approach is that the new design may need many more transistors and ultimately more area.

  The argument for static and fully complementary design is that such overhead, if there is any at all, is often outweighed by the benefits [300]. In Sandia Laboratories, a static and fully complementary version of National Semiconductor's 32C016 16/32 bit NMOS microprocessor, named SA3300, was designed. SA3300 had about 65,000 FETS which is about the same number as the number of FETs in 32C016. However, there were major

Table 3.8 Control signal specification for the bus

C0	C1	CS	Source
0	0	1	S0
0	1	1	S1
1	0	1	S2
1	1	1	None
X	X	0	None

advantages in using SA3300. It was possible to have an I_{DDQ} test set of less that 3,000 vectors that had very high stuck-at fault coverage. In addition, it was possible to determine the quality of the vendor supplied test set, assuming that it was an I_{DDQ} test set. On the other hand, it was estimated that ascertaining the quality of the same test set, assuming it to be for logic testing, using a MACH 1000 hardware accelerator would require about 21 years!! [300]

- **Avoid High Current State Designs:**

  There are a number of design practices that lead to high current states. In addition to improving I_{DDQ} testability, reducing the possibility of high current states also reduces power consumption and improves reliability.

  – **Bus Design:** Many ICs have internal buses and there are states during which they are not used. These are logical don't care states for the bus. Consequently, often as a result of reducing the complexity of the bus control logic, more than one bus driver drives the bus (**bus conflict**) or no bus driver drives it (**floating bus**). Both states result in high I_{DDQ}. Simple design techniques can be used to avoid these states.

  Consider the following example design. It calls for sourcing data from 3 sources $S0, S1, S2$ to a common bus. The definition of the control signals are given in Table 3.8. Note that there are several don't care states. A design requiring near optimal logic is shown in Figure 3.11(a). However, when $CS = 0$ or when $C0 = C1 = CS = 1$ the bus floats leading to high fault-free I_{DDQ}. The circuit can be redesigned, as shown in Figure 3.11(b), so that the bus is always driven. Source $S3$ is a dummy source added to ensure that the bus is always driven.

  – **Embedded RAMs:** Any kind of processor, microprocessors or special purpose DSP processors, usually have embedded RAMs. Care

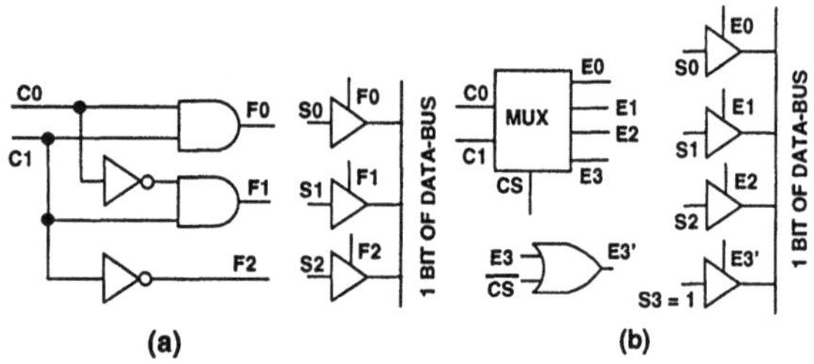

Figure 3.11 Redesigning to avoid floating bus/bus conflict

must be taken for I_{DDQ} testing of such processors because, depending on the RAM design, they may or may not draw large I_{DDQ} in the fault-free state. To understand this refer to Figure 3.12 which shows the organization of an SRAM. For a complete description of how such SRAMs store information see [215] or any other standard text on VLSI design. Here we provide a brief description.

Figure 3.12(a) is the block diagram of SRAMs. It is a 2-dimensional array of cells with some control and precharge logic. The address lines are not shown. The **row-decoder** selects exactly one row of cells by turning on one of the **row-select lines**. The **column-decoder** selects one cell from the selected row.

Each column is organized as shown in Figure 3.12(b). The jth column has two lines BL_j and BLI_j. They are, respectively, the **bit-line** and the **inverted bit-line**. There is one sense amplifier per column and the number of cells per column equals the number of rows.

The memory operates in two phases: precharge and compute. During the precharge phase all **row-select lines** are disabled, **SENSE** is off and **precharge** is pulled low. Thus, all BL_j and BLI_j lines are pulled high. In this phase nFETs $n1$, and $n2$ are cut-off. Therefore, the cells are isolated and each of them form a closed-loop consisting of two back-to-back inverters. They therefore do not need refreshing to retain the 1 bit information.

During the compute phase, a cell is selected either for read or write. When **precharge** and **row-select** is high the two nFETs $n1$, and $n2$ are conducting. Consider a read operation and assume that the selected cell stores a 1 (i.e., $Q = 1$ and $\overline{Q} = 0$). BL_j is driven by V_{DD} through $p1$ and $n1$, and BLI_j starts discharging through $n2$. When **SENSE** is turned high it aids this process of discharging and also drives the output buffers. The write opertation is similar. Thus, a

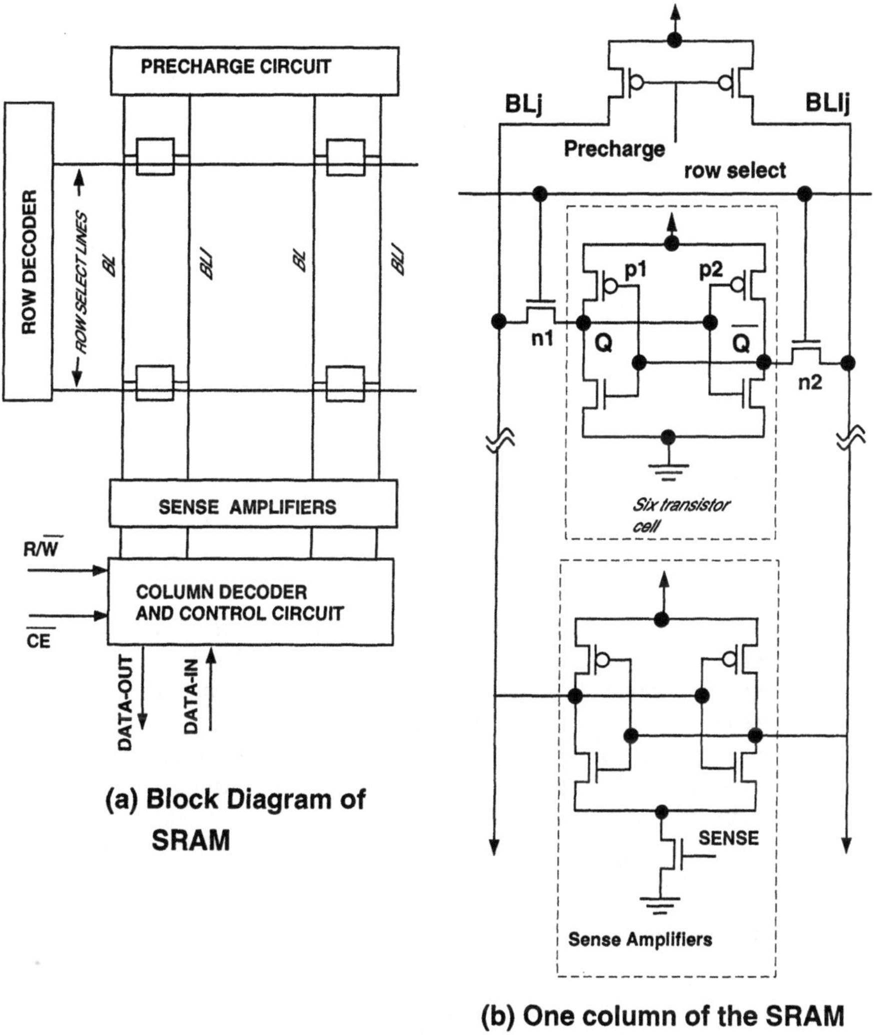

Figure 3.12 SRAM organization.

defect-free SRAM does not draw static current during the compute phase.

This was the SRAM design used in an experiment [39] of I_{DDQ} testing. In [279], circuits with embedded RAMs along with the core logic were tested using an I_{DDQ} threshold of 2μA. So, I_{DDQ} testing can and have been used for processors with embedded RAMs.

I_{DDQ} testing of processors with embedded RAMs is a problem when the memory cells do not conform to the design of Figure 3.12. In many designs, the two pFETs $p1$ and $p2$ are replaced either by two depletion mode n-channel load transistors or simply by two resistors. This creates a high current state as follows. When the cell is accessed the row select line is set to logic 1. This turns on the two pass transistors $n1$ and $n2$. Independent of what is stored in the cell, or what is written into it, one of the two resistors replacing $p1$ and $p2$ conducts for the entire duration of the time that the cell is accessed. To understand this assume that the cell stores logic 0 and a read is performed. Therefore, Q is at 0. Thus, there is a path from V_{DD}, through the resistor replacing $p1$, nFET $n1$ and the sense amplifier to ground. Thus, the fault-free I_{DDQ} is high and a high I_{DDQ} cannot be used as an indication of a faulty circuit. Note that this path does not exist in the design of Figure 3.12 since $p1$ is cut-off when the cell stores a 0.

One approach, used for I_{DDQ} testing of processors with such embedded RAMs, is to use a "standby mode" for the RAM. In this mode of operation none of the memory cells is accessed. Thus, nFETs $n1$ and $n2$ are cut-off. Therefore the RAM unit does not draw large I_{DDQ}. Facility must be provided to program the processor so as to put its internal RAM in the "standby mode" during testing. This approach assumes that the RAM is fault free. We have not addressed the issue of testing the RAM itself. I_{DDQ} testing of RAMs is discussed in Chapter 4.

- **Power Bus Separation:**

A number of other structures in the circuit can have high fault-free I_{DDQ}. Examples of these include the ring of I/O buffers on the periphery of the chip, analog blocks etc. I/O buffers, as required by the IEEE 1149.1 standard [235], have pullups which result in high I_{DDQ}.

For I_{DDQ} testing of circuits with such structures the ground and power bus are split. For example, in [279] the power and ground buses were split into two parts. One pair of power/ground buses supplied power to the core of the chip from one power source. This source is monitored during

I_{DDQ} testing of the core of the IC. The second pair supplied power to the I/O buffers. These buffers are sometimes tested by monitoring the second source [279]. The I_{DDQ} threshold for the second source was set at a much higher level (40μA) compared to the I_{DDQ} threshold for the first source (1μA).

3.5.2 Improving Controllability

In the previous subsection we saw that the presence of structures with high fault-free I_{DDQ} is a hindrance to I_{DDQ} testing. I_{DDQ} DFT techniques to address this either during design or during testing was also discussed. However, even if we can successfully design a circuit without high current states other issues pertaining to fault coverage, test application time and test generation need to be addressed.

In addition to the I_{DDQ} DFT techniques discussed in this section there are other I_{DDQ} DFT techniques in use. Once we have discussed defect detection mechanism using I_{DDQ} testing we will see that there are some commonly occurring defects that cannot be detected by I_{DDQ} testing. In order to detect such defects some design techniques have to be used. We will discuss this issue in Chapter 4.

Fault coverage is the percentage of targeted faults covered during testing. A test set is created to detect all target faults. However, if care is not taken during design, several things could happen: the test set could become very long and adversely affect the test application time; there might exist faults in the circuit for which the ATPG spends enormous amount of time to generate a test and often aborts without computing a test thereby reducing the coverage; etc. In this section we look at design techniques that address these issues.

Before we understand the techniques involved we need to understand two terms: **controllability** and **observability**. First consider combinational circuits. Controllability of a fault is the probability that a randomly selected input vector activates a fault. In Figure 3.5(a) the controllability of the fault *d s-a-0* is $\frac{1}{4}$. Observability of a fault is the probability of propagating the fault to an observable output. In Figure 3.5(a) the observability of fault *d s-a-0* is the probability of setting *e* to 1 which is $\frac{1}{4}$.

For sequential circuits the above definitions must be appropriately modified. However, the idea remains the same. If, for a fault, a long sequence of input vectors is required to activate, that sequence will be difficult and time consuming to compute. Therefore, the fault is said to have poor controllability. Similarly, if a fault requires a long sequence of vectors to propagate the fault effect to an observable output it is said to have poor observability.

For logic testing, a test must activate a fault and also propagate the fault effect to an observable output. In Figure 3.5(a), an input vector must activate *d s-a-0* by setting *d* to 1 and propagate it by setting *e* to 1. Therefore, in this context both controllability and observability of faults are important. For circuits in which the controllability and observability of faults are good ATPG usually spends less time generating tests and the test set is also small. It is therefore possible to compute acceptably small test set with high fault coverage in a reasonable amount of time.

For I_{DDQ} testing only controllability is important since the technique provides very good observability. For example, to activate some leakage faults in gate *d* of Figure 3.5(a) both nodes *a* and *b* have to be set to 1. That is the only requirement and the fault effect need not be propagated to an output. Although, the need for "good observability" is eliminated there is still the need for "good controllability".

The design for testability techniques proposed so far are very similar to those proposed for logic testing of stuck-at faults. The most basic technique is full-scan in which all memory elements of the circuit are connected together to form one long shift register. Inputs are scanned-in and outputs scanned out. This technique is described in most textbooks on testing [3]. The area overhead and the length of the test application time prevent its use in many situations. In these cases other techniques have to be used to address testability concerns during design. Here we look at the following techniques: Counter Design; and Partial Scan.

- **Counter Design:**

  Counters, if not properly designed, can be hazardous to test generators and ultimately to the effectiveness of the test set generated and the test application time. To understand this refer to Figure 3.5(b). Let A be a 16-bit counter with only a reset. To activate a leakage fault in gate G lines a_{15}, a_{14}, the two MSBs of A, must be set to 1. In the worst case, this may require a test of length $2^{15} + 2^{14}$. In order to get around this, counter A is designed as a register with parallel load so that it can be set to the appropriate value with a shorter test length.

Consider the situation shown in Figure 3.5(c). Here a 4-bit counter B is used to derive a clock for the D-Flip/Flop C. The MSB bit b_3 of B drives the clock line of D. If arbitrary logic states cannot be loaded into B then the ATPG will have great difficulty in controlling the output of C. The solution here too is to provide the counter with a parallel load capability so that arbitrary values can be loaded into register B.

- **Partial Scan:**

  In partial scan [311] only some of the system flip-flops are selected and chained into a shift register. Partial scan methods can be classified based on: whether they use a sequential ATPG [4, 29, 43, 68, 111, 171, 311] or a combinational ATPG [113, 162]; whether testability measures are used [4, 111, 311] or some topological analysis is used [43, 68, 113] to select the flip-flops to be chained; etc. The only reported work on partial scan pertaining to I_{DDQ} testing uses the topological method [43]. We therefore discuss that approach.

  The topological method is based on the observation that the key parameter of a circuit that makes sequential test generation difficult and time consuming are: **sequential depth** and **large feedback loops** [68] defined as follows. Let $G(V, E)$ be the graph of a sequential circuit such that: for every flip-flop in the circuit there is a node in V; and there is an edge from node v_1 to v_2 if and only if there is a path that passes only through combinational elements from the output of the flip-flop corresponding to v_1 to the input of the flip-flop corresponding to v_2. In Figure 3.13(a) we show the memory elements of an example circuit. The dependencies between memory elements are indicated using directed edges. These edges also represent the combinational circuit. The corresponding graph is shown in Figure 3.13(b). The **distance**, along a directed path, between two vertices is the number of vertices along that path. The is the length of the longest path in the graph of the sequential circuit. The sequential depth of the circuit of Figure 3.13(a) is 6 (from node 1 to node 10). Let D be the sequential depth of a circuit and n the number of primary inputs. Then, the following observations, valid for logic testing [68], are also valid for I_{DDQ} testing:
  (i) If the graph is a directed acyclic graph (**pipelined structure**) the test length for a fault is bounded by D. Clearly, this type of structure is "easy to test".
  (ii) The second topological category of sequential circuits is that where the graph contain only cycles of length 1. In Figure 3.13(a) flip-flop 8 has a self loop. Self loops correspond to finite state machines with two states. If we eliminate self-loops in the graph of a sequential circuit then each node represents either a flip-flop or a generalized two-state machine. In this case

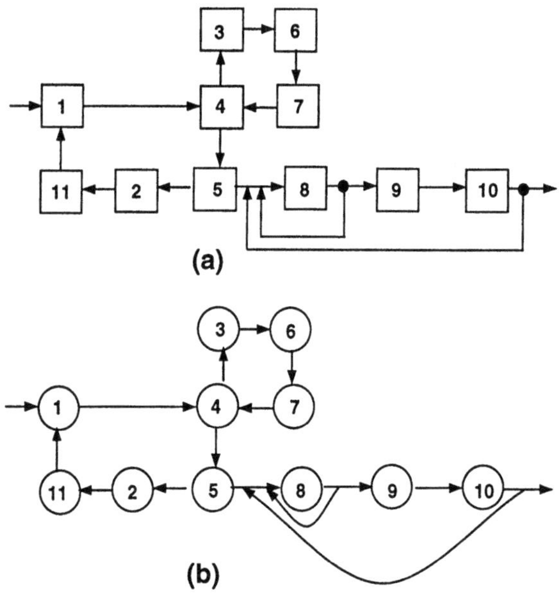

Figure 3.13 Graph of synchronous sequential circuit

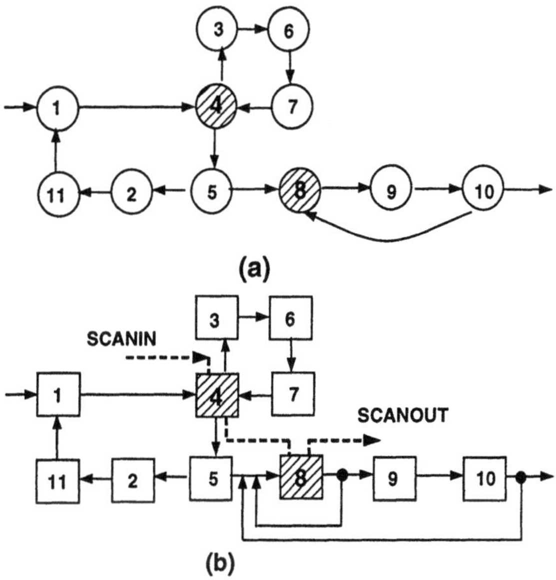

Figure 3.14 Selecting FFs to break cycles

an upper bound on the test length for a fault is $2 \times D$. Such structures are also not difficult to generate tests for.
(iii) The third topological category is sequential circuits whose graph contain cycles of length greater than 1. Let L be the length of the longest cycle. Then, an upperbound on the length of a test sequence for a fault is $D \times 2^L$.
These results point to the fact that the test length, and hence the complexity of test generation, is a function of the cycle lengths in a sequential circuit. The method used in [43], derived from the work in [68], is based on the following idea: absorb all self loops into vertices; and select a set of vertices whose removal makes the graph acyclic. This set of vertices comprise a **feedback vertex set**. The flip-flops corresponding to the vertices in the feedback vertex set are included in the partial scan chain. The graph of Figure 3.13(b), after absorbing the self loops, is shown in Figure 3.14(a). The shaded nodes comprise one vertex set. The corresponding circuit, with scan flip-flops as shaded boxes, is shown in Figure 3.14(b). Computing a feedback vertex set of minimum cardinality is NP-Hard [10]. Therefore heuristics are used to compute such a set. We present one such heuristic [43]. The **strongly connected component** (SCC) of a directed graph $G = (V, E)$ is defined as follows. Nodes v_i and v_j are said to be equivalent

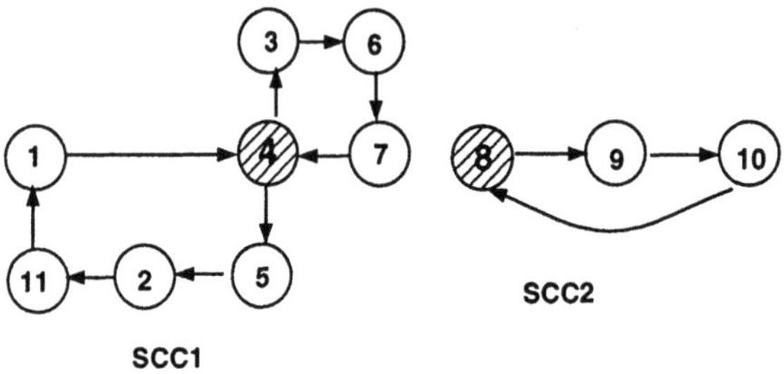

Figure 3.15 Illustration of heuristic BreakCycle

if and only if there is a directed path from v_i to v_j and vice-versa. We can therefore break up the set of vertices into disjoint sets of equivalent nodes $V_1, \ldots, V_k$. The subgraphs $G_1 = (V_1, E_1), \ldots, G_k = (V_k, E_k)$ induced by the graph G are the SCCs of G. All SCCs of a graph can be computed in time $O(n, e)$ where n is the number of vertices and e is the number of edges in G [10]. The SCCs of the graph of Figure 3.14(a) are shown in Figure 3.15(a). SCCs are important because removing a node from an SCC is guaranteed to eliminate at least one cycle. The reader can use the example of Figure 3.15(a) to verify this. In SCC1, picking vertices 5 and 6 breaks all cycles of that SCC. However, if we are careful and pick vertex 4 then all cycles will be broken by selecting only one vertex. BreakCycle, the heuristic given in Figure 3.16, attempts to select a small number of vertices to break all cycles in an SCC. The idea is to select a node with the largest value of "indegree $\times$ outdegree". The feedback vertex set of Figure 3.14(b) was computed from the SCCs of Figure 3.15 using the above heuristic. Experimental results, using the above heuristic for partial scan, are reproduced in Table 3.9 [43]. The results are for I_{DDQ} detection of single stuck-at faults. We are unaware of any such result for any other fault model.

In Table 3.9 for each circuit there are two rows. In the first row results after using the proposed partial scan methodology are presented. The second row contain results without using partial scan. The third column shows the number of FFs in the partial scan chain and the total number of FFs. The second to last column shows the number of faults proven to be untestable as well as the number of faults for which ATPG was aborted. The last column shows the fault coverage. There is a uniform and significant improvement in test generation time as well as fault coverage.

Table 3.9 Experimental results for I_{DDQ} testing of stuck-at faults using partial scan

Circuit Name	Total Gates + FFs	Selected / Total FFs	Scan FFs %	Test Gen. Time Seconds	Total Vectors	Total Faults	Faults Detected	Faults Untest Aborted	Fault Cov. %
510	217	5/6	83.3	11.6	39	584	584	0+0	100.0
		0/6	0	0.43	9	564	57	507+7	10.11
526	214	5/21	23.8	562.0	297	847	847	0+0	100.0
		0/21	0	1901.43	728	555	550	1+4	99.10
713	412	7/19	36.8	18.8	43	819	786	33+0	96.0
		0/19	0	22.41	28	581	534	47+0	91.91
838	422	15/32	47.0	7.08	110	1165	993	167+5	85.23
		0/32	0	71.40	41	857	597	260+0	69.66
953	424	5/29	17.2	37.7	72	1559	1559	0+0	100.00
		0/29	0	1.83	16	1079	461	618+0	42.72

BreakCycle(G)

/* Computes a minimal set of nodes P whose removal eliminates all cycles in G */

BEGIN

Step 1. Find all nontrivial SCCs , $G_1 = (V_1, E_1), \ldots, G_k = (V_k, E_k)$,in G.

Step 2. For each $G_i = (V_i, E_i)$

BEGIN

Select a node w in V_i s.t. $indegree(w) \times outdegree(w)$ is maximum among the nodes in V_i.

Delete w and all incoming and outgoing edges of w from G_i. Let the resulting graph be G'_i.

Add w to P.

If G'_i is not acyclic call BreakCycle(G'_i).

END

END

Figure 3.16 Pseudocode for algorithm BreakCycle

3.6 SUMMARY

In summary, we note that several issues have to be addressed before I_{DDQ} testing is used. In this chapter we addressed two of them. Determining the I_{DDQ} threshold that seperates the good ICs from the faulty ICs is a difficult problem. It also has a great impact on the yield and ultimately the total cost of the IC. Further work is necessary to improve the current techniques. There is concern that for future technologies a single threshold may not exist. To get around this the idea of using current signatures instead of single threshold I_{DDQ} testing has been suggested. Further investigation of this idea is required before we can draw any conclusion.

We saw that in order to make testing more effective some I_{DDQ} testing related issues have to be addressed during design. For I_{DDQ} testing designing circuits with good controllability is an important objective. I_{DDQ} testing assumes that a defect free circuit has low I_{DDQ}. This assumption is not valid for many real life designs. We saw some design techniques to get around this problem. We note that many components like RAMs, flip-flops, shift registers, etc., are not I_{DDQ} testable. Special care must be taken to design such components. We will discuss these techniques in the next chapter.

4

PHYSICAL DEFECTS

The dominant failure modes in VLSI circuits are opens and shorts [187, 188]. Gate-oxide shorts are a special type of shorts and are discussed separately. In addition to these, other less frequent defects are [298]: punch throughs, parasitic transistor leaks, defective *pn*-junctions, incorrect threshold voltages, etc. These defects also lead to abnormally high I_{DDQ} [298].

When test vectors are applied at a slow rate and the logic level of the primary outputs are monitored (**logic testing**), the static logic behavior of the circuit is tested. When test vectors are applied at a high speed they can detect degradation in the circuit's timing behavior (**delay testing** or **high speed testing**). In this case the dynamic behavior of the defective circuit is important. For I_{DDQ} testing we are interested in the effect defects have on the I_{DDQ} level.

In the next three subsections we study the behavior of VLSI circuits in the presence of gate-oxide shorts, shorts, and opens. The impact of these defects on the I_{DDQ} level, the static logic behavior, and the dynamic (timing) behavior of circuits is studied.

In Chapter 3 we saw that device scaling and some common system level circuit design practices tend to cause high fault-free I_{DDQ} levels. Possible techniques to make such circuits I_{DDQ} testable were also discussed. In Sections 4.4, 4.5, and 4.6, we discuss a number of other design issues necessary to make circuits I_{DDQ} testable. In the next two sections we will discuss the I_{DDQ} testability of flip-flops, shift-registers, scan chains and RAMs.

In the next chapter we present characteristics of tests used during I_{DDQ} testing, logic testing and delay testing. This leads us into a discussion of fault models

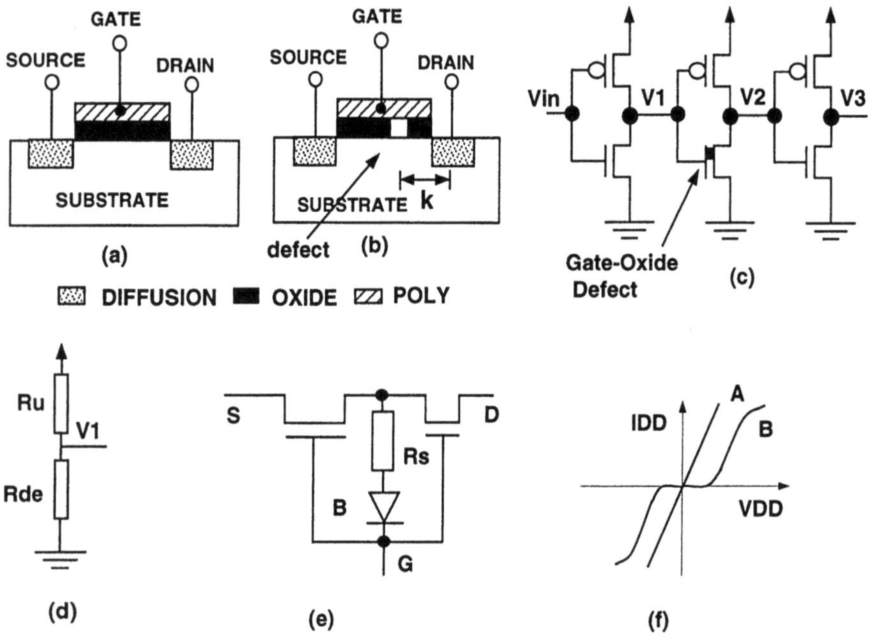

Figure 4.1 Illustration and modelling of gate-oxide shorts

used to compute these tests. We also analyze the defect detection capability of tests so computed. This will put into proper perspective the significant role that I_{DDQ} testing can play in detecting defects.

4.1 GATE-OXIDE SHORTS

Gate-oxide shorts have been recognized for some time to be a dominant mode of CMOS circuit failure [117, 118, 297]. The generic structure of an MOS device is shown in Figure 4.1(a). In a fault-free device there is no conducting path between the gate and the other two terminals of the device. The gate oxide beneath the poly-silicon layer of the gate isolates it from the other terminals.

As shown in Figure 4.1(b), when the oxide layer breaks down it results in a gate-oxide short. Gate-oxide shorts are caused either during or after fabrication. The fabrication related gate-oxide shorts are mainly due to imperfections in the silicon surface and presence of impurities in the oxide [297]. The gate-oxide defects that occur after fabrication are usually due to electrical overstress. Whatever the cause, these defects "magnify" when subjected to electrical stress for a long period of time.

To understand qualitatively the behaviour of CMOS circuits in the presence of gate-oxide defects, we refer to Figure 4.1(c). Assume that the gate-oxide defect is in the *n*MOS transistor of the second inverter and that $V1$ is driven to 1. This creates a conducting path from V_{DD}, through the pullup of the first inverter and the defect to grounnd. This conducting path from V_{DD} to ground results in an abnormally high I_{DDQ}. A "crude model" for this defect, when V_{in} is 0, is shown in Figure 4.1(d). *Ru* is the pullup resistance of the pullup network driving V_1 and *Rde* models the defect. The faulty I_{DDQ} increases with decrease in *Rde*. In addition V_{V1}, the voltage at $V1$, decreases as *Rde* decreases.

If V_{V1} is too low then the logic value at $V1$ is not recognized as 1 and the fault exhibits stuck-at 0 behavior. The probability of this fault being detected as a stuck-at 0 fault increases if $V1$ fans out to other gates. At the same time there is a range of values of *Rde* for which I_{DDQ} is elevated to a detectable level. So, in principle both I_{DDQ} testing and logic testing can detect such defects. To understand which of the two testing methods is better at detecting such defects, we abandon the "crude model" of Figure 4.1(d) and look at experimental [117, 298] and simulation studies [58, 266, 286, 288, 307].

As shown in Figure 4.1(e), to correctly model gate-oxide defects the defective transistor is represented by a pair of minor parasitic transistors, a rectifier barrier B and the effective resistance Rs of the defect [288]. The characteristics of the two defect parasitic transistors depend on the location of the defect since that determines the channel length of the minor transistors. As shown in Figure 4.1(b), k is the ratio of the distance of the defect from the drain to the total channel length.

The behavior of the three inverter chain of Figure 4.1(c) was simulated using SPICE assuming the above model. Figure 4.2 [288] shows how I_{DDQ} varies with the two parameters k and Rs. Note that I_{DDQ} increases with k and decreases with Rs. The shaded region in the chart denotes values of k and Rs for which logic testing detects the defect. If we assume that the I_{DDQ} threshold is 100 μA, defects corresponding to the shaded region, as well as the rest of the chart, are detected by I_{DDQ} testing. Plots of the voltage at the outputs of the three inverters are shown in Figure 4.3 for two different values of Rs. For a low value of Rs (50Ω) the output of the third inverter (V_3) behaves like a stuck-at 0 and

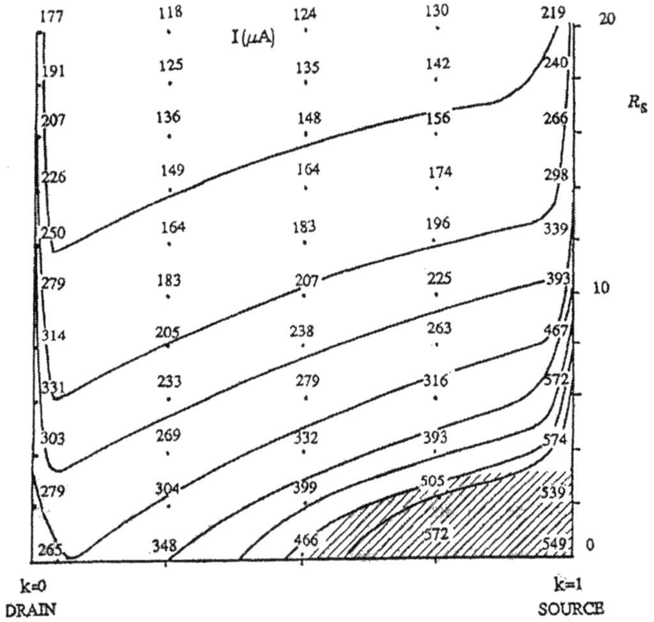

Figure 4.2 Logic and I_{DDQ} testing zones as functions of k, Rs

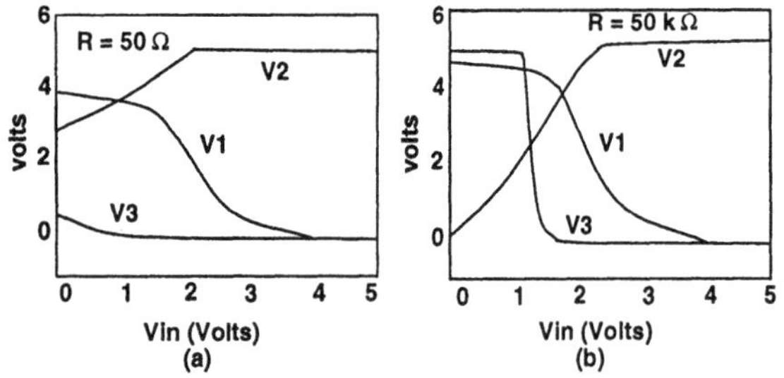

Figure 4.3 Logic behavior of the inverter chain

therefore can be detected by logic testing. When $Rs = 50\mathrm{k}\Omega$ the waveforms are degraded but the defect cannot be detected by logic testing. This clearly shows that a considerably larger variety of gate-oxide defects, compared to logic testing, are detected by I_{DDQ} testing.

The above simulation results confirm experimental findings [117]. Several defective three inverter chains, with intentionally created gate oxide defects, were fabricated. The input output voltage waveforms were observed to be similar to those in Figure 4.3(b). This implies that logic testing would not detect the defects. The nature of the I_{DDQ} plots obtained are similar to those in Figure 4.1(f). Two slightly different curves, marked A and B, are shown. A is a result of defects that show purely resistive effects. B is a result of defects that also show some non-linear effects. Both show a large increase in I_{DDQ} with increase in V_{DD}. In addition, sixty defective SA3000 (Sandia National Laboratories version of Intel's 8085 chip) with gate-oxide shorts were tested. All of them had I_{DDQ} above 100 μA but only one failed the functional test run at a number of speeds. These studies point to one fact: I_{DDQ} is better than logic testing at detecting gate-oxide shorts because it detects a wider variety of gate-oxide defects while covering all gate-oxide defects detected by logic testing.

A point often raised in this context is: if many gate-oxide defects do not exhibit logic failures, despite degraded electrical behavior, it is not advisable to reject ICs with such defects. Note however that many of these "weak ICs" degrade very rapidly, because gate-oxide shorts "degrade" with time when subjected to electrical stress, and ultimately cause logic errors. This was confirmed by experiments [297] where only one of the sixty SA3000 ICs having gate-oxide shorts failed functional tests. However, once this sample was subjected to burn-in (11V, 150C, 1 week) 15 additional ICs failed functional testing. This shows that gate-oxide shorts, not detected by logic testing, result in infant mortality and are a major reliability concern. This also explains the reliability data presented in Chapter 2.

Gate-oxide shorts, like other shorts, have an impact on the dynamic behavior of the circuit. We address this issue in the next section where other kinds of shorts are discussed.

4.2 SHORTS

Shorts, also referred to as bridges, occur when two or more electrically distinct nodes of the circuit get connected due to a defect. Some shorts occur when dust or extra material is deposited during fabrication. The result is that two distinct nets on the same layer get connected as shown in Figure 4.4(a). It can also occur due to imperfections in the mask, imperfections in the etching process, etc.

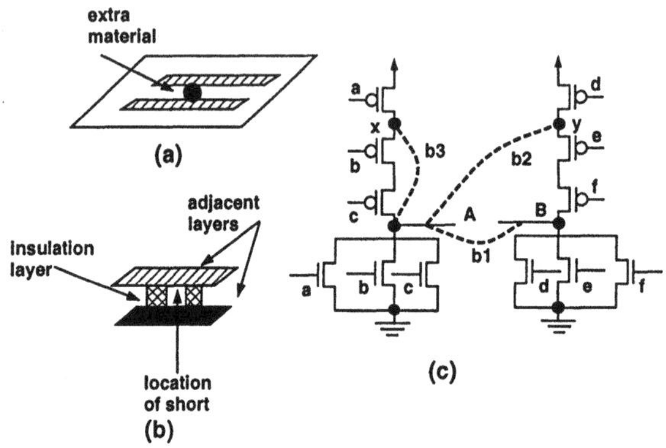

Figure 4.4 Illustration of different kinds of shorts

Circuit nodes having patterns on two adjacent layers can also get shorted. This happens, as shown in Figure 4.4(b), when the insulating layer between two adjacent layers break down or is not formed properly. This category of shorts deteriorates, when electrically stressed, over a period of time. This leads to reliability problems as discussed in the previous section.

We next define some terms. A node which is either an input or an output of a cell is an **external node** of the cell. In Figure 4.4(c), a, b and c are the external nodes of gate A. Other nodes of the cell, such as x of gate A in Figure 4.4(c), are **internal nodes**. Shorts between nodes of two different CMOS cells are **inter-cell shorts**. Inter-cell shorts can occur between the external nodes of two cells ($b1$ in Figure 4.4(c)) and are known as **external shorts**. Inter-cell shorts can also involve internal nodes of a gate ($b2$ in Figure 4.4(c)). Shorts involving only nodes of one cell are **intra-cell shorts** ($b3$ in Figure 4.4(c)). If the bridge resistance is small then it is a **hard short** else it is a **resistive short**. External bridges and intra-gate shorts are discussed in the next two subsections.

4.2.1 External Shorts

To understand how external shorts lead to high I_{DDQ} refer to Figure 4.5(a) [298]. Assume a hard short between f and g. Let $T1 = \langle a = b = c = 0, d = e = 1\rangle$ be applied to the circuit of Figure 4.5(a). As shown in Figure 4.5(b), a path from V_{DD} to ground is created. Thus there is a high I_{DDQ} on application of

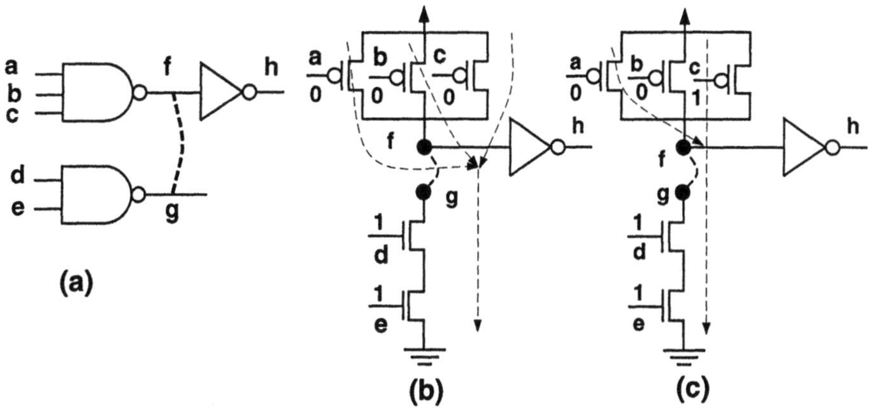

Figure 4.5 Detection mechanism for external shorts

$T1$. This is the primary source of high I_{DDQ} where the I_{DDQ} goes through the defect.

There is also a secondary source of I_{DDQ}. The activated path from V_{DD} to ground on application of $T2 = \langle a = b = 0, c = d = e = 1\rangle$ is shown in Figure 4.5(c). The pullup resistance in Figure 4.5(c) is larger than the pullup resistance in Figure 4.5(b) and the pulldown resistances are the same. The measured I_{DDQ} for $T1$ was 509 μA whereas for $T2$ it was 537 μA. We cannot explain this difference if we assume that a large I_{DDQ} goes only through the defect. The inverter also contributes to the total I_{DDQ}. The threshold for the nFETs (pFETs) used were 1.08 V(-1.3 V). On application of $T1$ ($T2$) V_f was 3.79 (2.83). Thus, for $T2$ both the pullup and pulldown of the inverter are on thereby contributing to the total I_{DDQ}. For $T1$ this effect is not as pronounced because the pFET was just below the threshold.

We present two simulation studies on the effect of such defects on the I_{DDQ} and the static and dynamic logic behavior of circuits. First consider the short between $V1$ and V_{DD} in the inverter chain of Figure 4.1(c) [287, 288]. The last two inverters were of the same size. The ratio of the channel width of the pFETs (nFETs) of the first and second inverters was K. The plot of I_{DDQ} as a function of K and R_b, the short resistance, is shown in Figure 4.6. For small values of R_b the defect behaves as $V1$ s-a-1 and also results in high I_{DDQ}. For larger values of R_b, the circuit does not show abnormal static logic behavior although I_{DDQ} is still abnormally high. This result point out that a

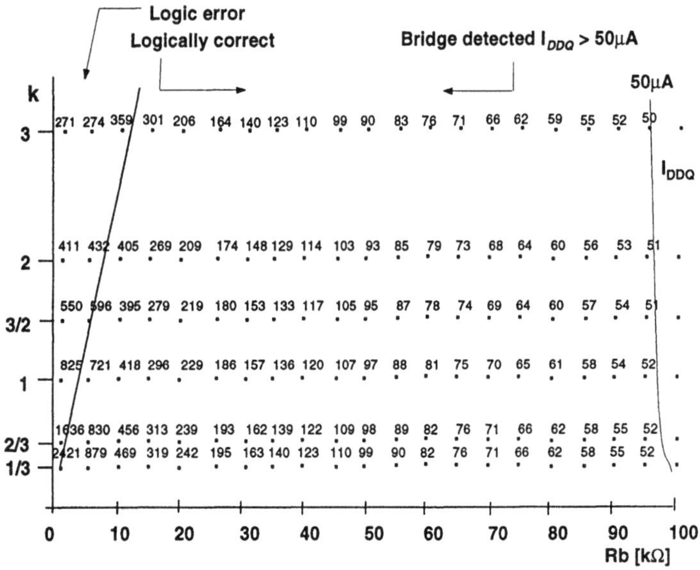

Figure 4.6 I_{DDQ} dependence on k and short resistance

considerably larger variety of external shorts, compared with logic testing, is detected by I_{DDQ} testing.

The second study simulated the behavior of the 2-NAND gate of Figure 4.7(a) for a short between its two inputs X and Y [321]. If $UV = 01$ and pulldown strength of N2 is larger than the pullup strength of P1 (n-dominant) then the static behavior is not affected. On the other hand if the pulldown strength of N2 is smaller than the pullup strength of P1 (p-dominant), for some values of the short resistance, the defective circuit exhibits a Y s-a-1 behavior.

If a node of a gate is slow-to-rise (slow-to-fall) then we have a **slow-to-rise (slow-to-fall) transition fault.** To understand how external bridges can lead to transition faults consider the short between $V1$ and V_{DD} in Figure 4.1(c). Assume that initially V_{in} is set to 0 which is followed by setting V_{in} to 1. This charges $V1$ and then attempts to discharge it. But, because of the short from $V1$ to V_{DD}, this discharge is delayed and leads to a slow-to-fall transition fault. If $V1$ fans out then this is equivalent to transition faults at inputs of more than one gate (**multiple transition fault**).

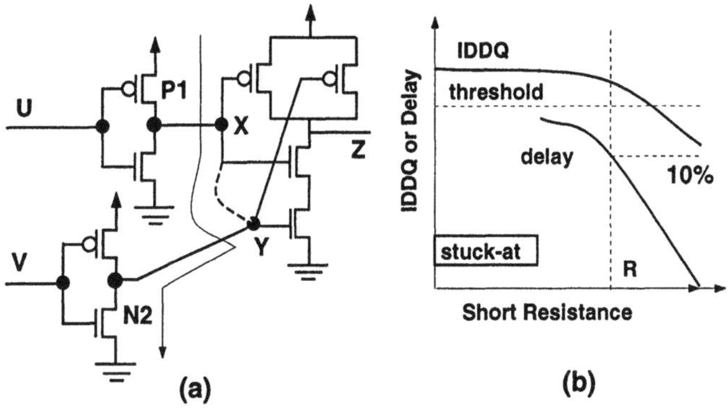

Figure 4.7 I_{DDQ} and delay variation versus short resistance

For another example refer to Figure 4.7(a). Assume that the input combination $UV = 11$ is applied followed by $UV = 01$. This attempts to create a rising transition at X. However, because of the short there is a conducting path from X to ground. This creates a slow-to-rise transition fault at X. The presence of this short can also lead to a slow-to-fall transition fault at Y. This transition fault is activated if the input vector $UV = 00$ is applied followed by the input vector $UV = 01$. The activated path from Y to V_{DD} through the short delays the fall of Y. If X or Y fans out this defect manifests itself as a multiple transition fault. It also maps onto different transition faults depending on the way the defect is activated.

In this example, the transition delays are either at the input or output of a gate involved in the defect. There are mechanisms that affect the delay through gates not involving the defect. We will see examples of such "spilling over" in the context of intra-gate shorts. They also occur in the presence of external bridges.

To get some sense of the usefulness of using I_{DDQ} testing versus delay testing for such defects we look at simulation results [321]. The nature of the plot of I_{DDQ} and delay versus the short resistance, for the circuit in Figure 4.7(a), assuming p-dominance, is shown in Figure 4.7(b). Observe that there are resistance values for which the delay increase is less than 10% but I_{DDQ} remains above the threshold. This happens around 10 kΩ. However, shorts having resistances in the range of 100 kΩ can also have elevated I_{DDQ} but have no perceptable effect on delay. Similar results were obtained for complex CMOS gates.

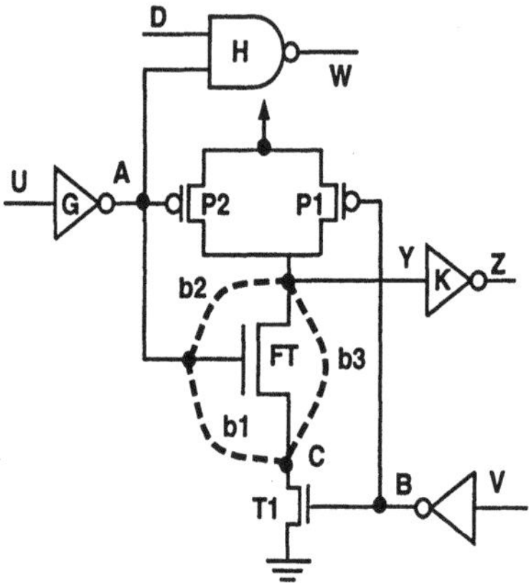

Figure 4.8 Illustration of intra-gate shorts

In conclusion we note that I_{DDQ} testing, logic testing and delay testing can detect external shorts. However, I_{DDQ} testing can detect external shorts for a wider variation of design parameters and a wider range of short resistances. It was observed [6] that external shorts either have very low resistance which either is of the order of 500 Ω or is above 50 kΩ. Logic and delay testing can only detect the 500 Ω short and not the 50 kΩs short. I_{DDQ} detects both of them.

4.2.2 Intra-Gate Shorts

The intra-gate shorts that we discuss exist between the three terminals of an individual transistor. Thus we have gate-source (GS), gate-drain (GD) and source-drain (SD) shorts for each transistor. Shorts labeled $b1, b2$ and $b3$ in Figure 4.8 are examples of, respectively, GS, GD and SD shorts. Note that GS and GD are different from gate-oxide shorts. When we consider GS or GD, unlike a gate-oxide short, we assume that the transistor is fully functional and equivalent to the fault-free transistor. This, as shown in Figure 4.1(e), is not the case with gate-oxide shorts. Our discussion here follows [115]. The three example shorts of Figure 4.8 are studied in more details to understand the effect of such defects on the circuit performance.

Gate-source short $b1$**.** For $b1$ application of $UV = 00$ sets C to 0. This raises I_{DDQ} to an abnormal level because of a conducting path from V_{DD} to ground, through the pullup of G, the defect and $T1$. This is the primary source of the elevated I_{DDQ}. In addition, there is a **secondary source** of high I_{DDQ}. Depending on the bridge resistance, the voltage at A may be at an intermediate level. Thus, the transistor pairs driven by A in the gate with output Y and, if D is set to 1 then, gate H conduct weakly thereby contributing to the total I_{DDQ}. Thus, such shorts can be detected by I_{DDQ} testing.

Such defects also affect the static and dynamic logic behavior of the circuit. Assume that the *short resistance is very small.* Then V_{AC} is very small (below the cut-off voltage for FT) and FT does not conduct. The output Y of the gate cannot discharge and FT is said to be **stuck-open**. So tests for stuck-open faults can detect such defects. If the pullup strength of gate G is small and the pulldown strength of transistor $T1$ is large then, for some input vectors, the defect manifests itself as A stuck-at 0.

If the pullup strength of gate G is comparable to the pulldown strength of $T1$ then node A has an intermediate voltage. This delays a transition propagating from D to W because of the following reason. Both transistors in gate W driven by A are turned on very weakly. Assume that the input vector $DUV = 000$ is applied followed by the input vector $DUV = 100$. This rising transition at D results in a slow-to-fall transition at W because of the weakly conducting transistor in gate W driven by A. This example is interesting because it shows that the effect of the defect can "spill over" onto a gate in which the defect is not located.

As the short resistance increases, V_{AC} can vary over a larger range. Therefore, instead of being cut-off FT conducts weakly. This leads to a slow-to-fall transition fault at Y when UV is changed from 10 to 00. Since A remains at an intermediate level, as explained above, a slow-to-fall transition fault could also appear at W.

Gate-drain short. Consider $b2$ in Figure 4.8. Such defects lead to high I_{DDQ}. If $UV = 11$ is applied then there is a conducting path from V_{DD} to ground through $P1, b2$ and the pulldown of G. This is the primary source of I_{DDQ}. As in the case of $b1$ secondary source of I_{DDQ} is also possible because A has a degraded voltage level.

This defect also affects the static and dynamic logic behavior of the circuit. First assume that the *short resistance is small.* If $UV = 00$ is applied then, depending on the relative strengths of the pullup of A and the pulldown of $T1$, one of two cases occurs: (i) $V_A = V_Y$ is well below V_{DD} but not recognized as 0; or (ii) $V_A = V_Y$ is above ground but not recognized as 1. Case (i) may be recognized as Y s-a-1 and case (ii) may be recognized as A s-a-0. The weakened levels at A manifest itself, as explained for $b1$, as transition faults at W. In addition, for some transitions, like UV from 10 to 11, Y may also experience a slow-to-fall transition. Similar analysis can be carried out when the short resistance is increased.

Source-drain shorts. Refer to defect $b3$ in Figure 4.8. Such defects can lead to high I_{DDQ}. The input assignment $UV = 10$ causes a path from V_{DD} to ground via $P2, b3$ and $T1$, thereby leading to a high I_{DDQ}.

This defect can also affect the static and dynamic behavior of circuits. The output Y of the gate may be at an intermediate level for some input combination and a range of values of the short resistance. This results in propagation delays through gates driven by Y. Depending on the resistance of the short and the pulldown strength of $T1$, the defect may manifest itself as Y stuck-at 0 for some input vectors.

Thus we see that an intra-gate short, in addition to producing abnormal I_{DDQ}, depending on the strength of other transistors and the short resistance can manifest itself as a stuck-at, stuck-open or transition fault. An important point to note is that the propagation delay effects "spill over" onto other gates. The delay effects were observed only for small short resistance of below 1 kΩ [115]. Abnormal I_{DDQ} can be observed for a much larger range of values of the short resistance. This adds to the versatility of I_{DDQ} testing in detecting such defects.

4.3 OPENS

Open circuit defects, or *opens* as we refer to them, are unintentional discontinuities. Due to defects a node splits up into two or more distinct nodes. Such defects are caused by missing metal, poly or diffusion. This could happen due to improper etching, masking error, electro-migration, etc. There are two types of opens - **narrow** and **large**.

Large opens. No current can go between the two ends of the open when a voltage is applied across it. To verify this an inverter with a large open at its input was created [123]. It resulted in the gates of the two transistors as connected together and floating. When the input was not driven and power was applied to the inverter the output was near 5V. Thus the voltage at the gate of the floating transistors was below the threshold of the n-transistor (0.5V). Next the input of the inverter was driven progressively to higher voltages in steps of 1V. Even after a delay of more than 50 seconds the output of the inverter remained high indicating that there was no increase in voltage at the floating gate. Similar results were also reported in [190].

They show that no leakage current goes through the rather large open created in the experiment.

Narrow opens. The opening is usually < 100 nm. In this case a small leakage current, due to tunneling effect, does go across the open [123]. Once again an inverter with a narrow open at the input was created. The leakage current through the open, when the circuit was powered by 5V and input set to 5V, was measured to be about 894 pA. If the circuit was driven at 100 Hz it functioned properly. At 100 kHz it did not respond to the input waveform properly. This was because in the former case there was sufficient time to discharge and charge the floating gate due to the leakage current across the open. Note that the current is small but is significant enough to cause a difference between the performances of the circuit in the presence of these two types of opens.

The impact of open defects on the performance of a circuit is a function of the location and size of the defect, as well as the electrical parameters of the circuits. In our discussion we use the opens marked $d1 \ldots d9$ in Figure 4.9(a). Defects $d1, d2, d7, d8$ are respectively **open-gate** defects in transistors $P1, P2, P3, N3$. We refer to $d4(d3)$ as **open-$n$net (open-$p$net)** defect. Defect $d5$ is referred to as **open-output** defect. Defect $d6(d9)$ is an **open-source** defect for $N1(N3)$. Defect $N2$ is also an **open-drain** defect for $N2$. Each of these defects has a different impact on the circuit performance.

Open-source: We consider two cases.

- The **first case** includes opens in a chain of transistors that is the only path to ground or V_{DD} from the output (defect class 3 [295]). Open $d9$ in Figure 4.9(a) is an example. If this *open is large* then there is no discharge path from the output H to ground. If we apply the sequence of input vectors that sets GF to 00 followed by 11 then the following events occur. H is charged to 1 by 00. On applying 11 H cannot discharge and remains at 1. Thus, this kind of defect manifests itself as a **stuck-open fault** [323] as was first predicted in [190]. We will discuss stuck-open faults in more detail in Chapter 5.

  If the *open is narrow* then the application of the same two input vectors results in H being charged to 1 followed by a slow decay of V_H, the voltage at H. There will therefore be two kinds of effects. Firstly, it will manifest itself as a slow-to-fall transition fault. Secondly, as shown in Figure 4.9(b), there is a time interval $t1 - t2$ during which V_H turns on both $P4$ and $N4$ of the load inverter resulting in a high

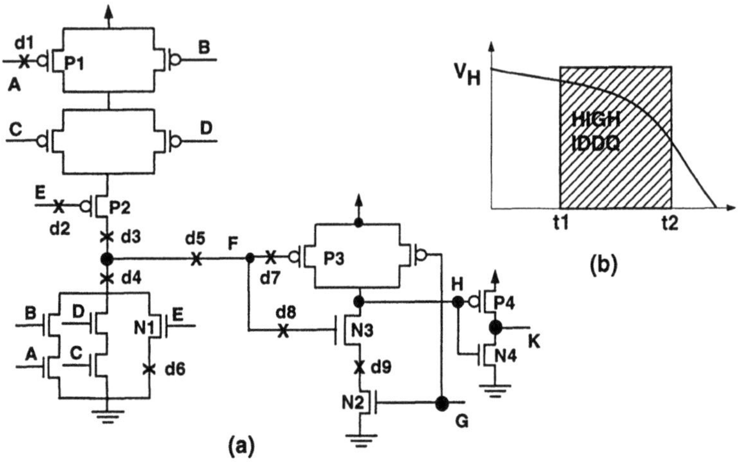

Figure 4.9 Illustration of opens

I_{DDQ}. The probability of such defects being detected by I_{DDQ} testing is rather small.

Results from two experimental chips show that all defect class 3 opens in two experimental chips were detected by logic testing but failed to be detected by high speed I_{DDQ} testing [295]. The opens created were rather large.

- The **second case** includes opens in a chain of transistors that is not the only path from the output to V_{DD} or ground (defect class 4 [295]). Open $d6$ in Figure 4.9(a) is an example. Note that like open $d9$, for large open, this defect behaves as a stuck-open fault. For a narrow open it may manifest itself as a delay defect. An input pair that activates this defect is $ABCDE$ = 10100 followed by $ABCDE$ = 10011. Experiment has shown that all 12 such opens in a fabricated chip were detected by logic testing [295]. Interestingly, unlike the first case, all these opens were detected by I_{DDQ} testing also [295]. This is explained as follows.

  Assume that the circuit of Figure 4.9(a) is embedded in a larger circuit and input $T2$ is applied after $T1$ such that: $T1$ sets $ABCDEG$ to 101001; and $T2$ sets $ABCDEG$ to 100111. Transitions at primary inputs, because of different delay paths they traverse, may result in glitches at the inputs to the subcircuit in Figure 4.9(a). Note that $ABCDE$ = 10100 i.e. $T1$ sets F to 1. Before reaching the final state

$ABCDE = 10011$ if the input state, due to glitches, is $ABCDE = 10111$ for a short duration then the output F gets partially discharged. This reduces the voltage at F and turns on both $P3, N3$ which results in a high I_{DDQ}. This hypothesis was verified experimentally [295].

Open-drain: Open-drains are very similar to open-sources and the above discussion about open-source apply.

Open-*n*net: Open *d*4 in Figure 4.9(a) is an example of such an open. Such faults behave similar to open-source/drain defects. Note that if the open is large there is no discharge path from F to ground and can be detected as a stuck-open fault [190]. However, if it is a narrow open then the defect manifests itself as a slow-to-fall transition fault. These malfunctions can be activated by setting $ABCDE$ to 00000 followed by setting $ABCDE$ to 00001. The slow-to-fall transition may also result in a high I_{DDQ} if G is also set to 1. However, the likelihood of detecting such defects using I_{DDQ} testing is rather small.

Open-*p*net: The behavior of circuits with open-*p*nets is similar to those of circuits with open-*n*nets [190]. They can be detected as either a stuck-open or a slow-to-rise transition fault. In addition, the probability of detecting such defects using I_{DDQ} testing is rather low.

Open-gate: Open-gate defects result in a transistor whose gate is not driven by either a primary input or the output of a gate. Such transistors are known as **floating-gate transistors**. The behavior of transistors with floating gates have been studied extensively [59, 123, 147, 261, 288]. It was found that floating gates acquired voltages below the cut-off of the *n*-transistor [123]. This implies that floating gate transistors do not conduct. Subsequently, measured gate voltages in the range 0.38 to 2.3V have been reported [147]. This suggests that a floating gate transistor could either be conducting or not conducting at all. Experimental results in [295] conclude that the floating gate transistor is cut-off.

The above results, at a first glance, appear confusing. These results can be explained by the fact that the voltage at a floating gate is a complex function of the "environment" it is in. Detailed analysis of the effect of the "environment" on the floating gate voltage can be found in [59, 261, 288]. We present the main findings.

A floating gate, although it appears to be so, is not isolated from the rest of the circuit. This is because of the capacitive coupling which, for an nMOS transistor, is shown in Figure 4.10(a) where [261]:
V_m, V_g, V_s and V_d are respectively the voltages at the metal line overlapping

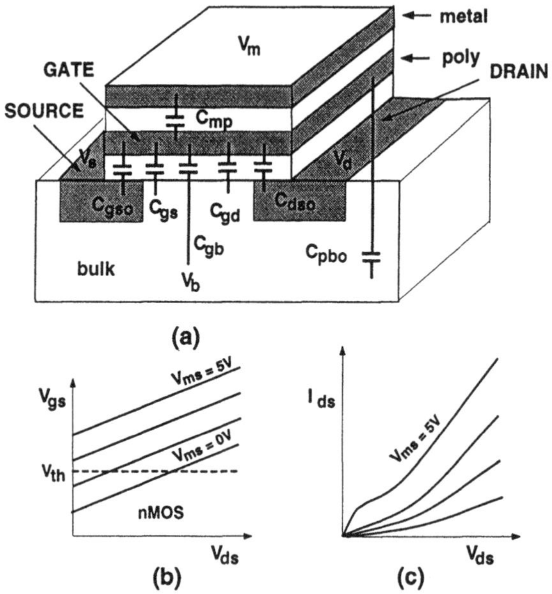

Figure 4.10 Modeling of floating gates

the poly-region, the gate, the source and the drain;
C_{mp} is metal-poly overlap capacitance;
C_{gso} is gate-source overlap capacitance;
C_{gs}, C_{gb} and C_{gd} are gate to source, bulk and drain capacitances; and
C_{pbo} is polysilicon on thick oxide to bulk overlap capacitance.
The gate region has some trapped charge. This could occur due to hot-carrier effects in which some of the electrons traveling from source to drain acquire enough energy to cross over the potential barrier and penetrate the oxide. There are a number of other reasons why charge could be trapped in the gate [59]. Let Q_0 denote this initial charge. If we write the charge conservation equation and simplify we have the following expression:

$$\begin{aligned}(C_{gs} + C_{gb} + C_{gd} + C_{gso} + C_{gdo} + C_{mp} + C_{pbo})V_{gs} &= \\ (C_{gd} + C_{gdo})V_{ds} + C_{mp}V_{ms} + Q_0 & \qquad (4.1)\end{aligned}$$

Equation 4.1 is used to derive expressions for V_{gs}, the gate-source voltage, for the three operating regions of the transistors, namely, cut-off, linear and saturation. The main conclusion is that the $V_{ds} - V_{gs}$ curves are of the form shown in Figure 4.10(b). If V_{ms}, the metal-source voltage, is held constant then V_{gs} grows linearly with V_{ds}. In addition, V_{gs} increases as V_{ms} is increased. Assuming that V_{th} is the threshold voltage for the transistor we note that depending on the voltage at the overlapping metal-layer the transistor can be operating in the cut-off or conducting region. Figure 4.10(c) shows the dependence of I_{ds} on V_{ds}, for a given value of V_{ms}. The magnitude of I_{DDQ} increases both with the increase in V_{ms} and V_{ds}. These results were confirmed by simulation experiments [59].

The above study did not investigate the effect of the location of the defect on the behavior of the floating gate transistor. From Figure 4.11(a) it should be clear that C_{mp} is a function of the metal poly overlap. C_{pb} is a function of the length of the break and hence depends on the location of the break. The I_{DDQ} curves, obtained by simulation, are plotted in the $C_{pb} - C_{mp}$ plane of Figure 4.11(b) [288]. The analytical model to analyze the effect of the location of the defect is discussed in [59]. These results were verified experimentally using the three-inverter chain of Figure 4.1(c). The gate of the n-transistor of the second inverter was assumed to be floating. Defects at different lengths from the gate were created for measurements. VMET is the voltage in the overlapping metal layer. This overlap was also varied in the chips that were fabricated. Note that if the length of the poly-break varies from 0 to 20 μm then C_{pb} varies from 0.69 fF to 6.39 fF [59]. In addition, if the overlap area varies from 0 to 28 μm^2 then C_{mp} varies from 0.97 to about 4.52 fF [59]. This gives the range of the overlap and poly-break lengths used in the experiment 4.10(c).

The plot in Figure 4.11(b) points out a number of important points. Firstly, if the break is close to the gate then C_{pb} is small. In addition, the overlap with the metal layer will be small. In this case the n-transistor operates in the sub-threshold region. To maintain a given level of I_{ds}, as the length of the polybreak is increased, the overlap with the metal layer must also increase. Even if the transistor conducts, it could be conducting rather weakly - with the current being in the tens of μA.

With the above background on the behavior of floating gate transistors we next look at how open-gate defects can be detected. Defect $d8$ of Figure 4.9(a) is an example of a defect such that $N3$ becomes a floating gate transistor. $N3$ is also a transistor on the only conducting path from H to the ground (defect class 2 [295]). If the environment of $N3$ is such that $N3$ is cut-off then this defect manifests itself as a stuck-open fault. Experimental results have confirmed that such defects cannot be detected by I_{DDQ} testing [295]. If $N3$ conducts, even very weakly such that a small I_{ds} of the order of 10 to 20 μA goes through $N3$, and measurements are taken at a realtively slow rate then the initially high value at H (presumably H is set to 1 by a previous vector) decays to an intermediate value. This turns on both transistors in the load inverter resulting in a high I_{DDQ}. Hence such defects can be detected by slow speed I_{DDQ} testing.

The open defect $D7$ which results in $P3$ becoming a floating gate transistor behaves very similarly to $d8$ except that it can be detected by high speed I_{DDQ} testing with a higher probability. The reason is that, if we set H to 0 and then try to pull it up to 1 by setting $FG = 01$ then, because of glitches, the values of GH can temporarily be 00 or 10. This will charge H to an intermediate level and result in the load inverter to draw significant I_{DDQ}. This was observed in the experiments reported in [295].

Since transistors with floating gates can either be cut-off, or conduct strongly, or conduct weakly, such defects can also lead to stuck-open faults and degraded voltage levels. The degraded voltage levels, as we have seen in the case of shorts, can either lead to a change in the functionality or affect the timing behavior. We discuss this in more details in the next chapter.

Open-output: Defects such as $d5$ in Figure 4.9(a) result in the complementary pair of transistors, $P3$ and $N3$, to float. Since the defect lengths are generally large and the overlap with metal is also high such transistor pairs conduct. Thus, they are readily detected by I_{DDQ} testing [295]. For open $d5$ of Figure 4.9(a) G has to be set to 1 for detection using I_{DDQ} measurement.

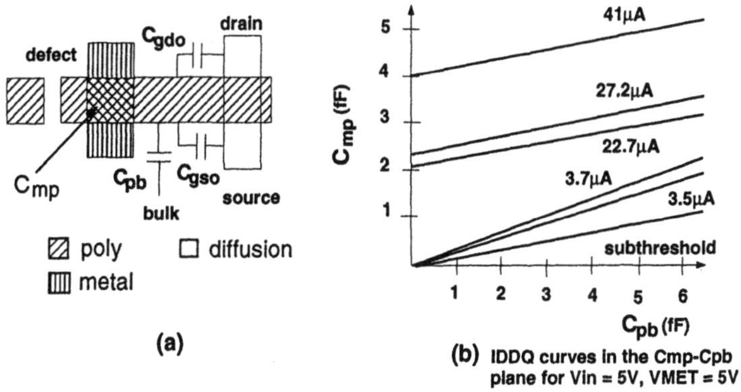

Figure 4.11 Dependence of I_{DDQ} on location of open-gate defect

4.4 ANOMALIES IN DETECTING BRIDGES USING I_{DDQ}

The discussion on detecting bridges using I_{DDQ} testing leads to the following conclusions:

Property 1. An I_{DDQ} test for a bridge $\langle x, y \rangle$ detects the bridge if it sets the two nodes to complementary values.
Property 2. In combinational circuits, if a bridge is I_{DDQ} detectable then there exists a single input vector that detects the bridge.

The next two properties of I_{DDQ} are usually assumed to be true.

Property 3. A test for a single bridge is not masked by multiple bridges.
Property 4. High I_{DDQ} signifies the presence of a defect.

Several examples [167] show that there are exceptions to each of these properties. We present design and test rules that guarantee the above properties [167].

4.4.1 Examples of anomalous behavior

Counterexamples for properties 1-4 are presented below.

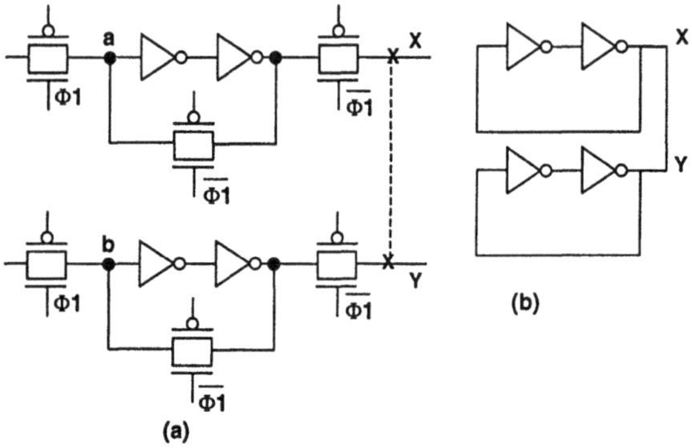

Figure 4.12 Bridges that cannot be detected by setting the two nodes to complementary values.

Property 1. Refer to Figure 4.12(a) [167]. In the absence of the bridge, x and y can be set to complementary values by setting nodes a and b to complementary values during ϕ_1 and propagating these values to x and y during $\overline{\phi_1}$. Figure 4.12(b) shows the equivalent circuit during $\overline{\phi_1}$. Note that there are no external sources driving the two closed loops that are shorted together. If the bridge resistance is very small then, depending on the device parameters, the voltages at x and y will settle to one of the two stable values V_{DD} or ground. Thus, I_{DDQ} testing cannot detect it. Note however that logic testing can potentially detect this fault since one of the two nodes settles to the incorrect value.

Property 2. Some bridges in combinational circuits cannot be detected by a single vector but can be detected by a sequence of vectors. Refer to Figure 4.13 [167]. Ignore bridge $\langle w, z \rangle$ and assume that only bridge $\langle x, y \rangle$ exists. Vectors that set z and A to 0 detect this bridge. However, both z and A cannot be set to 0 simultaneously and therefore the bridge $\langle x, y \rangle$ cannot be detected by a single I_{DDQ} test. It can however be detected by a sequence of two tests. The first vector sets z to 0 and the second vector leaves z floating. If we assume that z retains the value 0 then by setting A to 0 the bridge can be detected. A pair of vectors that can detect $\langle x, y \rangle$ is $ABC = 111$ followed by $ABC = 010$.

Property 3. The presence of multiple bridging faults (BFs) can invalidate I_{DDQ} tests. Refer to Figure 4.13 [167]. Assume that both bridges $\langle x, y \rangle$ and $\langle w, z \rangle$ exist and $ABC = 111$ is applied. This sets z to 0. However,

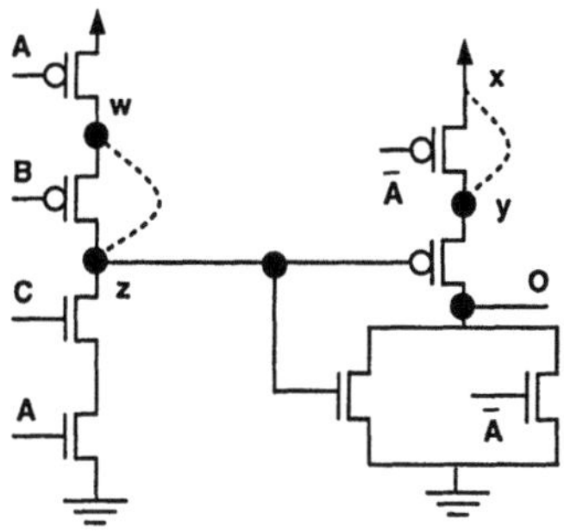

Figure 4.13 Multiple vector tests and fault masking.

when $ABC = 010$ is applied, because of the presence of $\langle w, z \rangle$, z is pulled up to 1. Thus, no path is created from V_{DD} to ground through $\langle x, y \rangle$. Note that since w and z cannot be set to complementary values a path from V_{DD} to ground cannot be set up through the bridge $\langle w, z \rangle$. Thus, this multiple fault cannot be detected by I_{DDQ} although, as we discussed before, the single bridge $\langle x, y \rangle$ is I_{DDQ} testable. Examples of fault masking in sequential circuits are known [167].

Property 4. I_{DDQ} testing sometimes identifies a circuit to be faulty even if it is fault-free. Refer to Figure 4.14. The MUX output is set to O_1 (O_2) if Sel is set to 1(0). When $ABC = 100$ nodes w and z are floating resulting in charge sharing between nodes w and z (since $B = 0$). If vector $ABC = 011$ is applied prior to $ABC = 100$ then w would have been set to 1 and z to 0. Thus V_z, the voltage at z, on application of $ABC = 011$ followed by $ABC = 100$ will be at an intermediate value (above the threshold for N but below the cutoff for P). This creates a conducting path from V_{DD} to ground through P and N, thereby resulting in a high I_{DDQ}.

If we design the circuit such that Sel is 0 whenever $ABC = 100$ then the circuit will function correctly, even though I_{DDQ} is high in the absence of any defect.

4.4.2 Design rules to avoid anomalies

We next present a set of design rules to guarantee that none of the anomalies discussed above occurs. To understand them we need a few definitions.

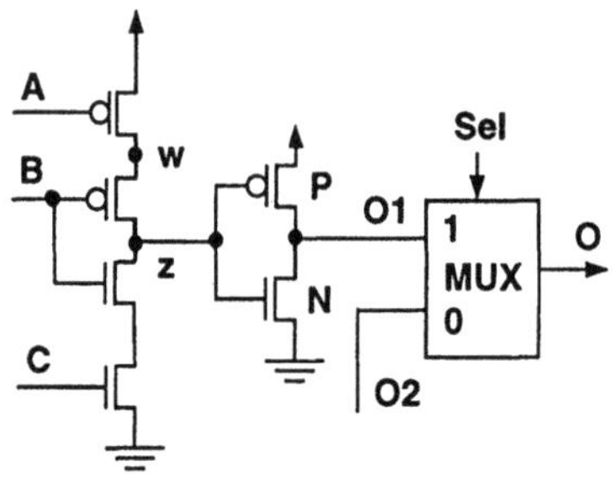

Figure 4.14 High I_{DDQ} in the absence of a defect

A MOS circuit is mapped into an undirected graph whose vertices are nodes in the circuit and edges correspond to transistor channels [33]. The V_{DD} and ground nodes are replicated so that each transistor connected to V_{DD} or ground has its own copy. The connected components of the graph forms "channel connected" components of the circuit. By merging the V_{DD} and ground nodes of all transistors of a connected component we get a **transistor group** (TG). In Figure 4.15(a) the dotted polygons demarcate transistor groups.

A TG TG_1 is said to have a **direct control** over TG TG_2 if an output of TG_1 is an input of TG_2. This relation can be represented by a graph. The graph for the TGs of Figure 4.15(a) is shown in Figure 4.15(b). The nodes in this graph represent TGs and the direct control relationship is represented by the directed edges. A TG TG_1 is said to be **related** to a TG TG_2 if and only if there exists a path either from TG_1 to TG_2 or vice-versa. Else, the two TGs are **unrelated**. In Figure 4.15(b), TG G_6 is unrelated to G_4 and G_3 is related to G_5. If there is cycle in the graph then there are pairs of TGs that can control each other and form part of a **control loop**. Next we state six design rules [167].

Rule 1. The gate and drain (or source) nodes of a transistor cannot be in the same TG.

Rule 2. During steady state operation each output of a TG must be connected either to V_{DD} or ground. In essence, no node driving other gates should be floating. This precludes the use of charge sharing or charge retention effects for correct circuit operation. Note that all our examples used circuits that violate this property.

Rule 3. There are no control loops among TGs.

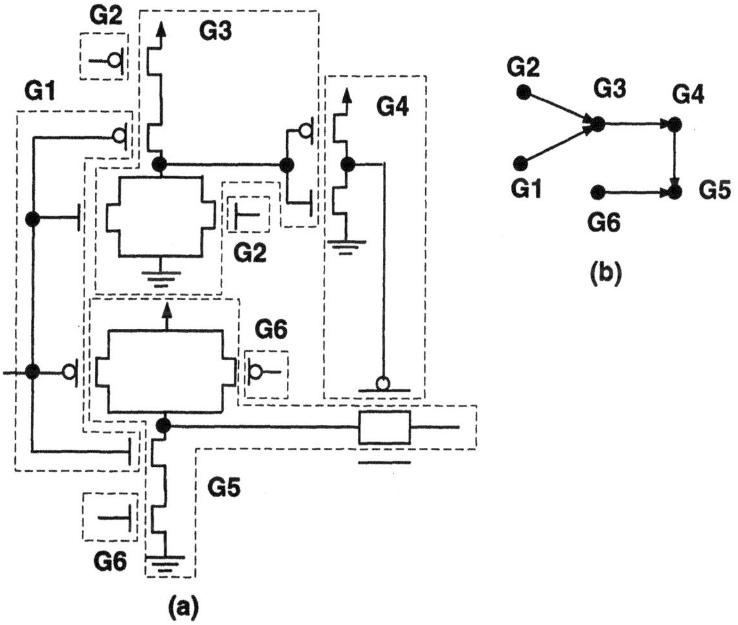

Figure 4.15 TGs of a CMOS circuit.

Rule 4. During testing, each primary input is controlled by a strong power source whose current is also monitored.

Rule 5. During steady state operation there should be no conducting path from V_{DD} to ground.

Rule 6. The bulk (or well) of an n-type (p-type) transistor is connected to ground (V_{DD}).

This set of rules is minimal and sufficient, but not necessary [167].

4.5 I_{DDQ} DFT FOR FLIP-FLOPS AND SCAN-CHAINS

Flip-flops are an integral part of all digital circuits. Unfortunately, implementation of this important component does not necessarily conform to the design rules of Section 4.4. In the next three subsections we will see: when and why bridges involving nodes in flip-flops cannot be detected by I_{DDQ} testing; alternative flip-flop designs that make them I_{DDQ} testable; and an I_{DDQ} design for testability (DFT) technique for scan chains.

4.5.1 I_{DDQ} undetectable bridges in sequential circuits

To understand why some bridges involving flip-flop nodes are not detected by I_{DDQ} testing, refer to Figure 4.16(a). It shows a standard master-slave D flip-flop design. CL and CLB are, respectively, the clock and complement of the clock signals, D is the data input and Q and QB are, respectively, the output and the complemented output of the flip-flop.

This flip-flop functions as follows. When CL = 1, TG1, TG4 are on, the current value of D sets the state of nodes 1, 2 and 3, and the floating control loop consisting of nodes 4, 5 and 6 retains the old state of the flip-flop. This is depicted in Figure 4.16(b). A control loop is said to be in a **floating state** if it is memorizing the previous value forced by its driver [167]. Else, it is in a **forced state**. As can be seen from Figure 4.16(b), when CL = 1, the slave forms a floating control loop. When CL = 0, TG2, TG3 conducts. This cuts-off D from driving nodes 1, 2 and 3 forming a feedback loop and retaining the value

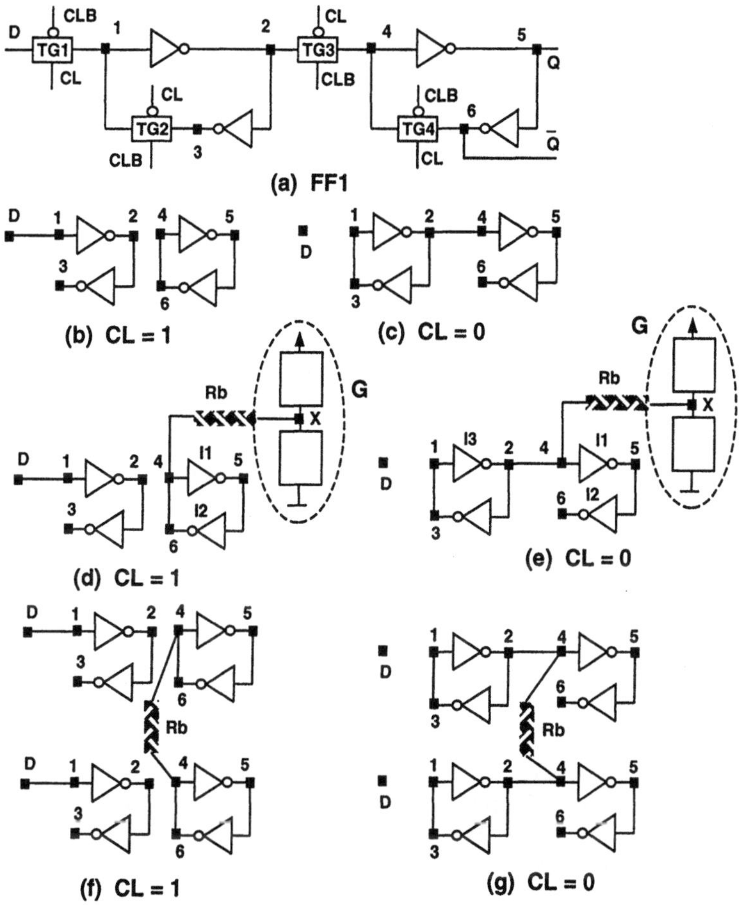

Figure 4.16 Difficulty of detecting bridges involving a master-slave flip-flop using I_{DDQ} testing

written into it. During this phase, node 2 drives node 4 and modifies the value of nodes 4, 5 and 6. Thus, data is transferred from the master to the slave. From Figure 4.16(c) we note that when CL = 0, the master forms a floating control loop.

The presence of floating control loops in sequential circuits causes difficulty in detecting bridges [264, 273]. There are two such classes of bridges. We discuss each of them.

- Only one bridged node is in a floating control loop.

  Example 1: Figures 4.16(d) and (e) show an example in which a node X in the combinational part of the circuit is bridged to node 4 of a master-slave flip-flop. Let R_b be the bridge resistance. We have two subcases:

  - R_b is small (i.e. the bridge is a hard short).
    When CL = 1, the flip-flop configuration, along with the bridge defect, is shown in Figure 4.16(d). Because of the presence of the floating control loop, nodes X and 4 always have the same logic value even if the current input vector sets them to different values in the fault-free case. The value attained is a function of the relative driving strengths of inverter $I2$ and gate G and is determined by the dominating gate. Hence, abnormal I_{DDQ} is not observed. However, since either X or node 4 (i.e, Q) changes from its fault-free value, this defect can potentially be detected as a stuck-at fault. The change in the value of Q due to a bridge is referred to as a **state change behavior**. Simulation studies confirm this hypothesis [264, 273]
    If $CL = 0$ then the flip-flop configuration, along with the defect, will be as shown in Figure 4.16(e). Even in this case node 4 forms part of a floating control loop. Therefore, as discussed above, abnormal I_{DDQ} will not be observed but the defect can be potentialy detected as a stuck-at fault. Thus, independent of when the I_{DDQ} measurements are made (during CL or during CLB), this bridge cannot be detected using I_{DDQ} testing.
  - R_b is reasonably large (i.e., the bridge is a resistive bridge). In this case, in both Figures 4.16(d) and (e), a bridge resistance nullifies the dominance of the gate with a higher driving strength. There is a voltage drop across the defect and a path is created from V_{DD} to ground through gates G and $I3$. Therefore, a steady state current flows through the defect making it I_{DDQ} testable. This hypothesis has been confirmed using simulation [264]. Thus, this defect can be detected by I_{DDQ} by setting the two nodes at different values in any of the two clock phases (CL or CLB).

- The two bridged nodes are in two different floating control loops.

  Example 2: An example of such a defect is shown in Figures 4.16(f) and (g). They show the state of the flip-flops in the presence of the bridge between node 4 of two master-slave flip-flops when CL = 1 and CL = 0, respectively. We have two subcases:

  - R_b is small. Then, it can be argued that, independent of the value of CL, even if the current input vector sets the two nodes to two different values in the fault-free case, the two nodes settle at the logic value of the dominating gate. Thus, this defect is not I_{DDQ} testable but can potentially be detected using logic testing.

- R_b is sufficiently large. The dominance of one gate over the other is nullified. Thus, there is a voltage drop across the defect. A path is created from V_{DD} to ground thereby making the defect I_{DDQ} testable. Simulation studies confirm this conjecture [264].

Thus, independent of when I_{DDQ} measurements are made (during CL or during CLB) this bridge cannot be detected using I_{DDQ} testing.

Note that the detectability or undetectability of the above classes of bridges is also a function of the driving strengths of the two inverters that form the closed loop. Simulation results studying this dependency can be found in [264].

Finally, we note that if both the bridged nodes lie in the same floating control loop and there is a state in which all the nodes of the loop are driven, then the fault can be detected by I_{DDQ} testing upon setting the two nodes to different values [264].

Simulation studies, for different kinds of flip-flops, have been published [342, 265]. In [265] bridging faults, from a master-slave flip-flop of Figure 4.16(a), were extracted using inductive fault analysis. These bridges were simulated using HSPICE. It was seen that for zero resistance bridges, 92% of extracted bridges were I_{DDQ} testable. The remaning 8% were detected by logic testing, at the output of the flip-flop. However, when the bridge resistance reached about 2KΩ all bridges were I_{DDQ} testable. The above discussion explains these findings. Thus, both I_{DDQ} testing and logic testing are required to detect bridges in sequential circuits.

4.5.2 I_{DDQ} testable flip-flop design

We have seen in Section 4.5.1 that some bridges involving flip-flop (ff) nodes cannot be detected using I_{DDQ} testing. This has lead to a search for D flip-flop designs that are 100% I_{DDQ} testable. We next discuss some of the research aimed at achieving this end.

Two kind of bridges are considered: ff-logic; and ff-ff. Table 4.1 summarizes the I_{DDQ} detectability of **ff-logic bridges**, which are bridges between a flip-flop node and a combinational node, for different flip-flop designs discussed below.

Table 4.1 Summary of I_{DDQ} detectability of bridges between flip-flop and combinational nodes.

Clock	FF1	FF-MS	FF-YM
CL = 1	1,2,3	1,2,3,4	1,2,3
CL=0	5,6	4, s1, 5, 6	4,5,6
Undetected	4	none	none

Table 4.2 Summary of I_{DDQ} detectability of bridges between flip-flops.

FF Type	FF1		FF-MS		FF-YM	
Detected						
A	1,2,3	5,6	1,2,3,4	4,s1,5,6	1,2,3	4,5,6
B	1,2,3	5,6	1,2,3,4	4,s1,5,6	1,2,3	4,5,6
Clock	CL=1	CL=0	CL=1	CL=0	CL=1	CL=0
Undetected						
A	4,5,6	4	1,2,3		1,2,3	
B	1,2,3	5,6	s1,5,6		4,5,6	

Example 1 of Section 4.5.1 illustrates such a bridge. Table 4.2 summarizes the I_{DDQ} detectability of **ff-ff bridges**, which are bridges between nodes of flip-flops A and B, for different flip-flop designs. Example 2 of Section 4.5.1 illustrates such a bridge. In deriving these tables we assume that node D is driven by a combinational circuit and nodes Q and $\overline{Q}$ drive combinational circuit. This rules out the possibility that these flip-flops are a part of a shift register.

Flip-flop type FF1 is the flip-flop of Figure 4.16(a). From Figure 4.16(b) we see that when CL=1, since nodes 1, 2 and 3 are driven by D, bridges between a logic node and any of the nodes 1, 2 and 3 are I_{DDQ} detectable. This is shown in the second row of Table 4.1 under the column marked FF1. The next row for FF1 is derived from Figure 4.16(c). We finally note that no ff-logic bridge involving node 4 is I_{DDQ} detectable. This is because, as can be seen from Figures 4.16(b) and (c), node 4 forms part of a floating control loop both when CL = 0 and CL = 1.

Let us next consider ff-ff bridges for FF1. From Figure 4.16(b) we note that, when CL=1, bridges between nodes 1, 2 and 3 of two flip-flops A and B are I_{DDQ} detectable. The next column of detected faults for FF1 was derived from

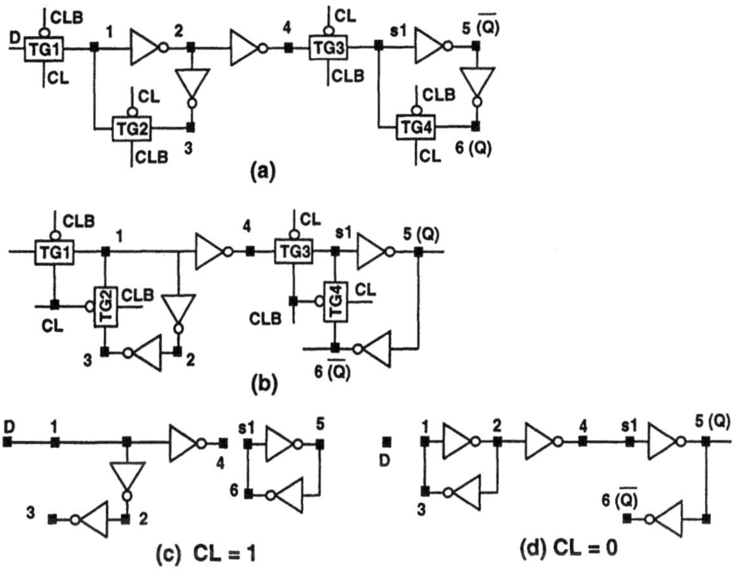

Figure 4.17 FF-MS, an improved flip-flop design for I_{DDQ} testing

Figure 4.16(c). From Figures 4.16(b) and (c) we can also see that the set of bridges where one node is in $\{1, 2, 3\}$ and the other node is in $\{4, 5, 6\}$, either in the same flip-flop or different flip-flops, is not I_{DDQ} testable. This is because, for the two clock phases CL $= 0$ and 1, one of the two nodes is a part of a floating control loop. For the same reason, a bridge between node 4 and one of 5 or 6 is not I_{DDQ} detectable. This is shown in the second column under undetectable faults for FF1.

To improve the I_{DDQ} detectability of sequential circuits various flip-flop designs have been studied. We discuss two of them. The first, which we refer to as FF-MS [274], is shown in Figure 4.17(b). Before discussing this flip-flop we first look at the flip-flop of Figure 4.17(a).

Note that in the design of Figure 4.17(a) a *unidirectional gate* has been inserted between nodes 2 and 4. As a result, unlike node 4 of FF1, no node of the flip-flop will be part of a floating closed loop for both CL $= 0$ and CL $= 1$. Since an inverter is inserted in the datapath, node 5 now becomes $\overline{Q}$ and node 6 becomes Q. Note that the path from D to Q is much longer than that in FF1. This is taken care of in the design of Figure 4.17(b) which we henceforth refer

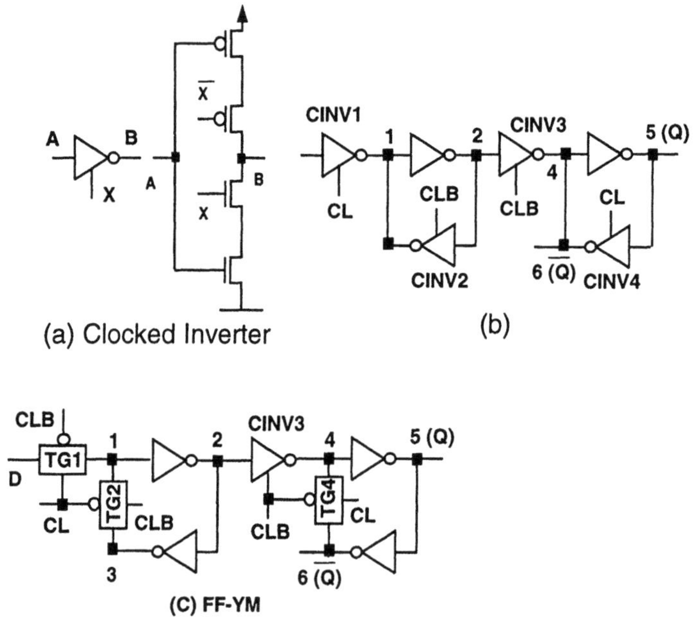

Figure 4.18 FF-YM, an improved flip-flop design for I_{DDQ} testing.

to as FF-MS. From Figures 4.17(c) and (d) the reader can verify that for each node, there is one clock phases when the node does not form part of a floating control loop. This eliminates the problem created by node 4 in FF1. The I_{DDQ} detectability of both ff-logic and ff-ff bridges are summarized in Tables 4.1 and 4.2. They can be derived from Figures 4.17(c) and (d) as was done for FF1.

Instead of using an additional inverter between nodes 2 and 4 of FF1, as was done for FF-MS, the same effect can be achieved by using clocked inverters [342]. This is shown in Figure 4.18(b). A clocked inverter implementations is shown in Figure 4.18(a). However, replacing all transmission gates by clocked inverters is an overkill. The same effect can be achieved by replacing TG3 of FF1 by CINV3. The resulting design is shown in Figure 4.18(c) [342]. We henceforth refer to this flip-flop as FF-YM. We leave it to the reader to verify the entries for FF-YM in Tables 4.1 and 4.2.

Note that FF-YM and FF-MS are improvements over FF1 in that: all ff-bridges are I_{DDQ} detectable; and fewer ff-ff bridges are undetected. The undetectable

Table 4.3 Summary of area/performance of flip-flops

FF Type	Prop Delay(ns)		Area
	High-Low	Low-High	Ratio
FF1	1.21	0.38	1.00
FF-YM	1.34	2.17	1.16
FF-YM(2)	0.08	1.87	1.16
FF-MS	1.06	1.46	1.19

ff-ff bridges are precisely those between nodes in the master and slave portions of the flip-flop. These defects have to be detected using logic testing.

A summary of the performance of the three flip-flops is presented in Table 4.3. FF-YM(2) is similar to FF-YM except that the width of the MOS devices is twice that of the MOS devices in FF-YM. Note that there is a marginal increase in the area of the cells for FF-YM and FF-MS compared to FF1. In addition the performance degradation, if any, is minimal.

4.5.3 I_{DDQ} DFT for shift registers, scan chains

In the previous subsection we had assumed that the D input of the flip-flops in Figures 4.16, 4.17 and 4.18 are driven by a combinational circuit and the outputs Q and $\overline{Q}$ drive combinational circuits. Therefore, the only possible floating control loop that a flip-flop node can be part of is one that consists of nodes in the flip-flop itself. This assumption is not valid when sequential circuits either drive flip-flop inputs or are driven by flip-flop outputs. Two very commonly occuring circuits of this type are shift registers and scan chains.

Figure 4.19 shows the configurations of a two-bit shift register made up of FF-MS master slave flip-flops. Figure 4.19(a), which is valid for CL = 1, was derived from Figure 4.17(c). Similarly, Figure 4.19(b), which is valid for CL = 0, was derived from Figure 4.17(d). Note that, like node 4 of FF1 (Figure 4.16), node 1 of the second flip-flop forms a part of a floating control loop for both CL = 0 and CL = 1. Consequently, none of the bridges involving this node is I_{DDQ} detectable. If we have an n-bit shift register there will be n-1 such nodes even if we use FF-MS and FF-YM. A similar problem arises in scan-chains.

To get around this problem a DFT technique, that configures flip-flops in a transparent mode of operation during testing, has been proposed [273]. In this

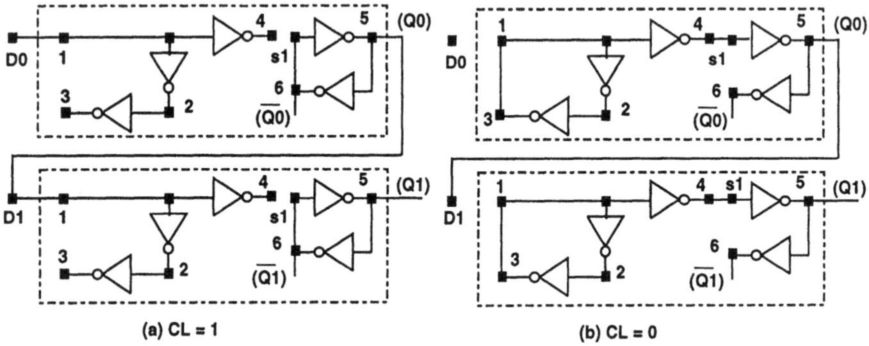

Figure 4.19 Two bit shift register configurations for the two clock phases

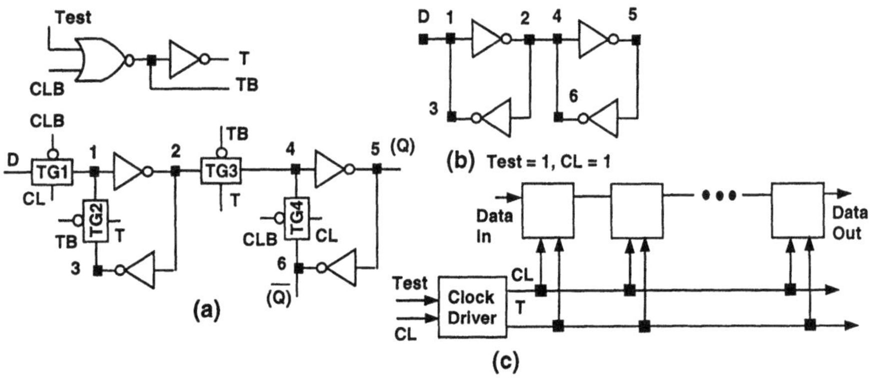

Figure 4.20 Shift register with transparent mode of operation

mode of operation every node in the flip-flop is a driven node. To understand this refer to Figure 4.20(a) which shows FF1 with a transparent mode of operation. When, Test = 0, T is equivalent to CLB and the latch operates exactly as FF1. When Test = 1, T = 1 and transmission gates TG2 and TG3 are closed. If we now set CL to 1 then TG1 and TG4 are also closed. Thus, as shown in Figure 4.20(b), all flip-flop nodes are driven when Test = 1 and CL = 1, provided node D is driven. In this mode of operation if D is driven, all bridges involving nodes in the flip-flop are I_{DDQ} detectable, provide the two bridged nodes are set to different values. This idea can also be used if the flip-flop is a scan flip-flop.

The block diagram of a shift register having a transparent mode of operation, is shown in Figure 4.20(c). The design of the transparent scan-chain will be very similar. Single phase clocking has been assumed. A similar idea can also be used for two-phase clocking. Note that, if Data-in is set to 1(0) then, when CL = 1 and Test = 1, nodes $D, 1, 3$ and 5 in all the flip-flops in the shift register are set to 1(0). Similarly, nodes $2, 4$ and 6 are set to 0(1). Therefore, bridges between node 1 of one flip-flop and node 3 of another flip-flop are not I_{DDQ} detectable using this mode of operation. There is a number of other such examples. On the other hand, recall that many bridges between the master and the slave of the same or two different flip-flops are not detectable when the flip-flops do not have a transparent mode of operation. An example would be the bridge between nodes 2 and 5 of two different flip-flops. In the transparent mode of operation all such bridges are now I_{DDQ} detectable, provided the bridged nodes can be set to two different values.

Another possible solution is to put unidirectional gates between shift register stages. If we use the flip-flop of Figure 4.18(b) then the problem stated above does not arise. Alternatively, one can modify FF-YM by using a clocked inverter instead of TG1 in Figure 4.18(c). Nodes Q and $\overline{Q}$ are interchanged. As seen in the previous subsection, if such flip-flops are used some bridges between nodes in the master and slave portions are not I_{DDQ} detectable.

To conclude, we note that careful design of flip-flops is necessary to make bridges involving flip-flop nodes I_{DDQ} detectable. This problem is more difficult for shift registers and scan chains. Solutions have been proposed that improve the I_{DDQ} detectability of such structures. However, we believe that there is considerable room for improvement.

4.6 I_{DDQ} TESTING OF RAMS

RAM sizes are increasing at a rapid rate. At the same time the feature size of such chips is decreasing. These trends have contributed to the difficulty of testing RAMs in several ways. Firstly, most known functional test procedures use a sequence of reads and writes. The number of reads and writes is linear in the number of memory locations. Consequently, the test time increases with the memory size. Secondly, the higher density of these chips implies that new defects must be considered in the test strategy. This also tends to increase the test time. It is estimated that the test cost of 64-Megabit DRAM will be about 240 times the cost of 1-Megabit DRAM [137]. In addition, the test cost for 64-Megabit DRAM is expected to be about 39% of the product cost [137]. The projected increase in the cost of SRAMs follows a similar trend. There is therefore a need to improve the effectiveness and reduce the complexity of RAM testing. Proposals for using I_{DDQ} testing for both testing and failure analysis of RAMs have been made [209, 217, 276].

Let us assume the RAM organization of Figure 4.21(a). This structure of the RAM is a little different from the one shown in Figure 3.12(a). Here we consider a one dimensional organization where each column represents one bit-slice of the memory. Each bit-slice is of the form shown in Figure 3.12(b) where the row select line is replaced by the word-select line W_j. W_j is enabled when location j is accessed. BL_i and BLB_i are respectively the ith bit line and its complement. For SRAMs, the memory cells are of the form shown in Figure 4.21(b). For DRAMs, the cells are different and require refresh circuitry. We will concentrate only on SRAMs.

The **fault model** used consists of several fault types. Some of them are marked as $d1 \ldots d6$ in Figure 4.21(a). They are defined as follows.

$d1(d2)$ is any defect that forces the cell to behave as if the cell is stuck-at 1(0);
$d3$ is a defect that bridges bit lines BLB_i and BL_{i+1};
$d4$ is a defect that causes a coupling between horizontally adjacent cells; and
$d5(d6)$ is a defect that causes a coupling between vertically adjacent cells (word-lines) and vice-versa.

In addition to these, two other faults are defined: a memory cell is stuck-open; and a memory cell can be changed from 0 to 1 but not from 1 to 0 (transition fault).

The standard functional test used to detect defective SRAMs is the augmented 13N march test (16N march test) where N is the number of memory locations [81]. This test set is defined in Table 4.4 where W0 (W1) implies writing 0 (1) into a given location; and R0 (R1) implies a read operation from a given location where the expected data is 0(1). It also shows two additional read write sequences to test for data retention. We assume that each memory access requires one unit of time. This basic march test does not detect horizontal coupling faults and additional passes of march tests are required with different patterns [81].

For SRAMs we observe that many defects cannot be detected by I_{DDQ} testing. The reason for this is two fold: the presence of floating control loops; and the short duration for which some of the lines are activated. Note that when a

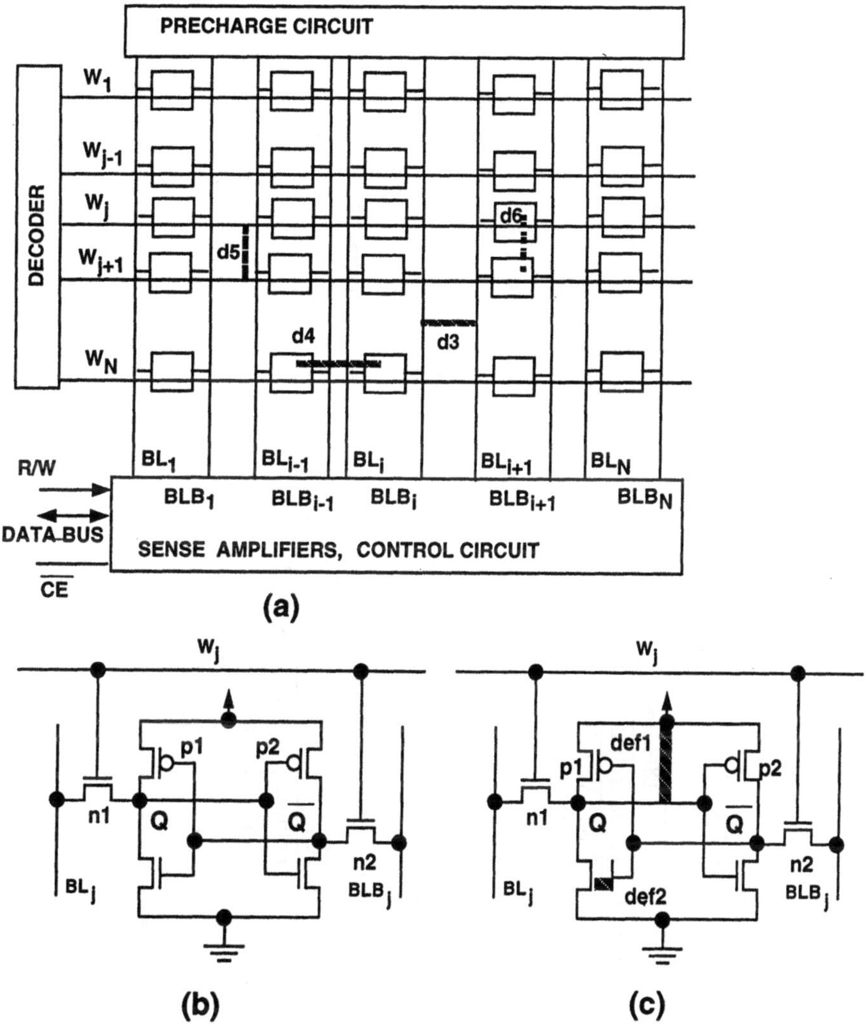

Figure 4.21 One dimensional organization of SRAM

Table 4.4 The 16N (augmented 13N) functional march test for SRAMs

	13N Test					Data Retention Test			
Addr	Initial	March 1	March 2	March 3	March 4	Wait	March 5	Wait	March 6
1	W0	R0, W1, R1	R1, W0, R0	R0, W1, R1	R1, W0, R0	Disable	R0, W1	Disable	R1
2	W0	R0, W1, R1	R1, W0, R0	R0, W1, R1	R1, W0, R0	RAM	R0, W1	RAM	R1
...	...	...	...	...	...	...	...	...	...
...	↓	↓	↓	↑	↑		↓		↓
...	...	...	...	...	...		...		...
N	W0	R0, W1, R1	R1, W0, R0	R0, W1, R1	R1, W0, R0		R0, W1		R1

Table 4.5 SRAM testing strategy using functional and I_{DDQ} testing

addr	IDDQ1	IDDQ2	IDDQ3	IDDQ4			March 1		March 2
1	All wordlines	All wordlines	All wordlines	Even Wordlines	W0	Disable	R0,W1,R1	Disable	R1,W0
2	activated	activated	activated	activated	W0	RAM	R0,W1,R1	RAM	R1,W0
	All bitlines	All bitlines	All bitlines	Odd wordlines	W0		↓		↑
...	set to 0	set to 1	101010...10	deactivated			...		...
N					W0		R0,W1		R1,W0

cell is not accessed, i.e., nFETs $n1$ and $n2$ are cut-off, the cell forms a floating control loop. In Figure 4.21(c) if node Q is bridged to V_{DD} then the content of the cell will be stuck-at-1 because, when the cell is not accessed, it forms a floating control loop and Q is pulled up. If we drive BL_j low and enable W_i long enough then this defect will be detected by I_{DDQ}. However, the cells are accessed for a very short duration (10s of ns) and therefore faults cannot be detected by I_{DDQ} testing. If we modify the design so as to increase the activation time of wordlines during testing the defect can be detected. Next, consider the gate-oxide defect in nFET $n3$. If the defect resistance R_b is very low then it manifests itself as a stuck-at-0 fault. Like the previous defect it cannot be detected using I_{DDQ} testing unless the access time is increased. If R_b is much larger then nodes Q and $\overline{Q}$ settle at an intermediate voltage. This results in a large static current and the defect is I_{DDQ} detectable.

Not withstanding these difficulties of using I_{DDQ} testing to detect defects in SRAMs, empirical results show that I_{DDQ} testing can improve the quality of SRAM testing [209]. In an experiment on an 8k×8 SRAM it was found that out of a total of 1582 devices that had elevated I_{DDQ} only 1490 failed the functional tests. The remaining 92 passed the functional test. This shows that additional defective SRAMs were picked up by I_{DDQ} testing.

Even if we increase the cell access time during testing, note that at least two I_{DDQ} test per address are required - one that drives the bit line to 1 and another that drives the bit line to 0. Such a large number of I_{DDQ} measurements is clearly impractical. We next discuss a testing method, proposed for RAMs [276], that uses a combination of I_{DDQ} and logic testing. The test is described in Table 4.5 and is explained below.

- The first I_{DDQ} test sets all wordlines to 1 and all bit lines to 1. It detects faults $d2$ and $d3$.
- The second I_{DDQ} test sets all bit lines to 0 and activates all wordlines. This test detects faults $d1$ and $d3$. Note that the first two tests detect all leakage faults.
- The third I_{DDQ} test sets the bit lines to an alternating pattern of 0s and 1s and activates all wordlines. It detects fault $d5$.
- The fourth I_{DDQ} test sets adjacent wordlines to complementary values. All even-numbered wordlines are activated and the odd-numbered ones are not. This detects fault $d5$.

Table 4.6 Definition of Even and Odd control signals

Even	Odd	Mode
0	0	Normal mode
0	1	Odd wordlines activated
1	0	Even wordlines activated
1	1	All wordlines activated

- Fault *d*6, which is the coupling fault between vertically adjacent cells is not detected by I_{DDQ} measurements. This is detected by the two march sequences that follow.
- Finally, the data retention and transition faults are detected by March 1 and March 2.

Recall that the complexity of the functional test of Table 4.4 is 16N whereas the complexity of the hybrid test of Table 4.5 is 5N + 4 I_{DDQ} measurements. Hence, the test complexity has been drastically reduced.

This reduction in test complexity has a price. The test strategy assumes: (i) a mode of operation in which all wordlines of the SRAM can be activated simultaneously; (ii) the cells are activated for a sufficient time duration for I_{DDQ} measurements to be meaningful; and (iii) odd (even) numbered wordlines can be selectively activated. Assumption (ii) can be satisfied by providing for a test mode during which the activation time for wordlines is increased. Assumptions (i) and (iii) are satisfied by providing two additional control signals to the address decoders: *Even* and *Odd*. They are defined in Table 4.6. The implementation of the four modes requires simple modifications to the address decoder.

Activating all cells simultaneously poses a design problem. Firstly, each driver has to be large enough to drive a column of cells. In addition, when all the cells are accessed a large current will flow and result in electromigration. This could either damage the memory chip or lead to reliability problems. To circumvent this problem it was suggested that address lines be activated gradually. First only W_1 is activated. Once W_1 is set at 1 for time δ then wordlines W_2 and $W3$ are activated. After time δ the next four address lines are activated, and so on. It was verified using simulation for $N = 64$ that the bit line drivers need not be changed and the total transient current is very small [276]. To what extent this approach can be scaled is not clear. However, by using the above idea on smaller blocks of addresses, test complexity can still be reduced but not as drastically as claimed above. The number of I_{DDQ} measurements will have to be increased. In addition, area and performance penalty of about 3-4% must also be paid [276].

4.7 SUMMARY

In summary we note that circuits ought to be tested primarily for three kinds of defects: shorts, opens and resistive contacts. Resistive contacts lead to delay effects and little is known about detecting such defects by I_{DDQ} testing. Shorts can be detected by logic or delay testing and I_{DDQ} testing. However, I_{DDQ} testing can detect a much larger variety of shorts than a combination of logic and delay testing. Some shorts, which are not detected by logic or delay testing do not immediately affect the logic performance of the circuit or its timing behavior. However, many of them degrade with time and ultimately end up causing field-failures very early in their life cycle. I_{DDQ} testing can detect such defects.

Opens affect the static as well as dynamic logic behavior of circuits. Hence they can potentially be detected by logic testing or delay testing. Several classes of opens can be detected by I_{DDQ} testing whereas others simply cannot be detected. Several mechanisms have been identified to explain the fact that I_{DDQ} testing detects a good percentage of opens.

The above discussion explains many of the experimental findings that were presented in Chapter 2. It highlights the potential impact of I_{DDQ} on IC quality. However, for many defects, especially opens, I_{DDQ} testing needs to be used in conjunction with logic and delay testing. In addition to understanding how defects modify the static or dynamic behavior of circuits or raise the I_{DDQ}, it is equally important to know how to compute good test sets to detect defects during logic, delay and I_{DDQ} testing. In the next chapter we look at fault models used for computing tests for logic, delay and I_{DDQ} testing. This will give us a better understanding of the effectiveness of these tests to detect defects. It puts into a better perspective the role that I_{DDQ} testing can play in detecting defects.

We also saw that, unless we carefully design circuits, anomalies arise in detecting defects using I_{DDQ} testing. A set of design rules, to avoid such anomalies, was presented. However, many standard circuit components do not conform to these design rules. They include flip-flops, shift registers, scan chains and RAMs. Design techniques to improve the I_{DDQ} testability of these components were also discussed.

5

TEST SUITES, FAULT MODELS, TEST SETS AND DEFECTS

In Chapter 4 we saw that defects change the I_{DDQ} level, the static logic behavior and the dynamic logic behavior of circuits. We henceforth refer to logic testing to detect faulty static logic behavior as **slow speed testing** and logic testing to detect faulty dynamic logic behavior as **high speed testing**.

Some defects cause elevated I_{DDQ} but do not change the logic behavior of circuits; other defects do not cause any change in the I_{DDQ} level but change the static or dynamic logic behavior of circuits; and still other defects do not change the static logic behavior but affect the dynamic logic behavior. It thus appears that I_{DDQ} testing, slow speed testing, and high speed testing have significant roles in ensuring IC quality. However, the vast majority of ICs manufactured today are subjected only to slow speed testing. Is this the right approach? In Section 5.1, we will see that empirical data suggests that it is not. The data also suggests that I_{DDQ} and high speed testing be included in the test suite.

The effectiveness of any component of a test suite is a function of the quality of the test set used. Test sets are computed based on fault models. Therefore, the quality of a test set is a function of the correlation between defects and fault models. In other words, we need to know how good the tests are in detecting defects. To understand this we first look at fault models used for computing tests for slow speed testing (Section 5.2), high speed testing (Section 5.3), and I_{DDQ} testing (Section 5.4). In Section 5.5, we discuss how good these tests are in targeting defects.

We find that there is a high correlation between I_{DDQ} fault models and defects. Hence I_{DDQ} tests are very effective in detecting defects. On the other hand,

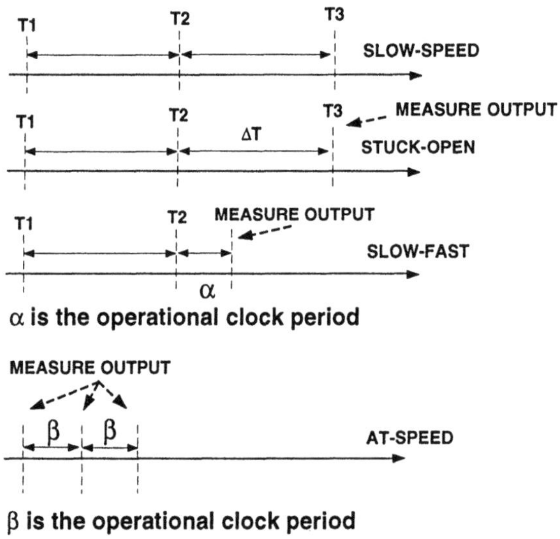

Figure 5.1 Timings for application of test-pairs

the correlation between high speed fault models and defects is not very good. This raises questions about the effectiveness of high speed testing in its current form. This is another very important advantage of I_{DDQ} testing.

5.1 TEST SUITES

In slow speed testing we assume that test vectors are applied at a slow speed. After application of an input vector, sufficient time is allowed for all transients and hazards to settle, before sampling the circuit outputs. This is shown in Figure 5.1. Slow speed testing attempts to detect defects that change the static logic behavior of circuits. Fault models to compute tests for slow speed testing are discussed in Section 5.2.

In order to detect faulty dynamic logic behavior of circuits, we need to know if the circuit under test responds to changes in the input when such changes occur at or near the operational speed of the circuit. We refer to this as high speed testing. Two varieties of high speed testing are considered here: **slow fast**; and **at-speed**. In slow fast testing, tests are applied in pairs as shown in

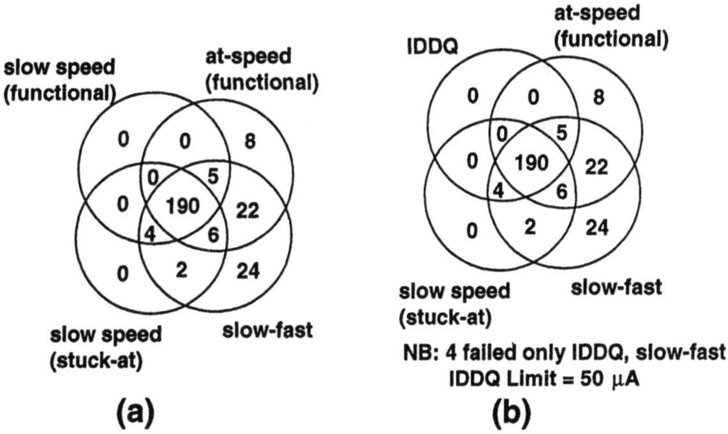

Figure 5.2 Data suggesting inclusion of I_{DDQ} and high speed testing in the test suite.

Figure 5.1. After application of $T1$, the circuit is allowed to settle down. This is followed by application of $T2$, a time period α is allowed to elapse before measurements are made. The value of α is the time period of the operating clock. The interval between the applications of $T1$ and $T2$ is much larger than α. Thus, two clocks are required to mark the time interval between $T1$ and $T2$ and the time interval between $T2$ and the measurement instant. In at-speed testing, as shown in Figure 5.1, vectors are applied at a constant rate. The time period β of the operating clock determines the rate at which the vectors are applied. The output is measured just prior to applying the next vector.

An important question is: what constitutes a good test suite? Most ICs today are subjected primarily to slow speed tests. However recent data suggests [203] that this needs to be augmented. We present some data in Figure 5.2. In Figure 5.2(a), the result of applying 4 different logic tests is shown. Two slow speed test sets and two high speed test sets were used. The two slow speed test sets used were a functional test set and a stuck-at test set. The functional test set was also applied at high speed in the at-speed mode. A specially constructed test set was applied in the slow fast mode. The results clearly show that high speed testing detects all defects detected by slow speed testing and many more. This suggests that high speed testing should replace slow speed testing.

Figure 5.2(b) shows the result when I_{DDQ} testing was added to the test suite and one of the slow speed testing, viz., slow speed testing using functional vectors, was removed from the test suite. Again, the results imply that slow speed testing, in this case, was redundant and I_{DDQ} testing picked up many defective ICs that were not detected by high speed testing. Similar results were obtained when the threshold was increased to $200\mu A$. This suggests that the test suite should consist of some form of high speed testing and I_{DDQ} testing.

5.2 FAULT MODELS FOR SLOW SPEED TESTING

Tests for slow speed testing are computed based on the **single stuck-at fault model.** We assume that this fault model and the associated test generation algorithms are known to our reader.

Stuck-open fault [323] is another fault model whose tests, in our opinion, fall into this category. In this fault model a transistor is assumed to be permanently open. For example, the switch level circuit for gate m of Figure 5.3(a) is shown in Figure 5.3(b). It has four transistors and each of them can be stuck-open. Thus, in the stuck-open fault model, there are four single transistor faults for this gate.

A test for a stuck-open fault consists of a pair of input vectors $\langle T1, T2\rangle$. First $T1$, the **initializing vector,** is applied and the circuit is allowed to settle. This is followed by the application of the **final vector** $T2$. Logic measurements are made after time ΔT has elapsed following the application of $T2$. This is shown in Figure 5.1. We assume that ΔT is large enough for all outputs to have reached the steady state and no hazards exist. As we will see shortly, tests for such faults are defined based on this assumption.

We next discuss properties that test-pairs must satisfy in order to detect a stuck-open fault. Assume that transistor $P1$, shown in Figure 5.3(b), of gate m of the circuit of Figure 5.3(a) is stuck-open. Then, the pair, $T1 = \langle X1 = 0, X2 = 0, X3 = 1, X4 = 1\rangle$ and $T2 = \langle X1 = 1, X2 = 1, X3 = 0, X4 = 1\rangle$, detects the fault. $T1$ sets both d and l to 1, thereby setting m to 0 in the presence of the fault; and $T2$ attempts to pull m to 1 through $P1$ only, by setting d to 0 and l to 1. If $P1$ is not stuck open then on application of $T2$, m is set to 1, else it remains at 0. If m is not a primary output then $T2$ must also propagate the erroneous value at m to a primary output. In fact, $T2$ has to be a test for m stuck-at-0. Methods for computing tests for stuck-open faults have been discussed [44, 61, 72, 83, 84, 143, 258, 259, 282, 331]. Such tests can be invalidated due to circuit delays and the notion of robust tests

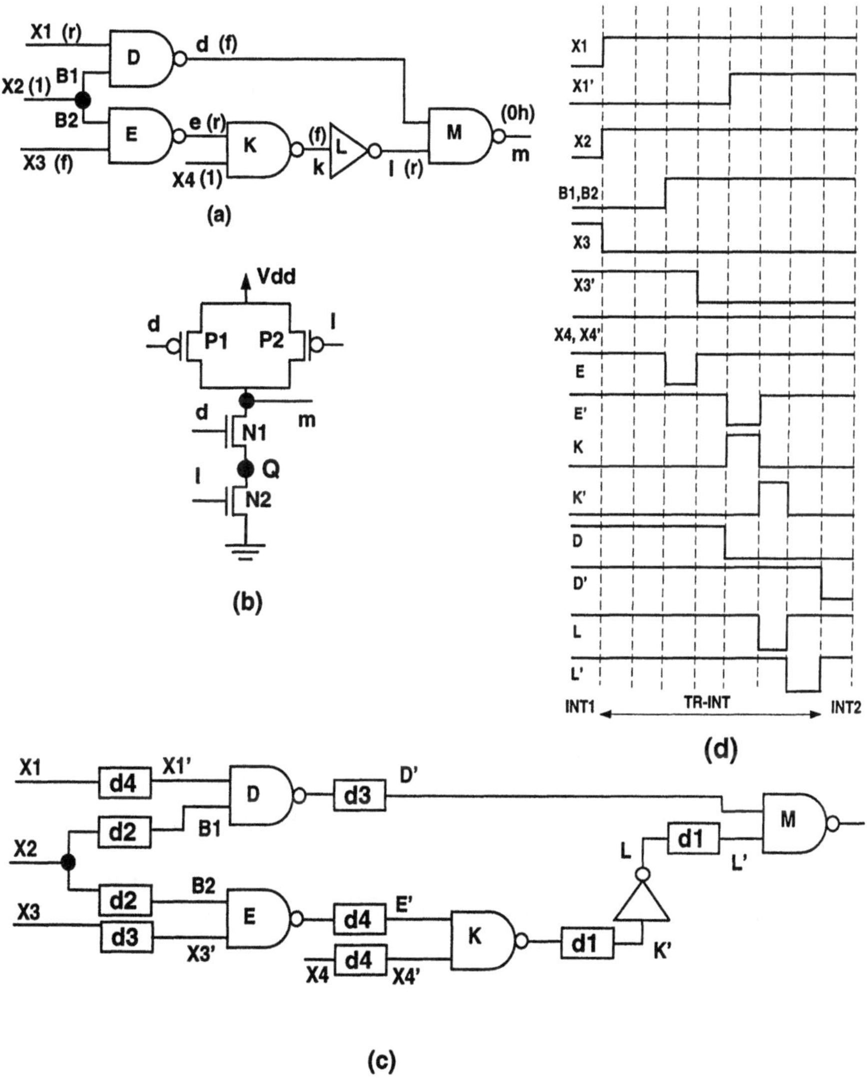

Figure 5.3 Robust test-pairs for stuck-open faults

was proposed [259]. That method [259] was subsequently refined [44]. The discussion to follow is from [44].

To see how delays invalidate tests for stuck-open faults, assume that the circuit has delays as shown within rectangular boxes in Figure 5.3(c), where $d1 = 1, d2 = 2, d4 = 4$ and $d3 = 3$. Let the input-pair $\langle T5, T6\rangle$ be applied at the primary inputs where $T5 = \langle X1 = 0, X2 = 0, X3 = 1, X4 = 1\rangle$ and $T6 = \langle X1 = 1, X2 = 1, X3 = 0, X4 = 1\rangle$.

The waveforms at each line of the circuit are shown in Figure 5.3(d) where INT1 and INT2 are the two steady state intervals and TR-INT is the transition interval. Note that there is a static 1-hazard at l. During this short interval of the static hazard when l is low $P2$ conducts and starts charging m towards 1. If m acquires a sufficiently high voltage then the test is invalidated. The test-pair was invalidated because there was a path from V_{DD} to the output that did not pass through the faulty transistor $P1$ and a hazard activated that path. Since there is only one path from the output to ground, no test-pair for $N1$ or $N2$ can be invalidated by hazards.

Since it is difficult to estimate circuit delays and delays vary from one manufacturing run to another we assume that we have no knowledge of circuit delays. Accordingly, a test-pair is said to be a **robust test-pair** if and only if it is not invalidated by arbitrary delays in the circuit [259].

The six-valued logic system $\Omega = \{0, 1, r, f, 0h, 1h\}$, defined in Table 5.1, is used to define the properties that must be satisfied by robust test-pairs. Assume that the two inputs of the NAND (NOR) gate are d and l. Since there is only one path in the pulldown (pullup) network of a NAND (NOR) gate, every test-pair for transistors in the pulldown (pullup) network is a robust test-pair. We present the robustness criteria for the rest of the transistors. Let $P1$ ($N1$) be the transistor driven by d and $P2(N2)$ the transistor driven by l. The robustness criteria for a test-pair is given in Table 5.2. A method to compute such tables for other complex gates has also been discussed [44]. This logic system can be used to modify standard gate-level stuck-at test generators to compute robust tests for stuck-open faults [44]. It has also been shown that a given stuck-at test set can be easily modified to compute robust test sets for a very large percentage of stuck-open faults [45].

Using Ω, a robust test-pair $\langle T3, T4\rangle$ for $P1$ can be computed, where $T3 = \langle X1 = 0, X2 = 1, X3 = 1, X4 = 1\rangle$ and $T4 = \langle X1 = 1, X2 = 1, X3 = 0, X4 = 1\rangle$. The values from Ω assigned to each line in the circuit are shown within "()"

Table 5.1 Definition of Ω

Logic Value	Initial Value	No Of Transitions	Final Value
0	0	0	0
1	1	0	1
r	0	≥ 1	1
f	1	≥ 1	0
$0h$	0	≥ 1	0
$1h$	1	≥ 1	1

Table 5.2 Properties of robust tests

NAND	d	l
$P1$	f	1
$P2$	1	f
NOR		
$N1$	r	0
$N2$	0	r

in Figure 5.3(a). Note that the output of the circuit has the value $0h$. Thus, a valid test-pair for a stuck-open fault can lead to hazards at a primary output.

Considering the time interval between the application of $T2$ and the measurement of logic values at outputs we note that ΔT should be at least one period of the operating clock of the circuit. This interval however should not be too large. If that is the case then the stuck open transistor, which may be conducting some leakage current, may actually charge or discharge the node and invalidate the test-pair.

As seen in the above example, a valid test-pair may result in hazards at the primary output. Thus, there is no requirement that hazards not be present at primary outputs. This is because it is assumed that ΔT is large enough for such transients to die out. This is the primary reason for classifying such test-pairs in slow speed testing. We will return to this issue again when we discuss test-pairs for delay faults.

All robust test generation algorithms for stuck-open faults neglect charge-sharing effects which can also invalidate test-pairs. To understand this once again refer to Figure 5.3(b). Assume that $N2$ is stuck-open and $T1$ and $T2$

are as defined above. $T2$ sets m to 1 and $T1$ attempts to discharge m to 0 through $N1$ and $N2$. But, because $N2$ is stuck-open m cannot discharge and m may remain high. This is detected by propagating the erroneous value by $T1$ to a primary output. Thus, $\langle T2, T1\rangle$ is a robust test pair for $N2$ stuck-open. However, this analysis neglects the fact that on application of $T1$, $N1$ is on so that the capacitances at nodes M and Q are connected. Thus, charge sharing between nodes M and Q may reduce the voltage at m so that it is not recognized as a logic 1. Hence, the test will be invalidated. This problem is especially acute for complex CMOS gates as well as for NAND and NOR gates with a large number of inputs.

For sequential circuits, during test application, only the primary inputs are changed. For full scan or partial scan circuits scan and hold latches [80] are used. This enables both $T1$ and $T2$ of the test-pair $\langle T1, T2\rangle$ to be scanned in prior to the application of $T1$. $T1$ is applied followed by $T2$ and no scan operation is required after application of $T1$. This ensures that after the application of $T1$ the system registers do not change state until $T2$ is applied. From the above discussion it should be clear that this requirement is important to prevent the invalidation of test-pairs during the scan operation.

5.3 FAULT MODELS FOR HIGH SPEED TESTING

Tests for high speed testing are computed in a variety of ways. An *ad hoc* method is to run the functional vectors at high speed. There are two other methods which compute two distinct classes of delay tests.

The original motivation for computing delay tests was to verify the timing specifications of circuits [14, 21, 155, 231, 278, 292]. Such tests are computed based on two different delay models: **path delay** [176, 292, 296] and **gate delay** [130, 141, 142, 247, 325, 326]. We first look at tests for gate delay faults, followed by tests for the path delay fault model.

In the **gate delay fault model** it is assumed that the fault is localized at one gate, either at the input or the output. Gate delay tests attempt to create a transition at a given line in the circuit and propagate the transition to any primary output. For an example, refer to Figure 5.3(a) and assume that a delay test for a rising transition at d is required. Let the test $T1$ set $X1$ and $X2$ to 1, and $T2$ be $\langle X1 = 0, X2 = 1, X3 = 0, X4 = 1\rangle$. Clearly, $T1$ sets d to 0, and $T2$

sets d to 1. Thus, it creates a rising transition at d. In addition, $T2$ propagates the value at d to a primary output. Thus, it is a gate delay test for d rising. Gate delay tests are often referred to as **transition tests** [203].

There are two problems with gate delay tests. Firstly, it is assumed that gates have zero delay and no hazards occur in the circuit. Secondly, the total delay may not be a result of a delay at a single gate but could be a cumulative effect of small additional delays at several gates. The second problem is important in the context of parametric faults where the excess delay is caused due to process variation. In this case, small additional delays occur at several gates along a path. These can add up to a large delay causing the circuit to malfunction. The first problem is more important for defect detection. We will see in the context of path delay tests that, due to non-zero gate delays, delay tests could be invalidated.

Not withstanding this problem, gate delay tests continue to be used [203] because path delay faults are difficult to compute and, in many cases, very few paths have robust test-pairs. However, the path delay fault model, which we study next, is more general than the gate delay fault model.

In the **path delay fault model**, along every physical path, from an input to an output of the combinational circuit, two distinct *transitions* - P_f, *input falling* and P_r, *input rising* - can propagate. For example, for the path $\langle e, n, q, s\rangle$ of Figure 5.4(a) there are two path delay faults: $\langle e, n, q, s\rangle_r$, input e rising transition propagated along path $\langle e, n, q, s\rangle$; and $\langle e, n, q, s\rangle_f$, input e falling transition propagated along path $\langle e, n, q, s\rangle$. Delay testing ascertains if, for every physical path P, both P_f and P_r can propagate within a predetermined time interval α.

A test for a delay fault consists of a pair of input vectors $\langle T1, T2\rangle$. As shown in Figure 5.1, first the **initialization vector** $T1$ is applied. All signals are allowed to stabilize under its influence. Then the **test vector** $T2$ is applied. The circuit outputs are monitored for the appropriate transition after a time interval α, where α is the time-period of the operational clock. Note the difference between this and the test application for stuck-open faults. For stuck-open faults, after application of the second vector it is assumed that measurements are made after a rather long interval ΔT. All transients are assumed to have died at that point. For delay faults, measurements are made after a very short duration following the application of $T2$. This makes a difference while defining the test-pairs for these two faults. As we have already seen, for stuck-open faults, the test-pairs could result in hazards at one or more primary outputs. On the

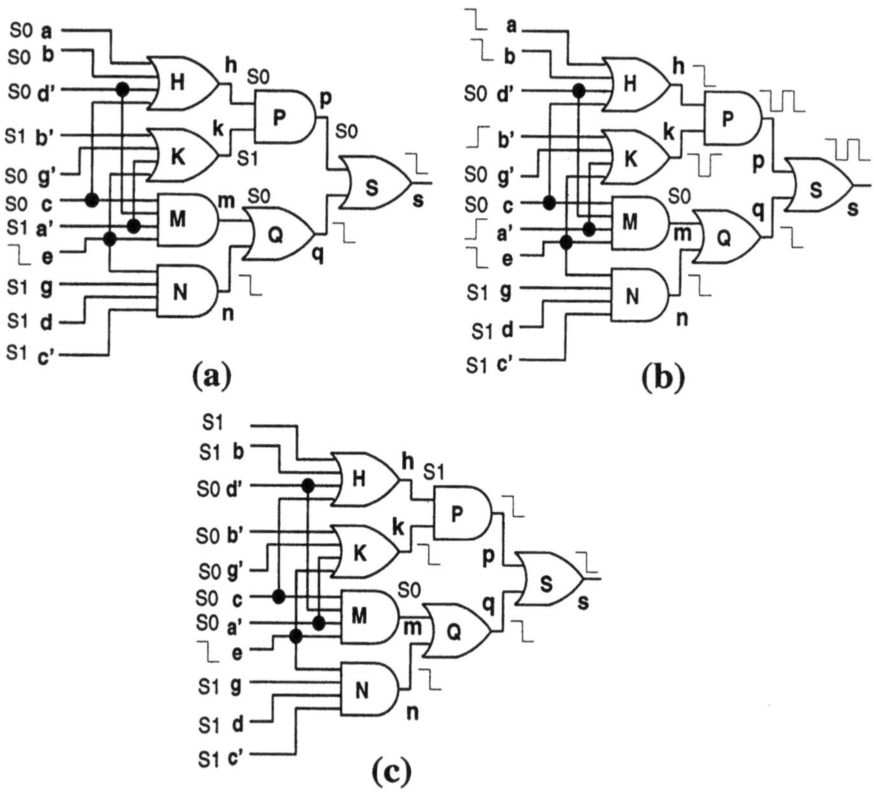

Figure 5.4 Illustration of different kinds of delay test-pairs

Table 5.3 Definition of Γ, the set of logic values for delay testing

Table 1								
Logic Value	S0	S1	h0	h1	r	f	hr	hf
$T1(X)$	0	1	0	1	0	1	0	1
$T2(X)$	0	1	0	1	1	0	1	0
Hazard ?	No	No	Yes	Yes	No	No	Yes	Yes

Table 5.4 AND operation over Γ

AND	$s0$	$s1$	$h0$	$h1$	r	f	hr	hf
$s0$	$s0$	$s0$	$s0$	$s0$	$s0$	$s0$	$s0$	$s0$
$s1$	$s0$	$s1$	$h0$	$h1$	r	f	hr	hf
$h0$	$s0$	$h0$	$h0$	$h0$	$h0$	$h0$	$h0$	$h0$
$h1$	$s0$	$h1$	$h0$	$h1$	hr	hf	hr	hf
r	$s0$	r	$h0$	hr	r	$h0$	hr	$h0$
f	$s0$	f	$h0$	hf	$h0$	f	$h0$	hf
hr	$s0$	hr	$h0$	hr	hr	$h0$	hr	$h0$
hf	$s0$	hf	$h0$	hf	$h0$	hf	$h0$	hf

other hand, delay faults test-pairs may or may not allow hazards at primary outputs.

To express the properties of the variety of delay tests we use the notation defined in Table 5.3. $T1(X)$ and $T2(X)$ are respectively the steady state values assigned to X by $T1$ and $T2$. Note that this notation differs from the notation in Table 5.1. Here we differentiate a hazard-free rising (falling) transition from a rising (falling) transition that allows for a hazard. Thus, there are two kinds of rising (falling) transitions r and hr (f and hf). This set of logic values is henceforth referred to as Γ. Logic operations over Γ are easy to define. The AND operation is shown in Table 5.4.

We can now define several classes of delay tests. For example, in Figure 5.4(a), assume that the initialization vector $T1 = \langle a = 0, b = 0, c = 0, d = 1, e = 1, g = 1\rangle$ is applied prior to the vector $T2 = \langle a = 0, b = 0, c = 0, d = 1, e = 0, g = 1\rangle$. A falling transition at e propagates along the path $\langle e, n, q, s\rangle$. The arrival time of the transition at s depends on the delay along the path and is *independent of delays in the rest of the circuit* [176]. This is an example of a **robust test**. Using the notation of Table 5.3, the logic values assigned by this test-pair to the lines in the circuit are shown in Figure 5.4(a). The test-pair is: $\langle a = S0, b = S0, c = S0, d = S1, e = f, g = S1 \rangle$. On applying this input pair, if the falling edge at the output does not arrive before measurements are made then a delay fault exists along $\langle e, n, q, s \rangle$. In Figure 5.4(b), the first falling edge at s can arrive only after the falling edge at q has arrived making the pair shown a robust test-pair for $\langle e, n, q, s \rangle_f$.

Robust test-pairs can be **multiple input change** or **single input change** [248]. In Figure 5.4(a) (Figure 5.4(b)) the two vectors differ in only one (more than one) bit position and is an example of a single input change (multiple input change) robust test pair.

Robust tests are classified as either **hazard-free** or **non-hazard-free** [248]. In Figure 5.4(a), lines g, d, c', m, p are inputs to gates that lie along $\langle e, n, q, s \rangle$ but do not lie on $\langle e, n, q, s \rangle$. These are the **off-path inputs** of $\langle e, n, q, s \rangle$. Under the arbitrary delay assumption if, on application of a robust test-pair T for path P no hazard occurs at the off-path inputs of P, as in Figure 5.4(a), then T is a **hazard-free robust test-pair** for P. Note that both hazard-free and non-hazard-free test-pairs can be used to detect delay faults. However, we have to be careful how they are used. If the output of the circuit is sampled after time α, following the application of $T2$, then only hazard-free robust test-pairs can be used. This is because, in Figure 5.4(b), the hazard could result in the sampling of an incorrect value. On the other hand, if edge detection is used to signify the arrival of a transition during the α interval, following the application of $T2$, then hazards can be allowed at the output. In this case, both types of test-pairs can be used.

There are two kinds of hazard-free robust test-pairs [248, 249]. In Figure 5.4(a) the falling transition at e propagates along only one path $\langle e, n, q, s \rangle$. This is an example of a **single path propagating hazard-free robust test-pair**. In Figure 5.4(c) the falling transition at e propagates along multiple paths $\langle e, n, q, s \rangle$ and $\langle e, k, p, s \rangle$. The falling edge at s can arrive only after both falling edges at p and q have arrived. Thus, it is an example of a **multiple path propagating hazard-free robust test-pair** for both $\langle e, n, q, s \rangle_f$ and $\langle e, k, p, s \rangle_f$.

Robust test-pairs without any restriction like the number of input changes, hazard free property, etc., are **general robust test-pairs** (Figure 5.4(b)). If we have a hazard-free robust test-pair then it is also a multiple path propagating hazard-free robust test-pair but not vice-versa. We next look at properties that must be satisfied by the two types of hazard-free robust test-pairs.

- An input pair $T = \langle T1, T2 \rangle$ is a **multiple path propagating hazard-free robust test** for input rising (falling) delay fault along path p if and only if T causes a rising (falling) transition at the input of p and T satisfies the constraints of Figure 5.5(a) [248].

  In Figure 5.5(a) the gate is assumed to be on the path p and A(B) is the on-path (off-path) input of this gate. For example, if the gate is an AND gate and there is a rising transition at A then each off-path input, that is also an input of this gate, must have either a rising transition or be hazard free $S1$ (steady 1).

- An input pair $T = \langle T1, T2 \rangle$ is a **single path propagating hazard-free robust test-pair** for input rising (falling) delay fault along path p if and only if T results in a rising (falling) transition at the input of p and T satisfies the constraints of Figure 5.5(b) [248]. Thus, if the gate is an AND gate and A has a rising transition then each off-path input that is also an input to this gate must have a hazard free $S1$ (steady 1).

- There are path delay faults for which there exist multiple path propagating robust test-pairs but there does not exist any single path propagating hazard-free robust test-pair [248]. For example [46], the reader can verify that, in the circuit of Figure 5.6(a), there exists no single path propagating hazard-free robust test-pair for b falling along either path $\langle b, d, g \rangle$ or path $\langle b, e, g \rangle$. However, the input pair $\langle a = S0, b = f, c = S1 \rangle$ is a multiple path propagating hazard-free robust test-pair for b falling for both these paths.

- Even if a circuit has a single path propagating hazard-free robust test-pair for all path delay faults there can exist a test set S consisting of one or more multiple path propagating hazard-free robust test-pair such that the size of S is smaller than all test sets consisting of only single path propagating hazard-free robust test-pairs [248]. For example [46], for the circuit of Figure 5.6(b), a multiple path propagating hazard-free robust test-pair for each path in the circuit is shown in Table 5.5. The corresponding test set consists of only 6 distinct pairs. Any test set consisting of only single path propagating hazard-free robust test-pairs will have 8 test-pairs.

The above constraints on test-pairs are henceforth referred to as **hazard-free robust propagation constraint**. In addition, robust test-pairs have to satisfy some additional constraints [241]. For the path delay fault $\langle c, e, g \rangle_r$ of Figure 5.7(a) there are a number of hazard-free robust test-pairs. Some of them are listed in Table 5.6. The question is: other than the fact that some of them are hazard-free robust test-pairs for a number of other paths, is there a difference in the capability of these tests to detect the path delay fault $\langle c, e, g \rangle_r$?

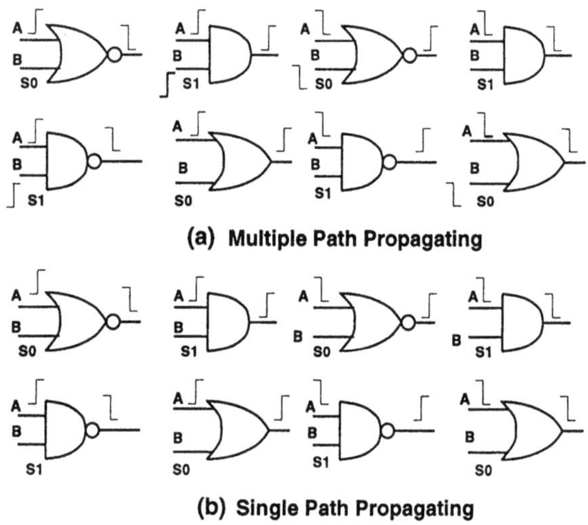

Figure 5.5 Definition of multiple, single path propagating hazard-free robust test-pairs

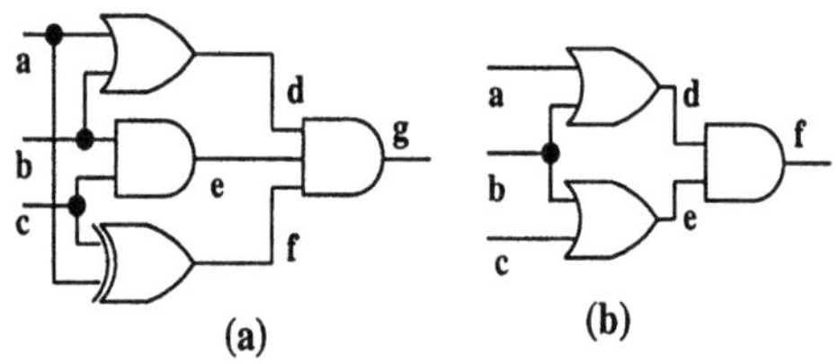

Figure 5.6 Example illustrating the advantage of using multiple path propagating robust test-pairs

Table 5.5 Multiple path propagating delay test for circuit of Figure 5.6(b)

	rising						falling					
	a	b	c	a	b	c	a	b	c	a	b	c
a,d,f	0	0	0	1	0	1	1	0	1	0	0	1
b,d,f	0	0	0	0	1	0	0	1	1	0	0	1
b,e,f	0	0	0	0	1	0	1	1	0	1	0	0
c,e,f	0	0	0	1	0	1	1	0	1	1	0	0

Table 5.6 Robust delay tests for paths $\langle c, e, g \rangle$ in Figure 5.7(a)

	a	b	c	d
1	S0	S1	r	S1
2	S0	S1	r	f
3	S0	S1	r	r
4	S0	r	r	S1
5	f	r	r	S1

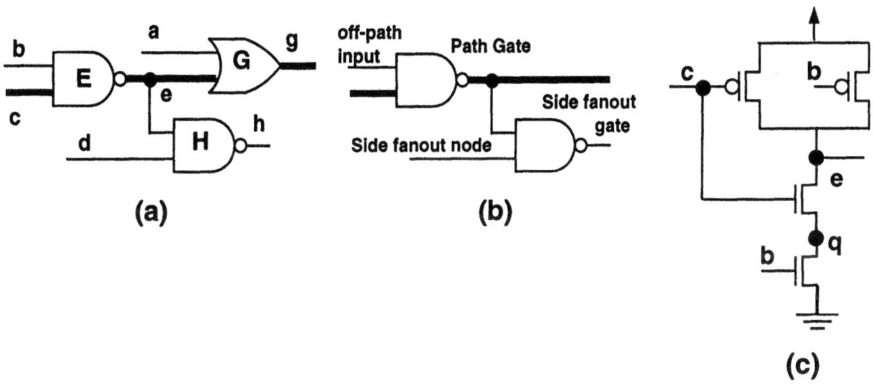

Figure 5.7 Example circuits

We say that a test-pair $p1$ is better than another test-pair $p2$, for the path delay fault $\langle c, e, g \rangle_r$, if and only if a wider range of delays along this path is detected by $p1$.

To understand why one of these test-pairs will be better than the other we need to define some terms. They are, **off-path input, side fanout gate** and **side fanout node**. Figure 5.7(b) defines them, assuming that the path under test is that shown in bold lines. The first important conclusion, based on simulation results [241], is that *the propagation delay along a path depends on the values at the side fanout node.* The first three test-pairs of Table 5.6 are single path propagating hazard-free robust test-pairs for the path delay fault $\langle c, e, g \rangle_r$ in the circuit of Figure 5.7(a). They however differ in the values they assign to the side fanout node d and will differ in the range of delay defects they can detect.

The second important conclusion is that the delay along a path **depends on the values assigned to the off-path inputs.** The 4th and 5th test-pairs of Table 5.6 are test-pairs that differ in the values they assign to the off-path inputs a and b. These two tests will therefore detect different ranges of delay variations for the path delay fault $\langle c, e, g \rangle_r$.

A third important conclusion is that the vector prior to applying the test-pair (**pre-initializing test**) is also important in ascertaining the effectiveness of the test-pair. This is based on the following observation [241]. The switch-level circuit for gate E is shown in Figure 5.7(c). Node q is an **internal node**. The propagation delay of a transition from b to e is affected by the voltage at the internal node q. Node q could be at 0 (b=1), floating (b=c=0) or a weak 1 value because of charge sharing between e and q (c=1,b=0). If $b = c = 0$ and q is floating then the voltage at q is a function of the pre-initializing test. To determine whether test-pair 1 or 4 of Table 5.6 is better we have to know what the pre-initializing test was.

Thus, to detect the smallest variation along the path, a test-pair for it, in addition to satisfying the hazard-free robust propagation constraints, has to satisfy some additional constraints. These constraints on the values assigned to off-path inputs, side fanout nodes and pre-initializing test, are henceforth referred to as **quality constraints**.

Subsequently, when discussing delay testing we assume that tests for delay testing are computed using the path delay fault model, test-pairs are robust and hazard-free, they can be either single path propagating or multiple path propagating type and satisfy both the quality and the hazard-free propagation constraints.

We will not be concerned here with methods for simulating or generating test-pairs for delay-faults. The following is a partial list of such work [28, 31, 41, 67, 69, 70, 97, 104, 105, 156, 173, 175, 179, 196, 232, 233, 234, 237, 256, 277, 281, 282, 283, 314, 316, 325, 326].

5.4 FAULT MODELS FOR I_{DDQ} TESTING

In this section, fault models used to compute I_{DDQ} test sets are discussed. Leakage faults target bridges between the terminals of a transistor as well as gate-oxide shorts. The notion of pseudo stuck-at patterns as a means to compute I_{DDQ} tests for leakage faults is discussed. This is followed by a discussion of the capability of I_{DDQ} tests for leakage faults to detect other intra-gate defects. Next, fault models that target bridges between nodes in two different gates are presented. This includes the extracted set of bridges and external bridges. Situations when these fault models are to be used is discussed.

All the above fault models target only bridge defects. The reason is that, until recently [295], it was not clear whether or not I_{DDQ} testing can detect opens. The only I_{DDQ} test tool that explicitly targets some kind of open defect uses the weak fault model [110, 197, 198] which we discuss.

5.4.1 Leakage Faults

leakage faults Every transistor has four terminals: gate (G), source (S), drain (D) and bulk (B). In the leakage fault model, for every transistor six shorts between these four terminals are assumed. Thus, for every transistor, we have gate-source (GS), gate-drain (GD), gate-bulk (GB), source-drain (SD), source-bulk (SB) and drain-bulk (DB). For example, in Figure 5.8(a), we show the switch level description of a 3-NAND gate. For each of $P1, P2, P3$, $N1, N2$ and $N3$ we have six leakage faults.

Leakage faults can be thought of as bridging faults between nodes of a gate. For example, for transistor $N2$ of Figure 5.8(a), the six equivalent bridging faults for GS, GD, SD, GB, SB and DB are respectively $\langle b, h \rangle$, $\langle b, e \rangle$, $\langle e, h \rangle$, $\langle b, 0 \rangle$, $\langle h, 0 \rangle$, $\langle e, 0 \rangle$ where 0 denotes the ground node. This is shown in the row labeled $N2$ in Table 5.7. These six faults target defects that short any two of the four terminals of $N2$ resulting either due to extra material deposition or insulation breakdown as in the case of gate-oxide shorts. The equivalent bridges for leakage faults corresponding to $P1, P2, P3, N1$ and $N3$ are shown in Table 5.7 in rows labeled $P1, P2, P3, N1$ and $N3$, respectively. Node 1 denotes the V_{DD} node.

Values assigned to each node of the gate of Figure 5.8(a), for all possible input patterns, is shown in Table 5.8. Column 1 lists the pattern number. Symbol f denotes a floating node which signifies that there is no conducting path from either V_{DD} or ground to that node. Assume that there is a short between the gate and drain of $N2$, i.e., nodes b and e, and the input pattern to the gate is I8. Then, there is a conducting path from V_{DD} to ground along the pullup of the gate driving b, through the short between b and e and transistors $N2$ and $N3$. Thus, any test that results in pattern I8 at the gate is an I_{DDQ} test for

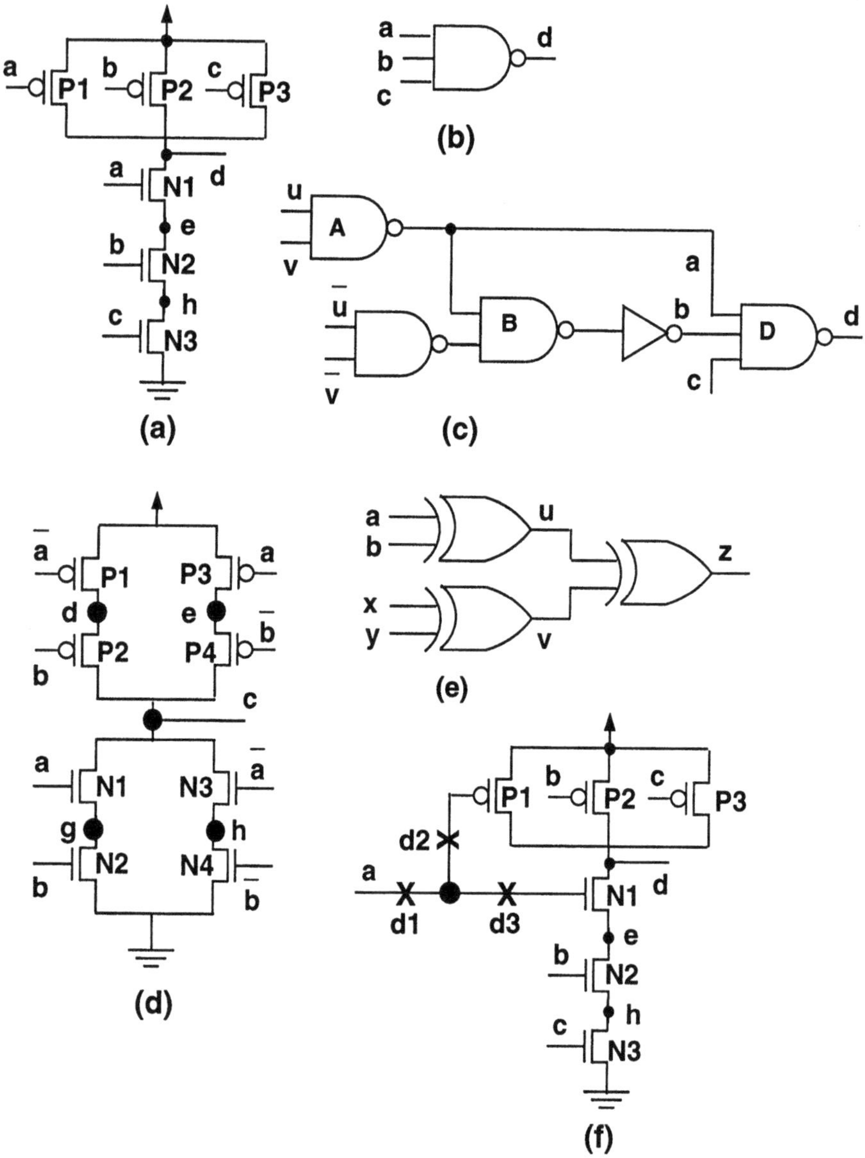

Figure 5.8 Example illustrating leakage and pseudo stuck-at faults

Table 5.7 Detection of leakage faults in a 3-NAND gate

	GS	GD	SD	GB	SB	DB
P1	(a,1) 1,2,3,4	(a,d) 1,2,3,4,8	(1,d) 8	(a,1) *	(1,1) NO	(d,1) *
P2	(b,1) 1,2,5,6	(b,d) 1,2,5,6,8	(1,d) 8	(b,1) *	(1,1) NO	(d,1) *
P3	(c,1) 1,3,5,7	(c,d) 1,3,5,7,8	(1,d) 8	(c,1) *	(1,1) NO	(d,1) *
N1	(a,e) 8	(a,d) *	(d,e) 4	(a,0) 5,6,7,8	(e,0) 5,6,7	(d,0) 1,2,3,4,5,6,7
N2	(b,h) 4,8	(b,e) 4,5,6,8	(e,h) 6	(b,0) 3,4,7,8	(h,0) 7	(e,0) 5,6,7
N3	(c,0) 2,4,6,8	(c,h) 2,4,6,7,8	(h,0) 7	(c,0) 2,4,6,8	(0,0) NO	(h,0) 7

Table 5.8 Possible combination of node values in a 3-NAND gate

	a	b	c	d	e	h
I1	0	0	0	1	f	f
I2	0	0	1	1	f	0
I3	0	1	0	1	f	f
I4	0	1	1	1	0	0
I5	1	0	0	1	1	f
I6	1	0	1	1	1	0
I7	1	1	0	1	1	1
I8	1	1	1	0	0	0

this fault. Similarly, it can be argued that if a test sets e to 1 and b to 0 then it is an I_{DDQ} test for the fault. Since pattern I1 sets e to f, it cannot cause a conducting path from V_{DD} to ground in the presence of the fault. Thus, no test resulting in pattern I1 at this gate is an I_{DDQ} test for the fault.

In Table 5.7, for each transistor of Figure 5.8(a) there are two rows. The first row lists, for each leakage fault, the equivalent bridging fault. The second row lists, for each leakage fault, the pattern numbers from Table 5.8 that detects the fault. Entries marked NO correspond to leakage faults in the gate that cannot be detected by I_{DDQ} testing independent of the circuit in which it is used. We differentiate such faults from undetectable leakage faults in a gate. We will come back to this issue later on. Entries marked * correspond to leakage faults that are equivalent to other leakage faults in the sense that they are shorts between identical nodes. For example, GD of N1 is equivalent to GD of P1.

5.4.2 Pseudo Stuck-at Tests

The idea of pseudo stuck-at tests has its root in the observation that a complete test set for stuck-at faults is very effective in detecting intra-node bridges [182]. The term pseudo stuck-at was however used much later [95, 202].

Pseudo stuck-at tests, as a vehicle for computing I_{DDQ} tests, are very popular with I_{DDQ} test tool developers. The reason is that they provide a simple approach to enhance existing stuck-at test generators to compute I_{DDQ} test sets. However, the approach has its limitations. We first discuss **pseudo stuck-at patterns** which is the basis for computing pseudo stuck-at tests. This is followed by a discussion of limitations of pseudo stuck-at test sets as I_{DDQ} test sets.

Consider the 3-NAND gate of Figure 5.8(b). Pattern I4 of Table 5.8 sensitizes and propagates a s-a-1 to d, the output of the gate. This is the **pseudo stuck-at pattern** for a s-a-1. Any input vector that causes pattern I4 at the input of the gate is a **pseudo stuck-at test** for a s-a-1 in a 3-NAND gate. Similarly, I6, I7 are pseudo stuck-at patterns respectively for b s-a-1 and c s-a-1. Pattern I8 is a pseudo stuck-at pattern for a s-a-0, b s-a-0 and c s-a-0. The four patterns I4, I6, I7 and I8 constitute an **exhaustive pseudo stuck-at pattern set** for the 3-NAND gate. This example can be generalized in several ways.

- The exhaustive pseudo stuck-at patterns for an n-NAND gate consists of $n+1$ patterns. Of these one pattern assigns 1 to all inputs. The rest of the patterns assign 0 to one and 1 to the rest of the inputs. The last n-vectors comprise a **walking 0 pattern**.
- The pseudo stuck-at patterns for an n-NOR gate consists of $n+1$ patterns. Of these patterns one assigns 0 to all its inputs. The rest of the patterns assign 1 to one of the inputs and 0 to the rest. The last n-vectors comprise a **walking 1 pattern**.
- Exhaustive pseudo stuck-at patterns for an arbitrary CMOS gate are a set of patterns such that one or more patterns sensitize each stuck-at fault at the output of the gate and, each stuck-at fault at the input of the gate is

Table 5.9 Detection of leakage faults in a 3 NAND gate by the exhaustive pseudo stuck-at test patterns

	GS	GD	SD	GB	SB	DB
P1	4	4,8	8	*	NO	*
P2	6	6,8	8	*	NO	*
P3	7	7,8	8	*	NO	*
N1	8	*	4	6,7,8	6,7	4,6,7
N2	4,8	4,6,8	6	4,8	7	6,7
N3	4,6,8	4,6,7,8	7	4,6,8	NO	7

sensitized by one or more patterns and propagated to the output of the gate (not the primary output of the circuit) [202].

Any set of input vectors that applies an exhaustive pseudo stuck-at patterns for a gate D in a circuit C is an **exhaustive pseudo stuck-at test set for D in C**. Any set of input vectors that is an exhaustive pseudo stuck-at test set for each gate in C is an **exhaustive pseudo stuck-at test set for C**.

Table 5.9 was derived from Table 5.7 by restricting the set of input vectors to the four exhaustive pseudo stuck-at patterns, I4, I6, I7 and I8. Note that these four patterns cover all leakage faults in the gate! Next, in Table 5.7, note that patterns I4, I6, I7 and I8 are such that for each of them there are leakage faults that are detected only by that pattern. For example, only I4 detects the leakage fault SD of N1. Thus, if any of the exhaustive pseudo stuck-at patterns of a gate is not applied by a test set it will not detect some leakage faults in that gate. The following, stated without a formal proof, is a generalization of the above example. A circuit consisting only of NAND and NOR gates is referred to as a **NAND/NOR circuit**.

Lemma 5.4.2.1. An I_{DDQ} test set detects all leakage faults in gate D in a NAND/NOR circuit C if and only if it is an exhaustive pseudo stuck-at test set for D in C.

Lemma 5.4.2.2. An I_{DDQ} test set detects all leakage faults in a NAND/NOR circuit C if and only if it is an exhaustive pseudo stuck-at test set for C.

The properties of exhaustive pseudo stuck-at test set summarized above and their close correlation with single stuck-at test set have motivated many computer aided test tool vendors to modify their stuck-at test generators to compute a pseudo stuck-at test set. The pseudo stuck-at test set is then used as an I_{DDQ} test set. There are two keywords the reader should make a note of. They are: NAND/NOR circuits and exhaustive pseudo stuck-at test sets.

First consider why exhaustive pseudo stuck-at test set imposes a constraint. Sometimes, when a gate is embedded in a circuit, no set of input vectors can apply the exhaustive set of pseudo stuck-at patterns to the gate. Consider the 3-NAND gate D in circuit C of Figure 5.8(c). The constraints on D are such that no input vector can cause its inputs to be set to $\langle a = 0,\ b = 1,\ c = 1\rangle$. However, there are input vectors that set its inputs to $\langle a = 1,\ b = 0,\ c = 1\rangle$, $\langle a = 1,\ b = 1,\ c = 0\rangle$, and $\langle a = 1,\ b = 1,\ c = 1\rangle$. These three patterns constitute the **complete set of pseudo stuck-at patterns** for D in this circuit.

A set of input vectors that applies a complete pseudo stuck-at pattern at a gate D in a circuit C is a **complete pseudo stuck-at test set for D in C.** Here, $\{\langle u,v,c\rangle\} = \{\langle 0,1,0\rangle, \langle 1,1,0\rangle, \langle 1,1,1\rangle\}$ constitutes a **complete pseudo stuck-at test set for D in C.** Any set of input vectors of C that is a complete pseudo stuck-at test set for each gate in C is a **complete pseudo stuck-at test set for C.** The test set $\{\langle u,v,c\rangle\} = \{\langle 0,0,1\rangle, \langle 0,1,0\rangle, \langle 1,0,1\rangle, \langle 1,1,0\rangle\}$ is a complete pseudo stuck-at test set for the circuit of Figure 5.8(c).

In the above example, since pattern I4 cannot be applied to D the leakage fault SD of N1 is not detectable. This is an example of an **undetectable leakage fault.** An I_{DDQ} test set that detects all detectable leakage faults is a **complete I_{DDQ} test** with respect to leakage faults.

With this background let us reconsider Lemmas 5.4.2.1 and 5.4.2.2. The best that a pseudo stuck-at test generator can do to is to compute a complete pseudo stuck-at test set for the circuit. The best I_{DDQ} test set on the other hand is a complete I_{DDQ} test set. The two lemmas state that if the complete pseudo stuck-at test set turns out to be an exhaustive pseudo stuck-at test set then it is also a complete I_{DDQ} test set. For many circuits, complete pseudo stuck-at test sets are not exhaustive pseudo stuck-at test sets. The question is: in such cases will a complete pseudo stuck-at test set for a circuit always be a complete I_{DDQ} test set? We do not know if this question has been resolved or not. We conjecture that the answer is no but do not have a proof for it.

We next consider what happens if the circuit is not a NAND/NOR circuit. Consider a CMOS cell implementing the two input XOR function. There are different switch level implementations for such a cell. One such XOR implemen-

Table 5.10 Possible combination of node values for XOR gate

	a	b	$\bar{a}$	$\bar{b}$	c	d	e	g	h
T1	0	0	1	1	0	0	1	f	0
T2	0	1	1	0	1	f	1	0	1
T3	1	0	0	1	1	1	f	1	0
T4	1	1	0	0	0	1	0	0	f

Table 5.11 Tests detecting leakage faults in XOR gate

	GS	GD	GB	SD	SB	DB
P1	T3,T4	T1,T3,T4	T3,T4	T1	NO	*
P2	T3	T3,T4	T1,T3	T4	*	T1,T4
P3	T1,T2	T1,T2,T4	T1,T2	T4	NO	*
P4	T2	T1,T2	T2,T4	T1	*	*
N1	T4	T2,T4	T3,T4	T2	*	T2,T3
N2	T2,T4	T2,T3,T4	T2,T4	T3	NO	*
N3	T1	T1,T3	T1,T2	T3	*	*
N4	T1,T3	T1,T2,T3	T1,T3	T2	NO	*

tation is shown in Figure 5.8(d). If complements of the inputs are generated within the cell itself, in addition to the switch level circuit of Figure 5.8(d), two inverters are needed.

Values assigned to the nodes of the gate by the four input patterns are shown in Table 5.10. Patterns from Table 5.10 detecting each leakage fault in the XOR gate are shown in Table 5.11. Entries marked NO and * are interpreted as for Table 5.7. The 34 leakage faults of Table 5.10 along with 7 leakage faults per inverter make the total number of leakage faults in an XOR gate as 48. Next, consider the following facts.

- For any two input XOR gate, any three of the four patterns constitute an exhaustive pseudo stuck-at test set. For example, $\{T1, T2, T3\}$ of Table 5.10 is one such pseudo stuck-at test set for the two input XOR gate.

Table 5.12 A pseudo stuck-at test set for 4 input parity checker

	a	b	x	y	u	v
T1	1	0	1	1	1	0
T2	1	1	1	0	0	1
T3	0	0	0	0	0	0
T4	0	1	0	1	1	1

- In Table 5.11, for each pattern P there are three leakage faults that are only detected by P. For example, for $T1$ these leakage faults are: GS of N3, SD of P1 and SD of P4.
- If a test set applies only three of the four input patterns at an XOR gate then it can detect only 45 out of the 48 leakage faults.

Next, consider the 4 input parity circuit of Figure 5.8(e). The reader can verify from Table 5.12 that the set of vectors $\{T1, T2, T3\}$ is a complete pseudo stuck-at test set. Observe that this complete pseudo stuck-at test set covers 45 of the 48 leakage faults in each XOR gate. Thus, this complete pseudo stuck-at test set covers 93.75% of all the detectable leakage faults. In fact, by adding one more vector, viz., T4 of Table 5.12, we get an I_{DDQ} test set that detects all leakage faults in the circuit. Thus, we conclude the following:

Lemma 5.4.2.4. A complete pseudo stuck-at test set for C need not be a complete I_{DDQ} test set.

5.4.3 Intra-Gate Defects

In addition to defects covered by leakage faults there are other defects that are confined to a gate. They include both opens and shorts. First we consider intra-gate shorts. For the 3-NAND gate of Figure 5.8(a) shorts that do not qualify as leakage faults are listed in row 1 of Table 5.13. Patterns from Table 5.8 that detect them are listed in the second row. In the third row patterns from the exhaustive pseudo stuck-at patterns that detect these faults are listed. Note that all these faults are detected by the exhaustive pseudo stuck-at patterns. This is another good property of pseudo stuck-at test sets. We conjecture that the result may be valid only for NAND/NOR circuit. We generalize as follows:

Table 5.13 Detection of intra-gate shorts

Faults	(a,b)	(a,c)	(a,h)	(b,c)	(c,e)	(d,h)
Vectors	3,4,5,6	2,4,5,7	6,8	2,3,6,7	4,5,7,8	2,4,6
Pseudo stuck-at	4,6	4,7	6,8	6,7	4,7,8	4,6

Lemma 5.4.3.1. An exhaustive pseudo stuck-at test set for a NAND/NOR circuit detects all intra-gate shorts.

Next, consider opens and use Figure 5.8(f) as an example. We have already seen that if the open is narrow then detection using I_{DDQ} testing is uncertain. Therefore only large opens are considered. First consider defect $d1$ which results in the gates of the complementary pairs $P1$ and $N1$ to float and the transistor pair conducts strongly. Any I_{DDQ} test that detects the SD leakage fault of either $P1$ or $N1$ also detects this defect. We refer to such defects as **large open input** defects.

Next consider defect $d2$. If the defect is relatively far away from the poly-diffusion overlap then $P1$ conducts strongly. This defect is equivalent to the SD leakage fault of $P1$. Similarly, it can be seen that defect $d3$, when located relatively far away from the poly-diffusion region, is equivalent to the SD leakage fault of $N1$. These defects are examples of **large open gate** defects. We generalize these examples as follows:

Lemma 5.4.3.2. A large open input defect at input X of gate G is detected by any I_{DDQ} test for the SD leakage fault in Q, where Q is any transistor in G driven by X.

Lemma 5.4.3.3. A large open gate defect in transistor Q is detected by any I_{DDQ} test for SD leakage fault in Q, provided the open is far away from the poly-diffusion overlap thereby causing Q to conduct strongly.

5.4.4 Intergate Bridges

We next look at fault models for computing I_{DDQ} test for bridges between nodes belonging to two different gates. Such defects are targeted by I_{DDQ} tests computed for bridging faults (**BFs**). A BF $\langle x, y\rangle$ is said to occur if two distinct nodes of the circuit are shorted. It is an inter-gate BF if nodes x and y belong to two different gates of the circuit.

On application of an input vector, a node can be set to either 0 or 1 or it could be floating. From the examples in Chapter 4 and the discussion on leakage faults, it should be clear that an input vector T is an I_{DDQ} test for the BF $\langle x, y \rangle$ if and only if T assigns 1 to one of the nodes and 0 to the other node. Note that none of the nodes can be floating.

An important question that I_{DDQ} test tool designers must answer is which set of BFs must the I_{DDQ} test tools target? The answer, we believe, depends on the context in which these tools are being used. We next discuss two different situations.

Extracted Bridges

The first situation is applicable when design flow I of Figure 5.9 is assumed. In this case it is assumed that everything about the design is complete and the layout of the circuit is known. Inductive fault analysis [89, 144] is done to determine a likely set of BFs. We refer to this set of BFs as **extracted BFs**. Test sets for I_{DDQ} testing are computed based on the BFs so determined.

Two kinds of defects can occur: local and global. Global defects tend to affect a large number of nodes and are often catastrophic in nature. Such defects are usually detected either during power-up or by tests that detect local defects. Local defects are those that occur due to point defects. Bridges due to such defects occur between nodes that are physically close.

Tools exist for extracting BFs resulting from point defects [144]. These tools assume that information about the defect data, after defect analysis of the process, is available. The likely point defects are assumed to be circular and the list of faults is extracted as follows [144].

Physically, every node in the circuit is a collection of rectilinear features on different layers of the circuit. Features on adjacent layers that belong to the same node are connected by a via. For most layouts rectilinear features on a layer are rectangles. Two types of BFs are possible: intra-layer and inter-layer BFs.

Intra-layer BFs occur when a defect connects features from two different nodes that lie on the same layer. In order to determine if two features on a layer can

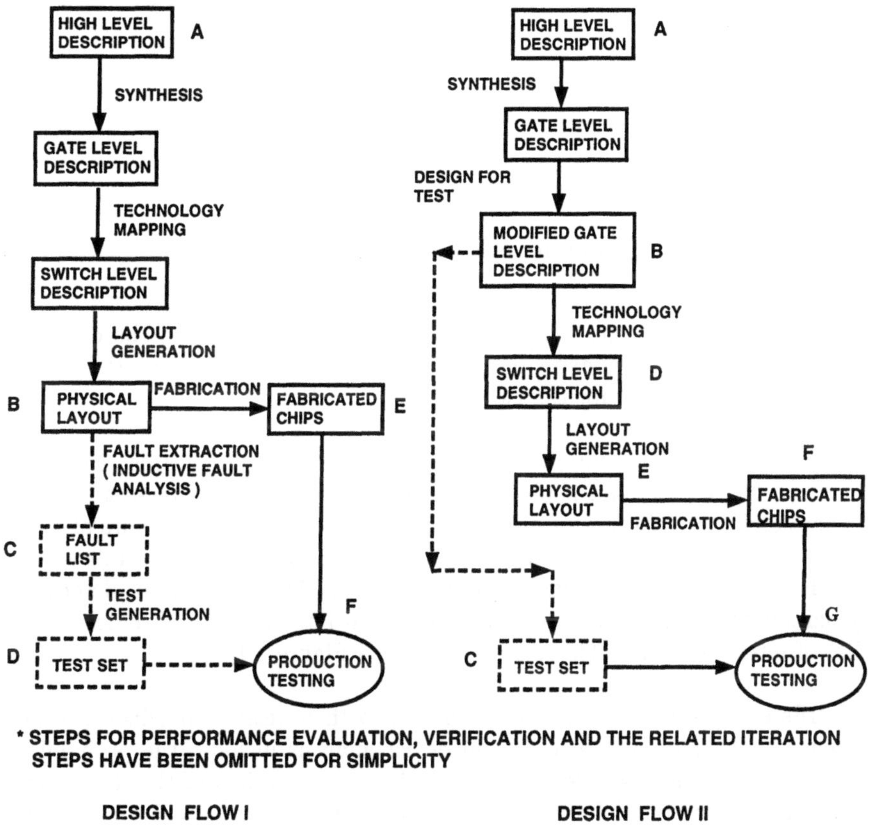

Figure 5.9 Design Flows

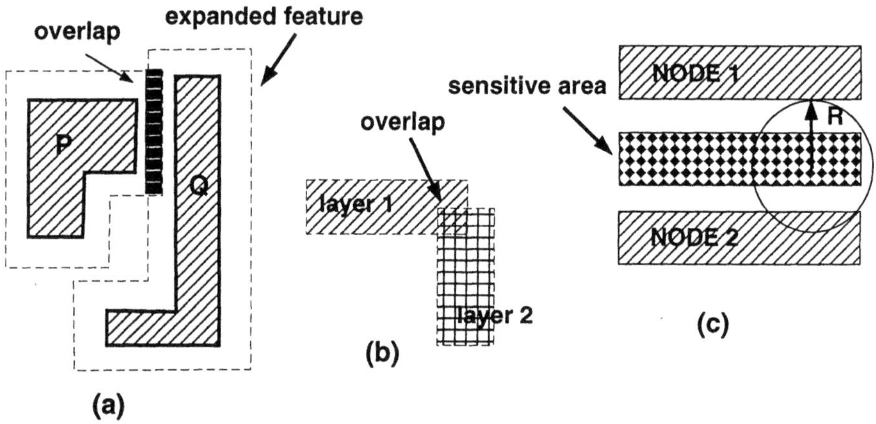

Figure 5.10 Extraction of Bridging Faults

be connected by a defect, consider the case where the radius of the defect is R. Then, each feature is expanded in all directions by an amount R. This is shown in Figure 5.10(a) for the two features P and Q.

The features outlined using dotted lines are the expanded features. Next, an algorithm for determining if pairs of the expanded features intersect is used. Efficient algorithms for performing this computation are known. Pairs of overlapping features belonging to different nodes determine the list of intra-layer BFs. For inter-layer BFs a similar method is used. An algorithm for determining pairs of features on the layers of interest that intersect is used to determine a list of inter layer BFs. This can be seen from Figure 5.10(b).

The BFs so collected are ranked according to their relative likelihood of occurrence. Information gathered from defect analysis includes the probability distribution function of defects as a function of the radius of the defect and the layer in which the defect occurs. This information, along with the relative likelihood of the BF occurring when the defect has radius R is used to rank the BFs. The relative likelihood of two BFs occurring, when the radius is R, can be measured by the overlap area of nodes in question. The method used in [144] is illustrated in Figure 5.10(c). For each BF the sensitive area of the BF is computed. The sensitive area for two features on the same plane is the area of the sensitive region. If the center of the defect with radius R is placed in the sensitive region then the two features will be shorted. If it is placed outside this region then there will be no short.

All and Metal Bridges

The extracted BFs should be used when test sets for production testing are required. In this case test sets are computed after the layout of the circuit is known. In addition to preparing tests for production testing, test tools are often used early in the design phase. These tasks are collectively referred to as **design for test** tasks. When design for test is used the design flow changes as shown in Figure 5.9. This new design flow is referred to as design flow II. In this case the layout of the circuit is not known.

There are many design for test tasks that are performed before the layout is known. We list a few of them here.

- Consider the task of determining which LFSR to be used as a test generator for a portion of the circuit. Fault simulation, to determine fault coverage and test length of the psuedo random test set, is performed.
- Weighted random patterns are used to reduce test length in a BIST environment [27, 152, 216, 243, 324, 339, 340]. In addition to finding an appropriate LFSR, the weight sets have to be computed. These weight sets determine the design of the on-chip test generator. The weight sets are computed using fault simulation.
- In one approach to designing on-chip test generators a test set is used as a guide [121, 122, 236, 319]. This approach improves fault coverage and reduces test application time.
- In a full-scan environment reduction of test application time is an important objective. Solutions to this problem use a combination of two techniques: finding a good ordering of flip-flops in the scan-chain; and finding a good order in which to apply the test vectors. The heuristics use extensive fault simulation and require a test set [112, 125, 172, 213, 244].
- In a partial scan environment, the primary objective is to select a subset of the flip flops and properly order the selected set of flip-flops. Flip-flop selection algorithms require extensive fault simulation [125].

The above examples suggest that fault simulation and test generation prior to layout are required. In many cases the difference between the circuit before and after the modifications done for ease of testing are small. For such cases it may sound attractive to generate the preliminary layout prior to modifying the design. Extract a fault list from the preliminary layout and use it during the design for test task. One needs to be careful in using such an approach.

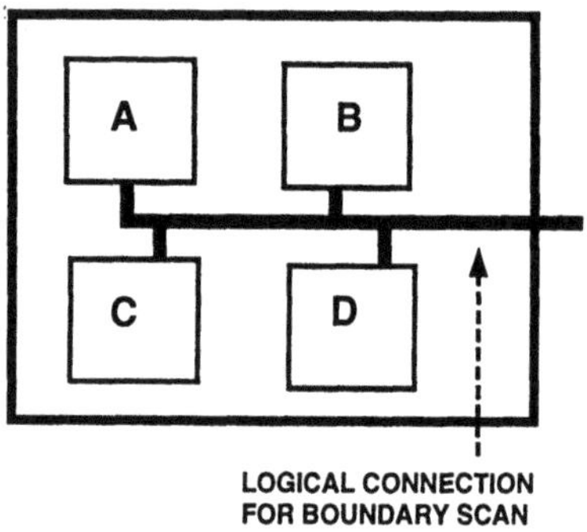

Figure 5.11 Board/MCM Assembly

The preliminary layout and final layout are generated using two different runs of the same layout generator. Layout generators use non-deterministic algorithms like simulated annealing, genetic-algorithm, simulated evolution, etc., for placement and routing [285]. Therefore, even if we have exactly the same circuit description and the same layout generator we would get vastly different layouts from two different runs. The proximity relationship between nets, which is the basis of fault extraction, even for identical circuits will not be preserved across two layouts generated by two different runs of the same layout generator.

We have so far considered the design flow for a single VLSI chip. Such chips are often integrated into larger systems. Two examples are board level and multchip module (MCM) level integration. As shown in Figure 5.11 the VLSI chips, or their unpackaged counterparts in MCMs, are mounted on a board/substrate.

For testing, the boundary scan test structure is built into the board/MCM. The assembled system is tested by first testing the boundary scan interconnect and then applying the test sets for A, B, C and D either in turn or in groups. In either case, test sets for A, B, C and D are required.

When such assemblies are manufactured in large volume, each module A, B, C or D may have several IC suppliers. In that case, if different test sets depending on the layouts chosen by suppliers are used, a large number of distinct test sets for the assembly will be required. During production testing, for each possible combination of suppliers for A, B, C and D, the proper test set will have to be selected. For example, assume that there are just two suppliers for each of A,

B, C and D. This implies that a total of 16 different tests for this assembly will have to be maintained. The number of such distinct tests grow exponentially with the number of chips. This is clearly not a feasible option.

One way to get around the above situation is to use a standard test set (independent of the supplier's layout) for each chip. If BFs are to be targeted then a standard test set obtained using layout independent analysis of BFs have to be computed.

Two different layout independent sets of BFs have been considered. In one case all possible BFs between all switch-level nodes are considered. This is the most comprehensive set of BFs and henceforth will be referred to as **all-BFs**. The set of all BFs between the lines of the circuit that are either inputs or outputs of a gate are referred to as **metal-BFs**. These nodes usually have features in the metal layer. This set of BFs is interesting for two reasons. Firstly, it has been observed that a good percentage of the more probable extracted BFs are due to overlap between features in the metal layer. This is because wiring consumes most of the layout area and all longer features tend to be part of metal wires. Secondly, in certain design styles like gate arrays, the only layer used for customizing a design is the metal layer. For such circuits the likelihood of a bridge between two metal wires is very high.

5.4.5 Weak Faults

I_{DDQ} tests for leakage faults detect some open defects although they only target bridges. Weak faults, on the other hand, target a class of open defects. The material on weak faults is taken from [197].

It is well known that NMOS transistors can pass logic 0 without degradation but degrade the voltage level for logic 1 [215]. If the threshold voltage of the device is 0.7 V then a 5 V gets degraded to 4.3 V. PMOS transistors, on the other hand, pass logic 1 without any degradation but degrade the logic level of 0. This was confirmed by circuit simulation where the circuit of Figure 5.12(a) was simulated [197]. From Table 5.14 we see that while the voltage level at A is degraded it is correctly interpreted by the succeeding gate. Therefore, such defects cannot be detected by logic testing. However, the degraded voltage at A leads to significantly high I_{DDQ} through the driven gate (the inverter of our example) as can be seen from Table 5.14.

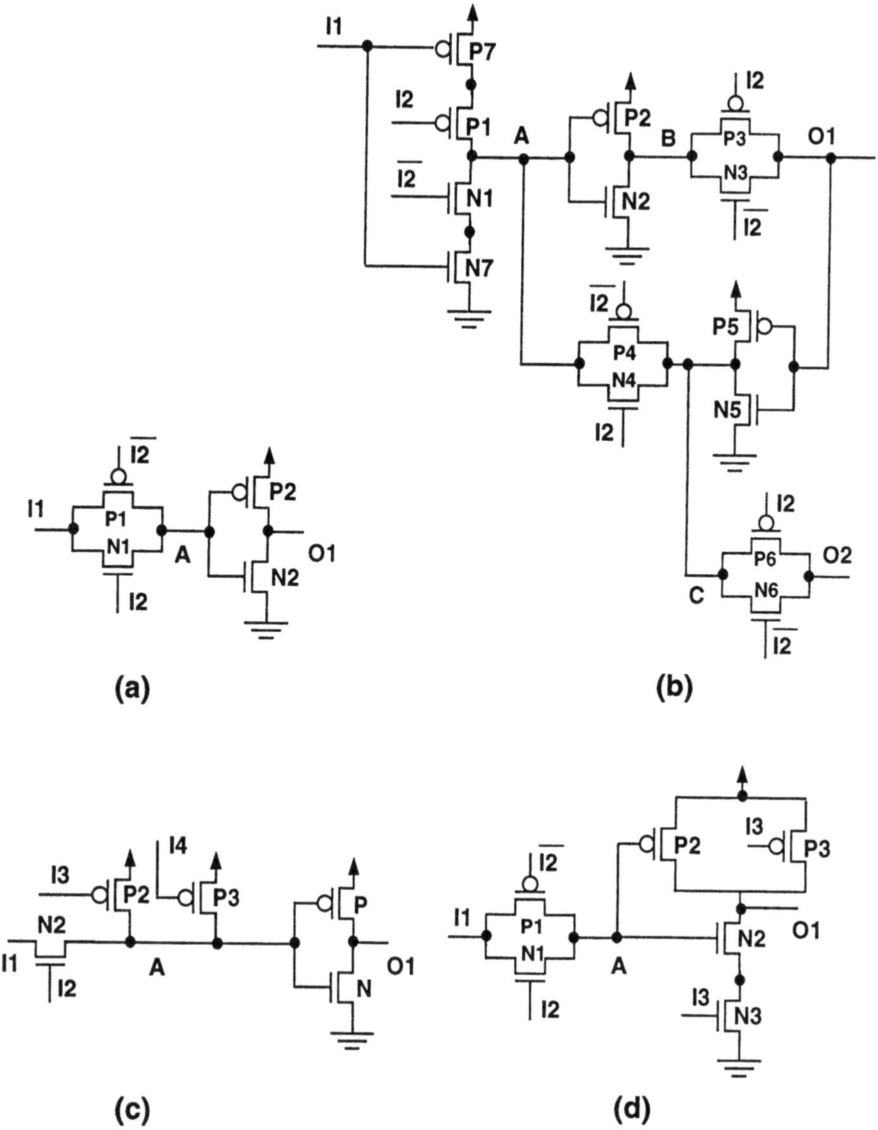

Figure 5.12 Weak faults

Table 5.14 Simulation results for the detection of weak faults

Defect	V_{I1}	V_A	V_{O1}	I_{DDQ}
P1 Open	5.0 V	3.6 V	0.1 V	56 μA
N1 Open	0.0 V	1.5 V	4.8 V	94 μA

If a node has a normal voltage level for logic 0 in a defect free circuit but has a degraded voltage level for logic 0 in a defective circuit the defect is modeled as a **weak-0 fault**. **Weak-1 fault** is similarly defined. A **transmission path** between nodes x and y is a chain of channel connected transistors. If all transistors in a transmission path are turned on then it is called a **conducting path**. A transmission path (conducting path) consisting only of PMOS transistors is a **normal-1 transmission path (normal-1 conducting path)**. **Normal-0 transmission paths (normal-0 conducting paths)** are transmission (conducting) paths consisting only of NMOS transistors. Conducting paths with at least one PMOS (NMOS) transistor is a **weak-0 (weak-1)** conducting path. In Figure 5.12(b): $N4$ and $N5$ form a normal-0 transmission path between A and ground; $P2$ and $P3$ form a normal-1 transmission path between V_{DD} and $O1$. If $I1 = 1, I2 = 0$: $P2, P3$ is a normal-1 conducting path between V_{DD} and $O1$; and $P2, N3$ is a weak-1 conducting path between V_{DD} and $O1$.

The necessary and sufficient condition for a defect to be mapped onto a weak-0 (1) fault at node x follows.

1. There should be at least one normal-0(1) conducting path between x and ground (V_{DD}) in the defect free circuit and all normal-0(1) conducting paths should be blocked in the defective circuit.

2. There should be at least one weak-0(1) conducting path between x and ground (V_{DD}) in the defective circuit.

For example, in Figure 5.12(b) if $I2 = 0$ and $I1 = 1$, then there is a normal-1 conducting path from V_{DD} to $O1$. Transistor $P3$ could be cut-off because of a number of reasons. Either the source or drain of $P3$ could be open. Any of these defects will result in a weak-1 fault at $O1$. Similarly, the source or drain of $N3$ could be open resulting in $N3$ being cut-off. This defect can result in a weak-0 at $O1$. Thus a node can have a weak-0 as well as a weak-1 fault.

Another inportant point to be noted is that different defects might cause the same weak-fault at a node but they are not equivalent. In Figure 5.12(c) if $N2$ and either $P2$ or $P3$ are on and I1 is set to 1 then A is at a normal 1. However, if any one among $P2$ and $P3$ is cut-off then there is a weak-1 at A. This is because, on application of $I1 = 1, I2 = 1, I3 = 0$ ($I1 = 1, I2 = 0, I3 = 1$) when $P3$ ($P2$) is open we get a weak-1 at A. Therefore, although both of them cause a weak-1 at A they are not equivalent and are therefore considered as two different faults.

All our examples so far were such that detection of the weak fault was guaranteed once it was sensitized. This is so because the weak faults were at nodes that drove inverters. Consider Figure 5.12(d) and assume that P1 is open. If we set $I1$ to 1 and $I2$ to 0 then there is a weak-1 at A. Although this assignment sensitizes the weak fault this is not guaranteed to detect the defect. In order to detect the defect a path from V_{DD} to ground has to be created through the weakly conducting transistors $P2$ and $N2$ by setting $I3$ to 1. This is referred to as **propagation of weak faults** . Thus, I_{DDQ} tests for weak faults, unlike all other I_{DDQ} tests that we have seen so far, must **sensitize and propagate** weak faults. The conditions are summarized below:

Sensitization of Weak-0(1) Faults. A weak-0(1) fault at node x, due to a defect, is sensitized by a test vector T if and only if T creates a normal-0(1) conducting path from ground (V_{DD}) to x in the fault free circuit, but creates only weak-0(1) paths with the defect in it.

Propagation of Weak-0(1) Faults. A node x can drive complementary pairs of transistors in a number of gates. Let G be one such gate such that x drives the complementary pair of transistors P and N. An input vector T is said to propagate the weak fault through G if and only if it creates a path in G from V_{DD} to ground such that the path passes through both P and N. T is said to propagate a weak fault at x if and only if it propagates the fault through any one of the gates it drives.

5.5 DEFECTS AND FAULT MODELS

We have so far studied different fault models and characteristics of tests computed using these fault models. In this section we look at a variety of defects and try to gauge the defect detection capability of tests so computed.

For slow speed testing it is assumed that tests for stuck-at and stuck-open faults are used. We will see that logic tests computed using stuck-at faults often fail to detect defects that change the function (static behavior) of the circuit.

For high speed testing we assume that hazard-free robust test-pairs for path delay faults are used. State-of-art delay test generators use only hazard-free propagation constraints to compute delay tests. We assume that future delay test generators will compute test-pairs that also satisfy the quality constraints. We will see that even if delay test-pairs satisfy both hazard-free propagation and quality constraints they often fail to "activate" the delay defect.

There is a very high correlation between the I_{DDQ} fault models and defects they target, especially for shorts. Therefore, I_{DDQ} tests are very effective in detecting bridges. For the three kinds of I_{DDQ} detectable opens, I_{DDQ} tests computed using leakage faults detect them. There are no known systematic methods for computing I_{DDQ} tests for source/drain opens. Opens in transmission gates are detected by tests computed for weak faults. Having noted this we only look at the defect detection capability of tests for stuck-at faults, stuck-open faults and delay faults. We first consider open defects followed by shorts. We will end with a brief summary.

5.5.1 Opens

Open *n*net.

(i) Large opens. These defects can be detected by any test-pair for a stuck-open fault in any transistor of the pulldown network. Tests for such opens cannot be invalidated by hazards at the input of the gate because no pulldown path conducts. Accordingly, tests for such defects are easier to compute than robust tests for stuck-open faults. Let x be the output of the gate and $T0$ ($T1$) any test for x stuck-at-0 (x stuck-at-1). Then, $\langle T0, T1 \rangle$ detects the fault. If x stuck-at-0 is redundant then any input vector that sets x to 1 can be used instead of $T0$.

(ii) Narrow opens. The detection of such defects by slow speed testing is questionable. Since they affect circuit delays using high speed testing is a possible approach to detect them. Any hazard-free robust test-pair for a path delay fault that results in a falling transition at the output of the gate will be a test for such defects.

Open *p*net.

(i) Large opens. These defects can be detected by any test-pair for a stuck-open fault in any of the transistors of the pullup network. Tests for such opens cannot be invalidated by hazards at the input of the gate because no path in the pullup conducts. Accordingly, test-pairs for such defects are easier to compute than robust tests for other stuck-open faults.

Let x be the output of the gate and $T1$ ($T0$) any test for x stuck-at-1 (x stuck-at-0). Then, $\langle T1, T0\rangle$ detects the fault. If x stuck-at-1 is redundant then any test that sets x to 0 can be used instead of $T1$.

(ii) Narrow opens. The detection of such defects by slow speed testing is questionable. Since they have an impact on circuit delay, high speed testing is a possible approach. Any hazard-free robust test-pair for a path delay fault that results in a rising transition at the output of the gate will be a test for such defects.

Open Source/Drain.

(i) Large Opens. Such defects are detected during slow speed testing by robust test-pairs for stuck-open fault at the transistor whose source/drain is open. If a test set has a test-pair for every stuck-open fault in a gate it will detect all such opens in the gate.

(ii) Narrow Opens. Such defects are not guaranteed to be detected by test-pairs for stuck-open faults. They have an impact on the circuit delay and can possibly be detected by high speed testing. However, as we show by an example, they cannot be detected by tests for path delay faults, unless we specifically target such defects.

Refer to Figure 5.13. The gate level circuit is shown in Figure 5.13(a). Let us consider two drain defects $d1$ and $d2$, shown in Figure 5.13(b), in gate E. Could these defects be modeled by the slow-to-rise fault at e? Consider the test-pair $\langle T1, T2\rangle$ where $T1 = \langle a = b = 1, c = d = 0\rangle$ and $T2 = \langle a = b = c = d = 0\rangle$. As can be verified from Figure 5.13(a) this test-pair detects a slow-to-rise fault at e. However, this test-pair will not activate the delay effect in the presence of either of the defects $d1$ and $d2$. In the presence of $d1$ ($d2$) e will be pulled up through transistor $P2$ ($P1$). Therefore, they cannot be modeled as delays at gate outputs.

The pair $\langle T1, T3\rangle$ ($\langle T1, T4\rangle$) activates $d1$ ($d2$) where: $T1 = \langle a = b = 1, c = d = 0\rangle$, $T3 = \langle a = 0, b = 1, c = d = 0\rangle$ and $T4 = \langle a = 1, b = c = d = 0\rangle$. Note that just like stuck-open faults we have to be concerned with hazards at the input of gate E because they can invalidate the test. For example, assume that there is a hazard at b when a test-pair $\langle X, Y\rangle$ for $d1$ is applied. If this hazard turns on $P2$ for a sufficiently long time then e is pulled high thereby invalidating the test. In addition, hazards at the primary output can also invalidate delay tests. We have already seen that test-pairs for stuck-open faults do not guarantee that primary outputs are hazard free. Hence, neither the test-pair for a stuck-open fault at the corresponding transistor (applied at the rate of high speed fault testing) nor the slow-to-rise delay fault at e is guaranteed to be a test for such defects. However, for our example, any single path propagating hazard-free robust test-pair

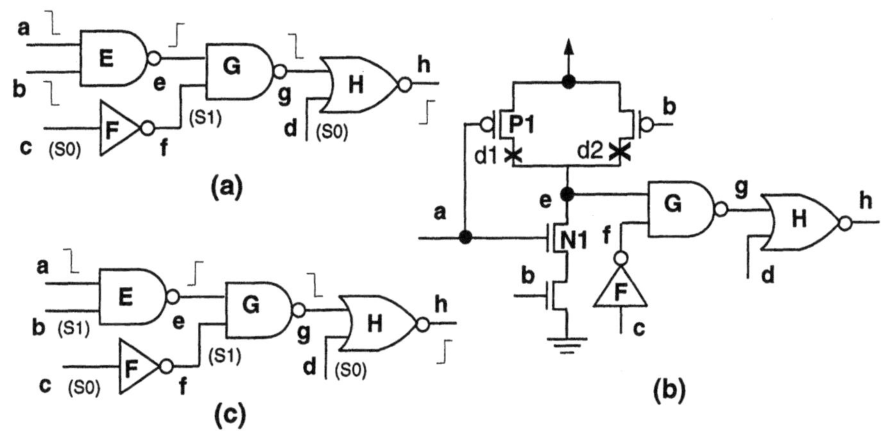

Figure 5.13 Delay tests to detect source/drain opens

for the path delay fault $\langle a, e, g, h\rangle_f$ ($\langle b, e, g, h\rangle_f$) detects $d1(d2)$. Using Figure 5.13(c) it can be verified that the test-pair $\langle a = f, b = S1, c = S0, d = S0\rangle$ detects defect $d1$. The test-pair is expressed using the notation of Table 5.3 which we will be using from here onwards.

Open gate

(i) Large Opens.

First consider *defect* $d1$ of Figure 5.14(a) where the gate of $N1$ is open. Then, $N1$ can either not conduct, conduct weakly or conduct strongly.

$N1$ does not conduct. In this case, during *slow speed testing*, a robust test-pair for $N1$ stuck-open detects the defect. During *high speed testing*, any hazard-free robust test-pair for $\langle a, k, p, s\rangle_r$ or $\langle b, k, p, s\rangle_r$ of Figure 5.14(b) detects it.

$N1$ conducts strongly. Depending on the characteristics of $P1, N1, N2$, an input vector that sets h and k to 1 and 0 respectively results in a weak 1 at p. Thus, the function of the circuit is not affected. However, the dynamic behavior is affected. The transistor pairs in gate S driven by p conducts weakly. This delays a falling transition, due to a rising transition at q, from propagating to s.

In order to activate the defect the second vector of the test-pair should set p to 1 by setting h and k to 1 and 0 respectively. Thus, a hazard-free robust test-pair for a path delay fault along a path passing through q and

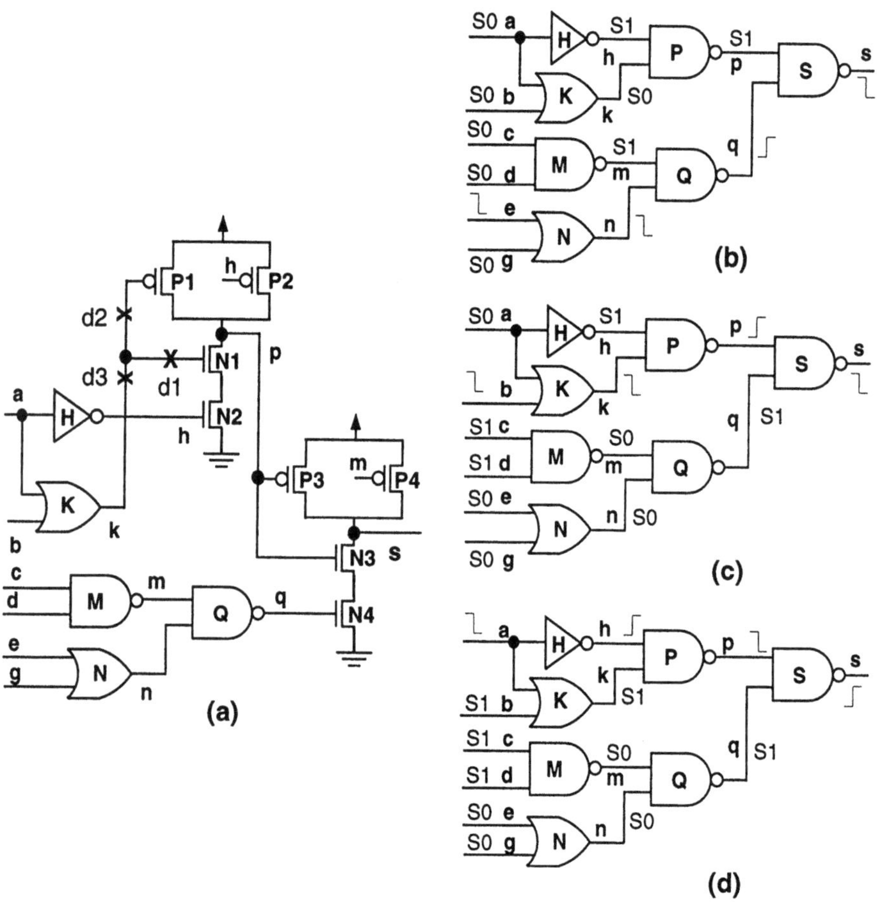

Figure 5.14 Delay tests to detect gate opens

causing a rising transition at q, detects the defect during high speed testing provided it also satisfies an additional constraint. That constraint is that the second vector sets h and k to 1 and 0 respectively. It can be verified using Figure 5.14(b) that $\langle a = S0, b = f, c = S0, d = S0\rangle$ is one such test-pair. On the other hand the test-pair $\langle a = S1, b = f, c = S0, d = S0\rangle$, which is a hazard-free robust test-pair for a path delay fault traversing q, that causes a rising transition at q, does not detect defect $d1$.

The additional constraint needed to "activate" the defect is interesting for the following reason. The constraints are on lines h and k which are neither off-path nor side fanout nodes of the path $\langle e, n, q, s\rangle$. Therefore, this does not qualify either as a hazard-free propagation or quality constraint. **Hence, even if we compute a delay test that satisfies both the constraints it is not guaranteed to detect the defect.**

$N1$ **conducts weakly.** In this case p falls slowly and any hazard-free robust test-pair for a path that passes through p and causes a falling transition at p detects this defect during high speed testing.

Next let us consider *defect* $d2$ of Figure 5.14(a) where the gate of $P1$ is open. Once again, $P1$ can either not conduct, conduct weakly or conduct strongly.

$P1$ **does not conduct.** In this case during slow speed testing, a robust test-pair for $P1$ stuck-open detects the defect. We have to be careful in creating a test-pair for this defect during high speed testing. Such a test-pair, in addition to causing a rising transition at p and propagating it to the output should also satisfy an additional constraint. The constraint is that the second vector of the test-pair should set h to 1 and k to 0. The reader can verify using Figure 5.14(c) that $\langle a = S0, b = f, c = S1, d = S1, e = S0, g = S0\rangle$ is one such test-pair.

$P1$ **conducts strongly.** In this case depending on the characteristics of $P1, N1$ and $N2$, any input vector that sets both h and k to 1 results in a weak 1 at p. Thus, unlike the defect $d1$, it may be detectable as a p stuck-at 1 fault during slow speed testing. It may also be possible to detect this defect during high speed testing. The weak 1 at p causes the transistor pair $P3$ and $N3$, in gate S driven by p, to conduct weakly. This, in turn, delays a falling transition due to a rising transition at q, to propagate to s. In order to activate the defect the second vector of the test-pair should set both h and k to 1 in the fault free case. Thus, analogous to defect $d1$, a hazard-free robust test-pair for the path delay fault that causes a rising transition at q detects the defect during high speed testing. However, the additional constraint that it must satisfy is different in that the second vector should set both h and k to 1.

Once again note that the constraints are on lines h and k which are neither off-path nor side fanout nodes of any path passing through q. Therefore, this does not qualify either as a hazard-free propagation or quality constraint. **Hence, like $d1$, even if we compute a delay test that satisfies the hazard-free propagation constraints and quality constraints it is not guaranteed to detect this defect.**

Next consider *defect* $d3$ of Figure 5.14(a) where the gates of both P1, N1 are joined and open from the rest of the circuit. In this case the two transistors are usually in a strongly conducting state. If pullup strength of $P1$ is much stronger than the pulldown strengths of $N1$ and $N2$ then setting both k and h to 1 will not set p to 0 and hence the fault can be detected as a stuck-at 1 fault during slow speed testing. Once again, p could be a weak 1 on application of a vector that sets h and k either both to 1 or to 1 and 0, respectively. This causes a slow-to-fall transition at s in response to a rising transition at q. Thus, it can be detected by any path delay fault test-pair that detects the defect $d1$. The test-pair of Figure 5.14(c) is one such test-pair. **Once again, the reader can verify that even if we compute a delay test that satisfies both the hazard-free propagation and the quality constraints it is not guaranteed to detect this defect.**

(ii) Narrow opens. For narrow opens a small amount of leakage current flows across the open. This slowly charges or discharges the floating gate.

First consider *defect* $d3$. Slow speed testing cannot detect this defect because, given sufficient time, the floating gates acquire the correct voltage and hence the static behavior is not affected.

This defect affects both rising and falling transitions at p. Will $d3$ be detected by a hazard-free robust test-pair for a path delay fault on a path passing through p and causing a falling transition at p? Such test-pairs detect the defect provided the first vector sets h to 0. This is necessary to ensure that p is in a known state. For our example, a test-pair for a path delay fault that propagates a rising transition along h is necessary. It can be verified, using Figure 5.14(d), that test-pair $\langle a = f, b = c = d = S1, e = g = S0\rangle$ detects the defect $d3$.

The reader can examine the other two defects using a similar argument. The main conclusion will be that we have to be careful during robust delay-fault test generation because additional defect detection constraints must be satisfied.

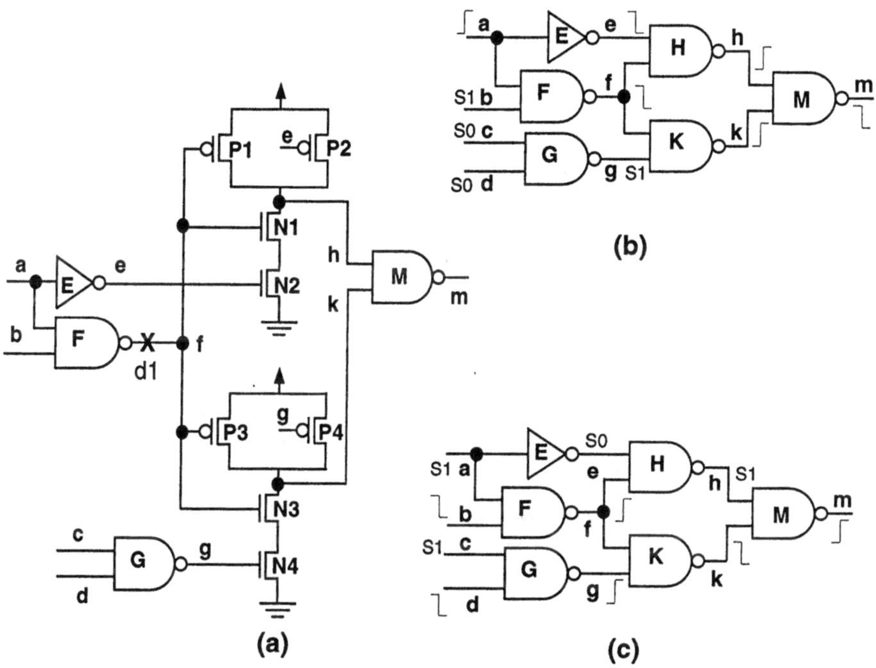

Figure 5.15 Delay tests to detect output opens

Open output.

(i) Large opens. Consider defect $d1$ of Figure 5.15(a) whose gate level circuit is shown in Figure 5.15(b). In this case all transistors $P1, N1, P3$ and $N3$ conduct strongly.

If $P1$ ($P3$) has a much larger pullup strength than $N1$ and $N2$ ($N3$ and $N4$) then slow speed testing will detect it as a h stuck-at-1 (k stuck-at-1). It may also manifest itself as both h stuck-at-1 and k stuck-at-1.

Note that if the pullup strength of $P1$ ($P3$) is not much larger than $N1$ and $N2$ ($N3$ and $N4$) then a degraded 1 could result at h (k). This, under certain conditions, delays the falling and rising transition at m. Let us again consider how a test-pair $\langle T1, T2\rangle$ can detect such a defect during high speed testing. Both $T1$ and $T2$ set g to 1 ensuring that k is set to a weak 1 during the entire interval in the presence of the defect. In addition, $T1$ and $T2$ should create a rising transition at h which causes a slow-to-fall transition at m. One such test-pair is shown in Figure 5.15(b).

(ii) Narrow opens. Slow speed testing cannot detect this defect because, given sufficient time, the floating gates will acquire the correct voltage. Hence the static behavior is not affected. They however affect the dynamic behavior.

Next, let us investigate how a test-pair $\langle T1, T2\rangle$ can detect this defect during high speed testing. Both $T1$ and $T2$ set e to 0 thereby ensuring that h is set to 1 for the entire duration. Let $T1$ set g to 0 thereby ensuring that the value at k after application of $T1$ is 1. Let $T1$ also set f to 0 so that the gates of $P3$ and $N3$ are discharged to a value close to 0. This will cut-off $N3$ and turn-on $P3$. Next, $T2$ switches both g and f to 1. Since the open conducts a small leakage current the falling transition at k will be delayed. It can be verified from Figure 5.15(c) that a test-pair that satisfies all these constraints is $\langle a = S1, b = f, c = S1, d = f\rangle$. It will therefore detect the defect.

This example is interesting because the constraint require transitions at two different inputs f and g of a gate. **Such constraints are not required by single path propagating delay tests but may be required required for a multiple path propagating delay test.**

5.5.2 Bridges

Two kinds of resistive bridges viz. external and intra-gate bridges are considered. Three kind of intra-gate defects viz. gate-source, gate-drain and source-drain bridges are considered.

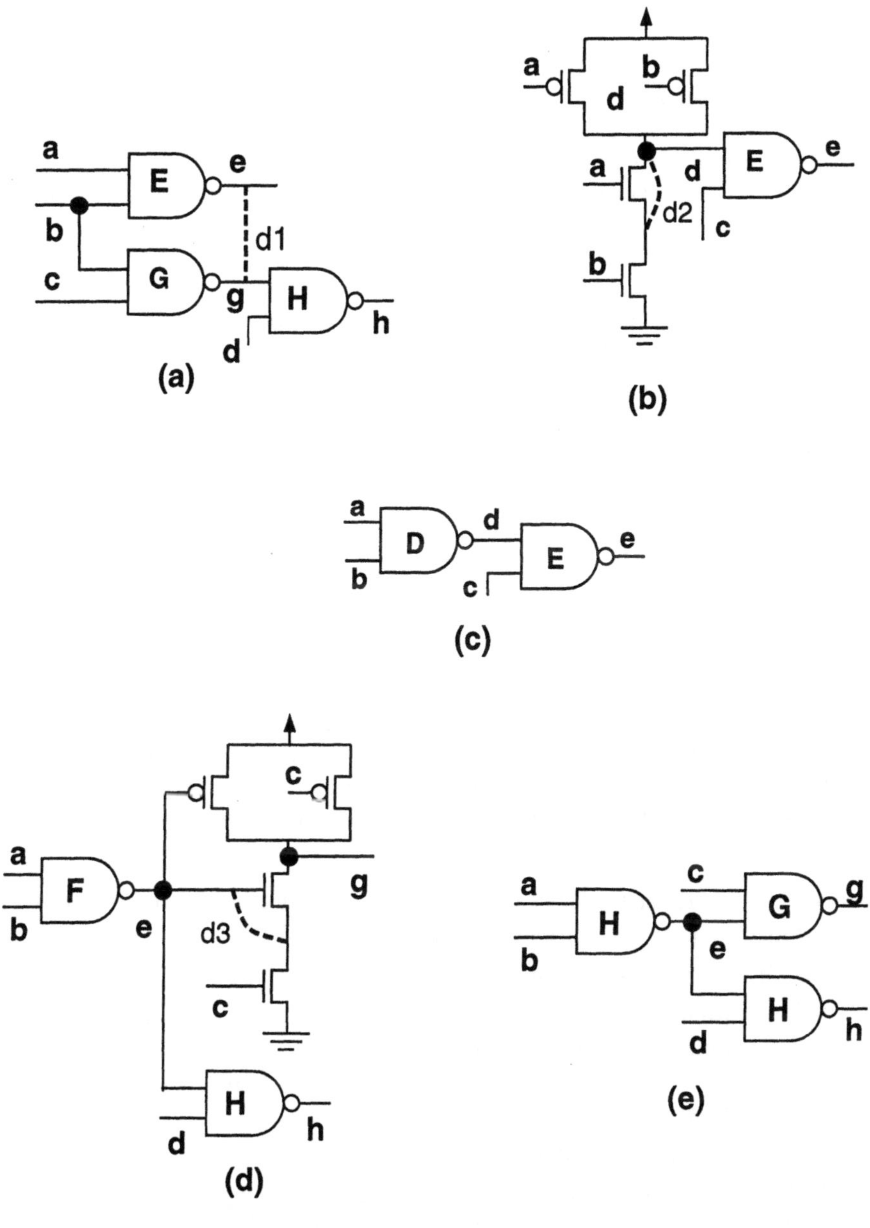

Figure 5.16 Delay tests to detect bridges

External. Refer to $d1$ of Figure 5.16(a) for an example. Two cases are considered.

Small bridge resistance.

In this case if T is such that in the fault-free case $T(e) = 0, T(g) = 1$ then the value at e in the faulty circuit will be a weak 1. Hence, it may be detectable during slow speed testing by a test for e stuck-at-1. However, note that not every test T for e stuck-at-1 detects this defect since T must satisfy the constraint that $T(a) = 1, T(b) = 1$ and $T(c) = 0$. This is an example of **defect sensitization.** There are a number of tests for e stuck-at-1, viz., $\langle a = 1, b = 1, c = 1, d = 1\rangle$ that will not activate this defect and hence will not detect it. The fault can also be activated by setting e to 1 and b to 0. The fault may be detectable as g stuck-at-1. The test $\langle b = c = 1, a = 0, d = 1\rangle$ for g stuck-at-1 detects it but $\langle a = b = c = d = 1\rangle$ does not. Thus, **such defects are not guaranteed to be detected by tests for stuck-at faults.**

Large bridge resistance.

In this case the defect is unlikely to be detected during slow speed testing. However, it may be detected during high speed testing. A test-pair for such a defect should activate the defect by setting g to an intermediate voltage. This causes the transistor-pair in gate H driven by g to conduct weakly and leads to the delay fault $\langle d, h\rangle_r$. To activate this path delay fault the test-pair $\langle T1, T2\rangle$ should be such that either: $T2(e) = 1, T2(g) = 0$; or $T2(e) = 0$, $T2(g) = 1$. This constraint is such that: e is neither an off-path input nor a side fanout node; and the test-pair cannot satisfy the hazard-free propagation property. Hence, **no hazard-free robust test-pair computed using present delay test generators detects this defect !!**

Source drain. Refer to defect $d2$ of Figure 5.16(b) for an example. The corresponding gate level circuit is shown in Figure 5.16(c).

Small bridge resistance.

In this case if T sets d to a weak 0 then it can detect the fault during slow speed testing provided it is also a test for d stuck-at-0. To do this T must set a and b respectively to 0 and 1. This implies that T should be a test for a stuck-at-1.

Large bridge resistance.

In this case the defect is unlikely to be detected during slow speed testing. However, it may be detected during high speed testing provided $T2$ of the test-pair $\langle T1, T2\rangle$ sets d to a weak 1. This activates the path delay fault $\langle c, e\rangle_r$. Thus, to activate the defect the test-pair for $\langle c, e\rangle_r$ must satisfy the constraints $T2(b) = 1$ and $T2(a) = 0$. Since a is neither an off-path input nor a side fanout node of $\langle c, e\rangle$ these constraints on the test-pair are neither hazard-free propagation nor quality constraints. **Hence, no hazard-free robust test-pair for path delay faults can guarantee the detection of this defect.**

Gate source. Refer to defect $d3$ of Figure 5.16(d) for an example. The corresponding gate level circuit is shown in Figure 5.16(e).

Small bridge resistance.

In this case a test T during slow speed testing can detect it as an e stuck-at-0 fault provided, in the fault-free case, $T(e) = T(c) = 1$. Thus, a test for $e1$ stuck-at-0, for our example, detects this defect.

Large bridge resistance.

In this case the defect is unlikely to be detected during slow speed testing. However, if e is set to an intermediate value due to the defect then it activates the $\langle d, h\rangle_r$ path delay fault. To activate the defect the test-pair should be such that the second vector should set both e and c to 1. The reader can verify that **this constraint is such that it is neither a hazard-free propagation constraint nor a quality constraint.** Hence, **hazard-free robust test-pairs for path delay faults are not guaranteed to detect such defects.**

Gate drain. Such defects are very similar to gate source defects. We leave it to the interested reader to construct examples showing that for small resistances such defects are often detected by stuck-at tests. On the other hand, for a large bridge resistance, even though it affects circuit delays, present delay test generators will not compute tests to guarantee detection.

5.5.3 Summary

From our discussion in Chapter 4 and this chapter the following conclusions can be drawn.

Bridges.

- I_{DDQ} testing is very effective in detecting bridges. A much wider range of bridge resistances cause appreciable elevation in I_{DDQ} when compared with the range of bridge resistances that cause a change in the functional or timing characteristics.

- Effective I_{DDQ} tests targeting bridges have been defined.
- Even if a bridge causes a change in the function such a defect is not guaranteed to be detected by tests for stuck-at faults.
- Even if a bridge causes a change in the timing behavior such a defect is not guaranteed to be detected by robust hazard-free test-pairs for path delay faults. In addition, there are some defects that require test-pairs that are neither robust nor hazard-free. Delay tests to detect such defects have not been defined.

Opens.

- Experiments have shown that I_{DDQ} testing detects a large percentage of opens.
- I_{DDQ} tests for all but one class of opens have been defined. A characterization of I_{DDQ} tests, that can be used by test tools, is required.
- Some opens cause changes in the function while others cause a change in the timing behavior of circuits. Hazard-free robust test-pairs for path delay faults are not guaranteed to detect such defects. A characterization of delay tests to detect such defects is required.

It is clear that I_{DDQ} testing has a major role to play in detecting defects even if stuck-at and delay test generation is rectified to guarantee defect detection. For the rest of the book we will be concerned with the generation and evaluation of I_{DDQ} tests for detection as well as diagnosis.

6

EVALUATING I_{DDQ} TESTS

The quality of a test set is defined by its **fault coverage** which is the percentage of targeted faults detected by the test set. Computing fault coverage is known as **fault simulation**. Fault simulation algorithms use fault models. The decision on the fault model(s) to be used is based on defects to be targeted, as well as the phase of the design cycle in which fault simulation is performed. Fault simulation algorithms for I_{DDQ} test sets using several fault models, defined in Chapter 5, are discussed. These include extracted BFs, all BFs, metal BFs, leakage faults and weak faults.

6.1 SIMULATION OF EXTRACTED BFS

Reported work [65, 180] on simulation of extracted BFs uses variants of methods for stuck-at faults. In this method to evaluate a given I_{DDQ} test set a table is created. Each row in this table corresponds to a target fault and there is a row for each fault in the fault list. Each column corresponds to test vector T_i and there is a column for each test vector. Thus, entry $\langle i, j \rangle$ corresponds to fault i and vector j. It is set to D if vector j detects fault i. Else it is set to U. The entries in this table are computed by logic simulation of each input vector and verifying if the condition for detecting the fault is satisfied. Fault coverage is computed from the table by determining which faults are detected by at least one vector.

The following example, which uses Figure 6.1(a) and the test set in Table 6.1, illustrates the above method. Figure 6.1(b) shows the switch-level representation of the circuit in Figure 6.1(a). Let the list L of extracted faults

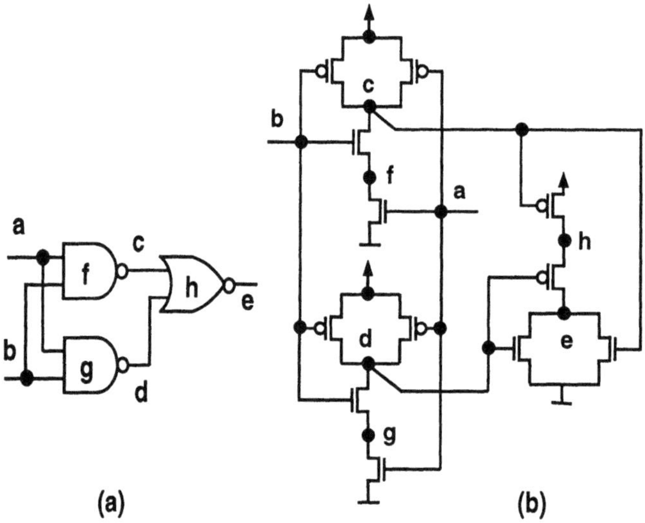

Figure 6.1 Switch-level description of example used to illustrate simulation of extracted BFs

be $\{\langle a,b\rangle,\langle a,c\rangle,\langle a,d\rangle,\langle a,f\rangle,\langle b,c\rangle,\ \langle b,d\rangle,\langle b,g\rangle,\ \langle c,e\rangle,\langle c,f\rangle,\langle d,e\rangle\}$. First, Table 6.2 described above is created. Entries marked "U" and "D" are interpreted as defined above. Column "Status" records the detection status of the faults. Thus, 9 out of the 10 faults are detected. Hence the fault coverage is 90%.

Table 6.1 Test set used to illustrate simulation of extracted faults

	a	b	c	d	e	f	g	h
T_1	0	0	1	1	0	f	f	f
T_2	0	1	1	1	0	1	1	f
T_3	1	1	0	0	1	0	0	1

Refinements using techniques from fault simulation of **logic** tests for stuck-at faults, such as parallel-pattern [180] and concurrent [65] fault simulation, have been used. In CURRENT [180], parallel pattern fault simulation was used with parallel pattern size (N) varied from $N = 32$ to $N = 256$. For a discussion on parallel pattern and concurrent fault simulation see [3].

Table 6.2 Table used to illustrate simulation of extracted faults

Fault	T_1	T_2	T_3	Status
$\langle a, b\rangle$	U	D	U	1
$\langle a, c\rangle$	D	D	D	1
$\langle a, d\rangle$	D	D	D	1
$\langle a, f\rangle$	U	D	D	1
$\langle b, c\rangle$	D	U	D	1
$\langle b, d\rangle$	D	U	D	1
$\langle b, g\rangle$	U	U	D	1
$\langle c, f\rangle$	U	U	U	0
$\langle c, e\rangle$	D	D	D	1
$\langle d, e\rangle$	D	D	D	1
Faults detected				9

An important point to be noted about this method is that faults are explicitly enumerated. The number of rows equals the number of faults. Hence, for large fault lists the size of the table explodes and the computational resource requirement increases dramatically. We will look at implicit enumeration techniques that are better as larger fault lists can be handled. They can also be used for extracted BFs.

6.2 SIMULATION OF METAL BFS AND ALL BFS

Fault simulation of I_{DDQ} test sets for metal BFs and all BFs requires that the detection status of each bridging fault (BF) be computed. In principle, the method in Section 6.1, which explicitly enumerates the list of target BFs, can be used. However, as pointed out earlier, it requires enormous computational resources. Therefore, for metal BFs and all BFs a different class of fault simulation algorithms has been proposed [55, 78, 79, 310]. The simulator implicitly enumerates the list of BFs. In this subsection we look at these fault simulation algorithms.

Background information to understand the algorithms is given in Section 6.2.1. Two fault simulation algorithms are discussed in Sections 6.2.2 and 6.2.3. In

Section 6.2.4 experimental results are presented. They attest to the computational efficiency of these methods.

6.2.1 Background Information

First, issues pertaining to potentially detected faults, state computation and the state change problem are discussed. Next, an implicit representation of the fault list using *set of ordered pairs of sets* (SOPS) [52], node partitions [54, 55, 78, 310] and indistinguishable pairs [310] are discussed. We then show how BF coverage is computed from node partitions and indistinguishable pairs.

The notion of *equivalence classes* of nodes, based on which node partitions were developed, was initially used in fault simulation of metal-BFs in combinational circuits [54, 78]. As we will see this by itself cannot be used for simulating BFs involving nodes that can have undefined values. In [78, 79], however, a method based on *compatibility classes* for simulating BFs involving nodes having unknown values was also presented. We will discuss the difference between the method presented here and the method in [78].

Potentially Detected Faults:

Let C be a sequential circuit with N flip-flops and S the initial state vector. For simulating sequential circuits, the initial state is assumed unknown and $S(1)$ through $S(N)$ is set to an unknown value designated by X. Thus, during simulation nodes in the circuit can be assigned the value X.

Consider BF $\langle a, b \rangle$ such that on simulating test vector T, $T(a) = 0$ and $T(b) = X$. During testing, $T(b) = 0$ or 1. If $T(b) = 0$ then $\langle a, b \rangle$ is not detected. However, if $T(b) = 1$ then BF $\langle a, b \rangle$ is detected. Thus, the BF may or may not be detected by T depending on the initial state. BF $\langle a, b \rangle$ is said to be potentially detected by T and referred to as a potentially detected fault (**PDF**) with respect to T. During fault simulation, if T potentially detects a fault then we assume that T will not detect it and hence PDFs are not dropped from the fault list during simulation of T.

Next State Computation:

Let F be the fault list obtained after fault simulation of tests $T_1, \ldots, T_{i-1}$ and Pi be the present state obtained by fault free logic simulation of $T_1, \ldots, T_{i-1}$. We assume that this is same as the present state obtained by simulating

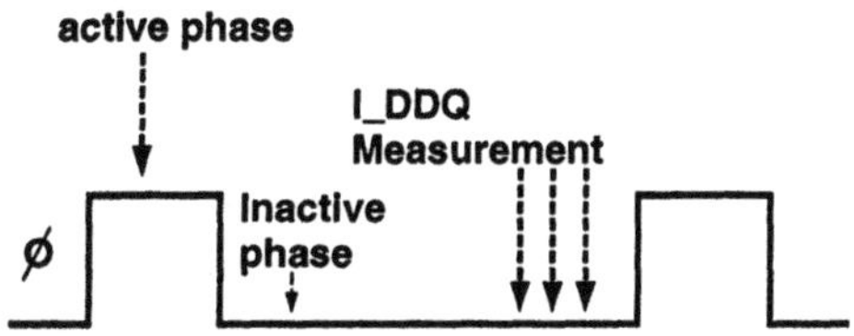

Figure 6.2 I_{DDQ} measurement point

$T_1, \ldots, T_{i-1}$ in presence of BFs in F. After processing T_i all BFs $\langle x, y \rangle$ such that $T_i(x) \neq T_i(y)$ and $T_i(x)$ and $T_i(y) \in \{0, 1\}$ are dropped from F to derive F'. Therefore, for all BFs $\langle a, b \rangle$ in F' such that $T_j(a) = T_j(b), 1 \leq j \leq i$, the present state P_{i+1}, obtained by simulating $T_1, \ldots, T_i$ in the presence of these BFs is the same as the present state obtained by fault free logic simulation of $T1, \ldots, T_i$.

Only other faults in F' are those that are PDF w.r.t. some $T_j, 1 \leq j \leq i$. The fault-free next state computation is also valid for all such BFs. Consider PDF $F = \langle a, b \rangle$ w.r.t. T such that during simulation of T, $T(a) = 0$ and $T(b) = X$. Let N be the fault free next state and N' the next state with F in the circuit. During testing, $T(b) = 0$ or 1. If $T(b) = 1$, the fault is detected and we stop. If $T(b) = 0$ then the BF is not activated and $T(a) = T(b)$. In this case $N' = N$.

State Change Problem:

Care should be taken to simulate BFs involving internal nodes of flip-flops. As shown in Figure 6.2, we assume that I_{DDQ} measurement is made during the inactive phase of the clock. In the circuit shown in Figure 6.3, x represents a node in the combinational block of a sequential circuit and nodes y and z represent the output node and an internal node, respectively, of a master slave flip-flop. On application of test T let the value at nodes x, y and z be as shown. Based on our discussion so far we would declare T to be a test for $\langle x, z \rangle$. However, as explained below, T may not detect $\langle x, z \rangle$ [167, 264]

If node x is stronger than node z then z is pulled up to 1. Since I_{DDQ} measurement is made during the inactive phase of the clock Φ, inverter A is enabled. Hence, in presence of BF $\langle x, z \rangle$, faulty value at z feeds back through loop $z - z' - z$. Consequently, it sets z' to 0 and z to 1. Thus, during the inactive phase of the clock the pulldown path from z to ground does not exist and I_{DDQ} will be at the fault-free level.

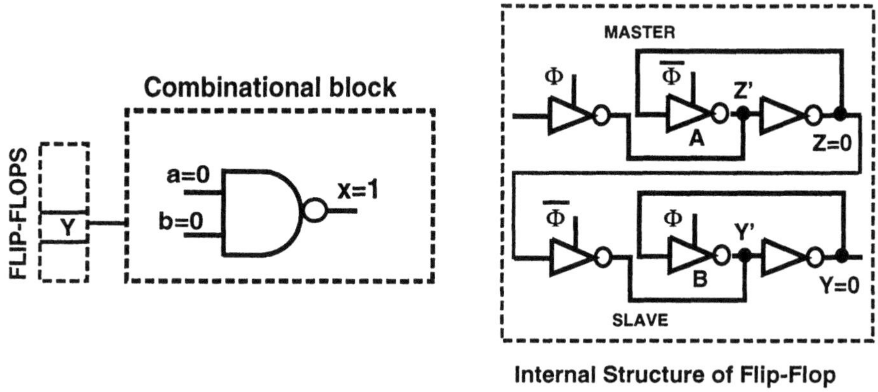

Figure 6.3 State change problem

The faulty value at z propagates through y' to y. The new value at y may propagate to change the value of the next state outputs and the circuit will function incorrectly. This is known as the **state change problem.**

The above problem does not arise for BFs involving the output node of the slave half of the flip flop. T detects $\langle x, y \rangle$ of Figure 6.3 for the following reason. Assume that node x is stronger than node y and x pulls y up to 1. During the inactive phase of the clock Φ inverter B is disabled and loop $y - y' - y$ is open. Therefore the faulty value at y cannot feedback through the loop $y - y' - y$. Thus, a path from V_{DD} to ground continues to exist and T will detect $\langle x, y \rangle$. The fault simulation algorithm does not simulate BFs involving internal nodes of flip-flops. Every flip-flop is represented by the external node of the slave half of the master-slave flip-flop.

SOPS Representation:

An implicit enumeration technique based on SOPS [52] (defined below) is used. Let A, B be two sets of nodes in the circuit. Then the ordered pair $\langle A, B \rangle$ form the set of BFs $\langle x, y \rangle : x \in A$ and $y \in B$. For example the SOPS $\{\langle \{a, b\}, \{c, d\} \rangle, \langle \{e\}, \{f, g\} \rangle\}$ represent the following set of BFs $\{\langle a, c \rangle, \langle a, d \rangle, \langle b, c \rangle, \langle b, d \rangle, \langle e, f \rangle, \langle e, g \rangle\}$.

Node Partitions:

Consider the circuit in Figure 6.4 and test set $T = \{T1, T2, T3\}$ of Table 6.3. The values assigned to nodes of the circuit by sequence T, during each clock

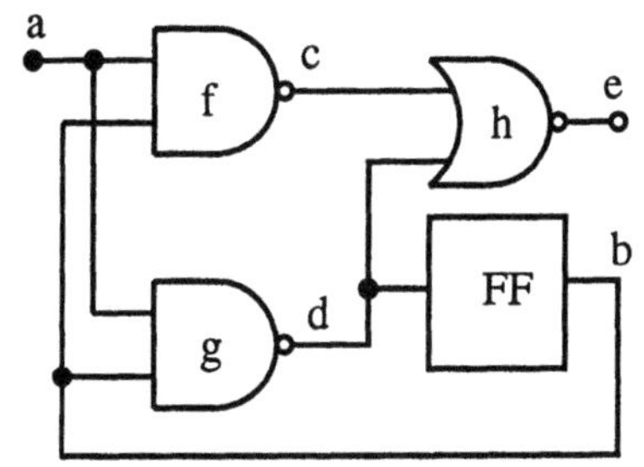

Figure 6.4 Circuit to illustrate fault simulation algorithm

Table 6.3 Test set used in all examples

	a	b	c	d	e	f	g	h
T_1	0	X	1	1	0	X	X	X
T_2	0	1	1	1	0	0	0	f
T_3	1	1	0	0	1	0	0	1

cycle, are shown in Table 6.3. Unknown initial state was assumed. Observe that nodes a and e have the same values for tests $T1, T2$ and $T3$. Such a pair of nodes is said to be equivalent with respect to T. In general, two nodes a and b are equivalent with respect to a test set T iff for all $T_i \in T$, $T_i(a) = T_i(b)$. This binary relation on the set of nodes is an equivalence relation and therefore partitions the set of nodes into equivalence classes. However, to avoid confusion with equivalence classes of BFs we refer to equivalence classes of nodes as **Node Partitions (NP)**. For the circuit in Figure 6.4 and test set $T = \{T1, T2, T3\}$ of Table 6.3 the set of NPs with respect to test set T are $\{\{a, e\}, \{c, d\}, \{f, g\}, \{b\}, \{h\}\}$. The issue of computing NPs is discussed later.

Indistinguishable Pairs:

Consider the circuit in Figure 6.4, test set $T = \{T1, T2, T3\}$ of Table 6.3 and nodes a and h. For input vector $T1 \in T$, $T1(a) \neq T1(h)$ and $T1(h) = X$. For test $T1$, BF $\langle a, h \rangle$ is a PDF. Thus, as discussed before, we assume $\langle a, h \rangle$ is not detected by $T1$. Similarly, for test $T2$, BF $\langle a, h \rangle$ is a PDF and assumed not detected by $T2$. Finally, for test $T3$ since $T3(a) = T3(h)$, BF $\langle a, h \rangle$ is not detected by $T3$.

We have seen that if nodes a and e are in the same NP then BF $\langle a, e\rangle$ is not detected by T. In addition, as seen above, although nodes a and h are in distinct NPs BF $\langle a, h\rangle$ is not detected by T. To keep track of undetected BFs between nodes in distinct NPs, the notion of indistinguishable pairs (**IP**) (defined below) is used.

Consider NPs $P1 = \{a, e\}$ and $P2 = \{h\}$ with respect to test set T. For the nodes in $P1(P2)$, the logic values assigned by $T_i \in T$ are given by response vectors $R1[i](R2[i])$. For example, $R1[1] = 0$ and $R1 = \langle 0, 0, 1\rangle$ and $R2 = \langle X, f, 1\rangle$. We say that $\langle P1, P2\rangle$ is an IP [310] iff for all $T_i \in T$, either: $R1[i] = R2[i]$; or $R1[i] = X$ or f; or $R2[i] = X$ or f. We will see in Sections 6.2.2 and 6.2.3 that it is not necessary to compute these response vectors.

Fault List Representation:

Fault simulation algorithms have to keep track of undetected BFs. This is done using an SOPS of the form $\{\langle A1, B1\rangle, \ldots, \langle Ak, Bk\rangle\}$. For the example from Table 6.3 and Figure 6.4, the undetected BFs is given by $\{\langle\{a, e\}, \{a, e\}\rangle, \langle\{c, d\}, \{c, d\}\rangle, \langle\{f, g\}, \{f, g\}\rangle, \langle\{a, e\}, \{h\}\rangle, \langle\{b\}, \{h\}\rangle\}$

This list consists of two sublists: $L1 = \{\langle\{a, e\}, \{a, e\}\rangle, \langle\{c, d\}, \{c, d\}\rangle, \langle\{f, g\}, \{f, g\}\rangle\}$, $L2 = \{\langle\{a, e\}, \{h\}\rangle, \langle\{b\}, \{h\}\rangle\}$. List $L1$ contains pairs of sets such that the two components of a pair are the same, i.e., both the sets are equal to an NP. In addition, $L1$ contains one such pair for each NP of size two or more. This information is implicitly available in the list of NPs as follows. A node partition P represents all BFs $\langle x, y\rangle$ where $x, y \in P$. Hence there is no need to explicitly store $L1$. Pairs $\langle\{a, e\}, \{h\}\rangle$ and $\langle\{b\}, \{h\}\rangle$ are IPs. Thus, at any given stage of fault simulation the fault list has two parts. The first part is a list of NPs and the second part is either an explicit or implicit representation of the list of IPs. Techniques to manipulate and process these two parts of the fault list are given in Sections 6.2.2 and 6.2.3.

Computing Coverage:

Fault coverage computation using NPs and IPs is explained with the help of Figure 6.4 and test set $T = \{T1, T2, T3\}$ of Table 6.3. The final list of NPs and IPs are as follows: $NPs = \{\{a, e\}, \{c, d\}, \{f, g\}, \{b\}, \{h\}\}$; $IPs = \{\langle\{a, e\}, \{h\}\rangle, \langle\{b\}, \{h\}\rangle\}$. From this the number of undetected faults is computed as follows. Total number of undetected BFs = number of undetected BFs in NPs + number of undetected BFs in IPs. Number of undetected BFs in NPs = $1 + 1 + 1 + 0 + 0 = 3$; number of undetected BFs in IPs = $2 + 1 = 3$; and total undetected BFs = 6.

Formally, the new coverage after processing tests $T_1, \ldots, T_i$ in the given test set $T = \{T_1, \ldots, T_N\}$ is computed as follows: Let $NP_i = \{E_1, \ldots, E_k\}$ and $IP_i = \{\langle A1, B1\rangle, \ldots, \langle An, Bn\rangle\}$ be the set of NPs and set of IPs respectively after processing tests $\{T_1, \ldots, T_i\}$. Total number of undetected BFs = NP_Size(NP_i) + IP_Size(IP_i), where NP_Size(NP_i) = $\sum_{j=1}^{k} \frac{|E_j|*(|E_j|-1)}{2}$ and IP_Size(IP_i) = $\sum_{j=1}^{n} | Aj | \times | Bj |$.

Other approaches:

The following fault simulation method [78] for metal BFs and all BFs involving nodes having values X/f has been discussed. Initially, let $S = \{S_0\}$, where S_0 is set of all nodes in the circuit. For each test T_i, do (i) and (ii):

(i) Using logic simulation compute values assigned by T_i to each node in the circuit;

(ii) For each set $S_j \in S$ that contains at least one node that has the value 0 *and* at least one node that has the value 1: replace S_j by two sets S_j^0 and S_j^1, where $S_j^0(S_j^1)$ contains all nodes in S_j that do not have the value 1(0). That is, S_j^0 contains all of the nodes in S_j that have the value $0, X$, or f, and S_j^1 contains all of the nodes in S_j that have the value $1, X$, or f.

(iii) Retain only one copy of any duplicated sets in S, and delete any set that is a proper subset of any other set in S.

An I_{DDQ} measurement needs be taken for test T_i only if it causes at least one set to be split on step (ii). Because of the ambiguity introduced by the X and f states, a pair of nodes (representing a BF) may appear in more than one set. The actual set of BFs remaining undetected after the application of tests $T_1, \ldots, T_i$ can be obtained by expanding the sets of nodes and keeping only the unique pairs. However, an upper bound on the number of undetected BFs is given by $\sum_{j=1}^{k} \frac{|S_j|*(|S_j|-1)}{2}$.

We illustrate this with the circuit of Figure 6.4 and test set T of Table 6.3. Initially, $S = \{S_0\}$, where $S_0 = \{a, b, c, d, e, f, g, h\}$. On evaluating T_1, from S_0, we get the set $S_1^1 = \{a, e, b, f, g, h\}$, the set of nodes that do not have logic value 1; and $S_1^0 = \{c, d, b, f, g, h\}$, the set of nodes that do not have nodes with logic value 0. Note that nodes b, f, g, h are repeated in sets S_1^0 and S_1^1. This is because these nodes have a logic value that is neither 0 or 1. On evaluating T_2, we get the following sets: $\{\{a, e, f, g, h\}, \{b, h\}, \{c, d, b, h\}, \{f, g, h\}\}$. Note that $\{b, h\} \subset \{c, d, b, h\}$; $\{f, g, h\} \subset \{a, e, f, g, h\}$. Therefore, we eliminate

$\{b, h\}$ and $\{f, g, h\}$ to get $\{\{a, e, f, g, h\}, \{c, d, b, h\}\}$. This step however is computationally very expensive. On evaluating T_3, we get the following final list of undetected BFs: $\{\{a, e, h\}, \{f, g\}, \{b, h\}, \{c, d\}\}$.

6.2.2 List-based algorithm

As shown earlier, NPs and IPs can be used to compute fault coverage. In the list-based algorithm [310] a list of NPs and a list that explicitly enumerates IPs are kept. For *every test* $T_i \in T$, both lists are updated.

NPs can be computed by first computing a table similar to Table 6.3 and then performing pairwise comparison of the values assigned to each node. This is an expensive algorithm. The basic idea of a faster algorithm [54, 56] is discussed with the help of an example. Let NP_0 be the initial set of NPs, NP_i the set of NPs obtained after processing input vectors $\{T_1, \ldots, T_i\}$. For the circuit of Figure 6.4 and test set $T = \{T1, T2, T3\}$ of Table 6.3, the following list of NPs is computed.

$NP_0 = \{\{a, b, c, d, e, f, g, h\}\}$.
$NP_1 = \{\{a, e\}, \{c, d\}, \{b, f, g, h\}\}$
$NP_2 = \{\{a, e\}, \{c, d\}, \{b\}, \{f, g\}, \{h\}\}$
$NP_3 = \{\{a, e\}, \{c, d\}, \{b\}, \{f, g\}, \{h\}\}$

Initially, there is only one NP. Next, each NP in the list of NPs is replaced by its 0, 1, f or X subsets. As explained before, since BFs involving nodes with value X or f are considered undetected, nodes with value X or f are considered equivalent and hence placed in the same NP. A new list NP_i can be computed from NP_{i-1} using $O(N)$ space and $O(N)$ time, where N is the number of nodes in the circuit [54, 56].

Computation of IPs is explained with an example. For the circuit of Figure 6.4 and test set $T = \{T1, T2\}$ of Table 6.3, let the initial set of NPs NP_0 be $\{A\}$, where $A = \{a, b, c, d, e, f, g, h\}$. Let IP_0, the initial set of IPs, be Φ. The values assigned by $T1$ to the nodes are shown in Table 6.3. Thus, $NP_1 = \{A0, A1, A2\}$ with respect to $T1$ (shown in Figure 6.5). Note that faults in $\langle A0, A2\rangle$, $\langle A1, A2\rangle$ are not detected by $T1$. These two IPs were obtained as a result of splitting NP A. Thus, as shown in Figure 6.5, the IPs are: $\langle A0, A2\rangle$ and $\langle A1, A2\rangle$.

Each NP of Figure 6.5 is split with respect to the values assigned by $T2$. The new set of NPs, with respect to $\{T1, T2\}$, is shown in Figure 6.6. Thus,

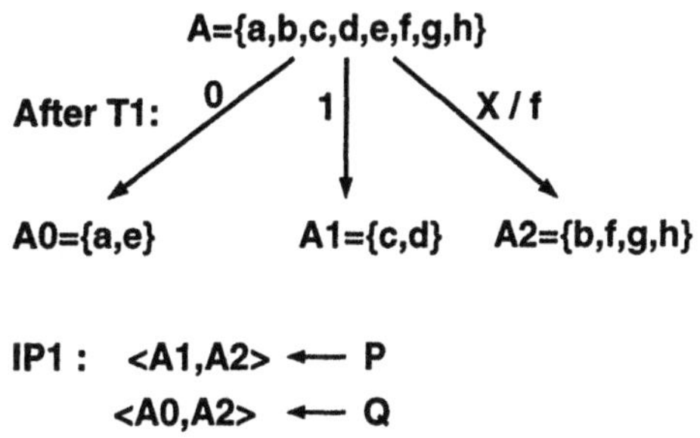

Figure 6.5 Example computation of NPs and IPs after T_1

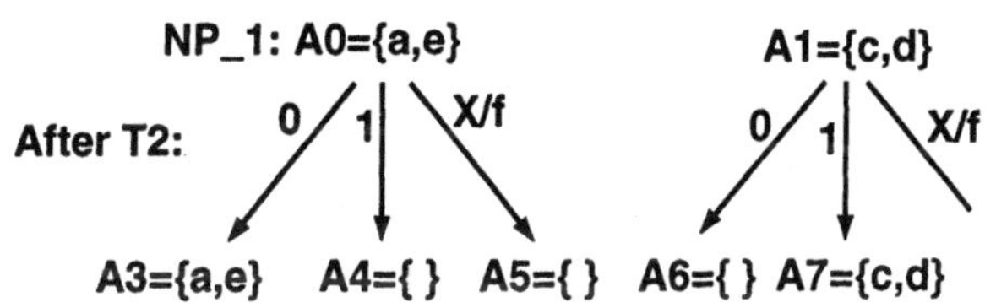

Figure 6.6 Example computation of NPs after T_2

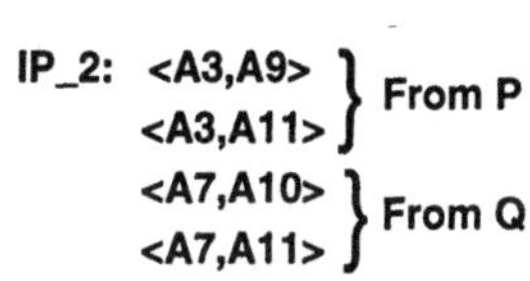

Figure 6.7 IPs obtained from IP_1

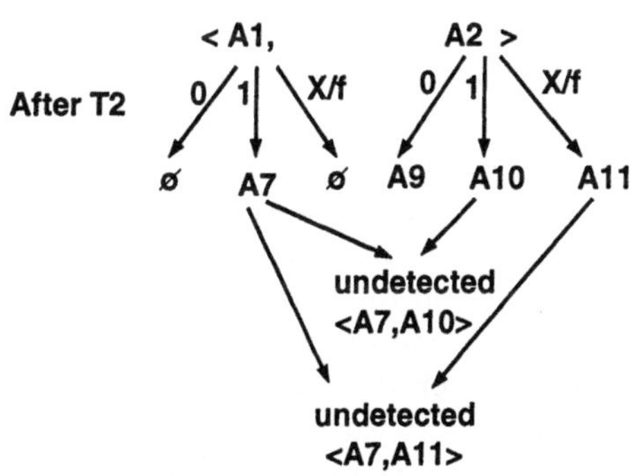

Figure 6.8 Computation of IPs from IP P

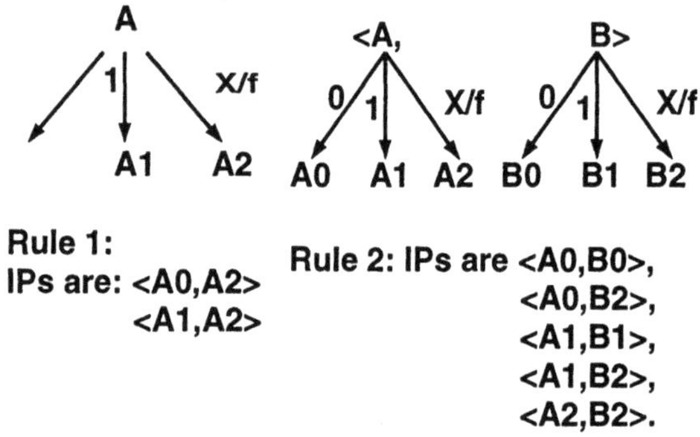

Figure 6.9 Rules to compute IPs

$NP_2 = \{A3, A7, A9, A10, A11\}$. The new set of IPs, viz. IP_2, is computed as follows. Splitting each NP in NP_1 gives rise to two new IPs. Thus, $A2$ gives rise to $\langle A9, A11\rangle$ and $\langle A10, A11\rangle$. The potential IPs resulting from splitting $A0$ and $A1$ i.e. $\langle A3, A5\rangle$, $\langle A4, A5\rangle$, $\langle A6, A8\rangle$ and $\langle A7, A8\rangle$ are empty and therefore not added to IP_2. The rest of the entries in IP_2 are a result of entries in IP_1. For example, for $P = \langle A0, A2\rangle$ in IP_1, we get the entries marked P in Figure 6.7.

The derivation of new IPs from the entry $\langle A1, A2\rangle$ in IP_1 is shown in Figure 6.8. To summarize, for every test T_i, there are two steps in updating the list of IPs:

Step 1. Generate new IPs resulting from splitting an NP in the list of NPs. This is shown as Rule 1 in Figure 6.9.

Step 2. Generate new IPs from IPs in IP_{i-1}, the previous list of IPs. Rule 2 in Figure 6.9 shows how the new IPs are generated.

When implementing the list-based algorithm, the following observation leads to reduction in space required. We use the example of Figure 6.4 and test set $T = \{T1, T2, T3\}$ of Table 6.3. The initial set of NPs NP_0 be $\{\{a, b, c, d, e, f, g, h\}\}$ and initial set of IPs IP_0 be Φ. After processing $T1$ we obtain $NP_1 = \{\{a, e\}, \{c, d\}, \{b, f, g, h\}\}$. The new set of IPs is given by $IP_1 = \{(\{a, e\}, \{b, f, g, h\}), (\{c, d\}, \{b, f, g, h\})\}$. Note that in the above example we have three copies of the set $\{b, f, g, h\}$. One in NP_1 and two in IP_1. This can lead to an enormous amount of space requirement for the algorithm [55]. Now, assume we had pointers $A0, A1$ and $A2$ pointing to each of the three NPs in NP_1. Using these pointers the new set of IPs IP_1 is as follows: $\{\langle A0, A2\rangle, \langle A1, A2\rangle\}$. This leads to a saving in space because now, we do not have any duplicates of NPs. This is what is used in the list-based scheme.

For step 1 all NPs in NP_i, the current list, have to be visited. This step takes $O(K)$ time and $O(K)$ space, where K is the number of NPs in NP_i. In the worst case, the number of NPs = number of nodes in the circuit. For step 2, each IP in IP_{i-1} is visited. This takes $O(M)$ space and $O(M)$ time, where M is the number of IPs in IP_{i-1}. In the worst case, $M = \frac{N \times (N-1)}{2}$. Hence, the worst case time required by the list-based algorithm to derive the new set of IPs from the previous set of IPs IP_{i-1} is $O(N^2)$ and the worst case space required is also $O(N^2)$. However, in practice although M is large, it never attains the worst case bound.

The reader can complete the example shown. On application of test $T3$, $NP_3 = \{A12, A13, A14, A15, A16\}$ where $A12 = \{a, e\}$, $A13 = \{c, d\}$, $A14 = \{f, g\}$, $A15 = \{b\}$, $A16 = \{h\}$ and $IP_3 = \{\langle A12, A16\rangle, \langle A15, A16\rangle\}$. The numbers of undetected BFs are: 3 from NP_3; and 3 from IP_3. Thus, the total number of undetected BFs is 6.

6.2.3 Tree-based algorithm

In the example of Section 6.2.2, prior to processing $T1$, the number of IPs was 0. After processing $T1$, the number of IPs increased to 3 and after processing $T2$ it increased further to 6. It then decreased to 2 after test $T3$ was processed. This trend of increase followed by a decrease in the number of IPs, as faults are dropped from the fault list, was observed for almost all circuits. The rise in the size of list of IPs has two implications. The space requirement and time to update the list of IPs increases with the size of the list. Thus, if IPs can be stored implicitly and the list computed after applying all the tests in the given test set, the space and time required can be reduced. This is the motivation behind the tree-based algorithm [310].

Recall that fault simulation reduces to computing NPs and IPs either explicitly or implicitly. In the tree-based algorithm NPs are computed exactly as in the list-based algorithm. Hence, for the example of Figure 6.4 and test set $T = \{T1, T2, T3\}$ of Table 6.3, $NP_3 = \{A12, A13, A14, A15, A16\}$, where $A12 = \{a, e\}$, $A13 = \{c, d\}$, $A14 = \{f, g\}$, $A15 = \{b\}$ and $A16 = \{h\}$.

The difference in the two algorithms lies in the way the final set of IPs is computed. In tree-based algorithm, no list of IPs exists and hence the expensive processing step of updating the list of IPs for each input test vector is eliminated. This is illustrated with the circuit of Figure 6.4 and test set $T = \{T1, T2, T3\}$ of Table 6.3. A labeled ternary tree is computed in which

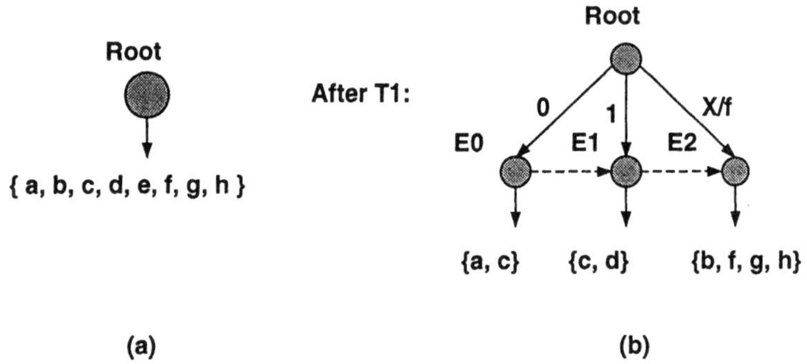

Figure 6.10 Growing the ternary tree

the leaves are linked together. Every node has three children pointing to subsets 0, 1 or f. If a subset is empty then the corresponding child node is not inserted in the tree. Let S_0 be the initial tree and S_i the tree obtained after processing $\{T_1, \ldots, T_i\}$. As shown in Figure 6.10(a) $S_0 = \{A\}$ is initialized to be the *Root* node. On application of test $T1$, the updated tree S_1 is as shown in Figure 6.10(b). The *Root* node has three children $E0, E1$ and $E2$. Note that each leaf node points to NPs corresponding to labeling on its incoming tree edges.

The list of IPs can be enumerated from implicit representation S_1 as follows. For each pair of leaf nodes Ei and Ej of S_1, traverse up the tree starting from the leaf nodes. Start the process with the pair Ei and Ej, where Ei is the leftmost leaf $E0$ and the neighbor leaf node of $E0$, obtained from the linked list of leaves. Let $Pi[k]$ be the kth edge, on the path from leaf node Ei to the *Root*, and $Label(Pi[k])$ be the label on $Pi[k]$. From leaf nodes Ei and Ej in tree S_N, $\langle Ei, Ej \rangle$ forms an IP iff for all $1 \leq k \leq N$, either $Pi[k] = Pj[k]$ (i.e., the two paths merge) or $Label(Pi[k]) = Label(Pj[k])$ or $Label(Pi[k]) = X/f$ or $Label(Pi[k]) = X/f$.

For our example, let $Ei = E0$ and $Ej = E1$. Note that $P0[1] \neq P1[1]$, $Label(P0[1]) \neq Label(P1[1])$, $Label(P0[1]) \neq X/f$ and $Label(P1[1]) \neq X/f$. Hence $\langle E0, E1 \rangle$ is not an IP. Next, let $Ej = E2$ and $Ei = E0$.
Since $Label(P2[1]) = X$, we get IP $\langle E0, E2 \rangle$. Next, let $Ei = E1$ and $Ej = E2$. Since $Label(P2[1]) = X$, we get IP $\langle E1, E2 \rangle$. Hence, the set of IPs IP_1 is implicitly represented in S_1.

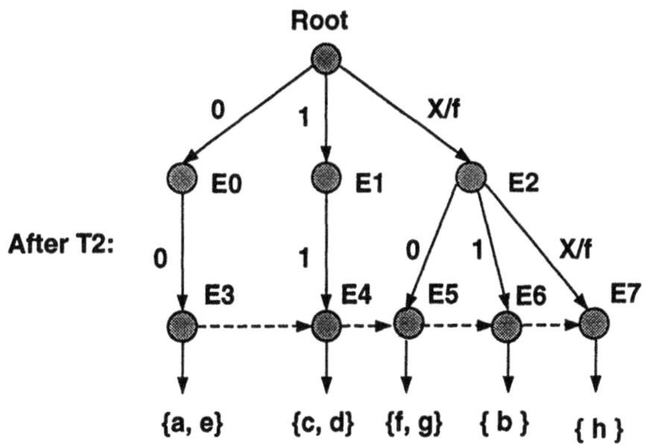

Figure 6.11 Tree after test T_2

On processing $T2$, we get S_2 shown in Figure 6.11. The set of IPs can be enumerated as before. The steps for enumerating IP $\langle E6, E7\rangle$ are illustrated. $Label(P7[1]) = X$, and by traversing one more tree edge up the tree, we see that $P6[2] = P7[2]$. Thus paths $P6$ and $P7$ merge, so we stop the upward traversal and conclude that $\langle E6, E7\rangle$ is an IP for the following reason. If paths from distinct leaf nodes Ei and Ej merge after k edges then for all edges further along the path to the *Root*, $Label(Pi[y] = Label(Pj[y])$ for all $y > k$. The reader can verify that the set of IPs is $\{\langle E3, E5\rangle, \langle E3, E7\rangle, \langle E4, E6\rangle, \langle E4, E7\rangle, \langle E5, E7\rangle, \langle E6, E7\rangle\}$.

Note that in the tree-based algorithm, IPs are not computed after processing every input test vector. It was done in this example to illustrate that IPs can be obtained after processing any test vector from the implicit representation. On application of $T3$, the tree S_3 is as shown in Figure 6.12. This completes **the first phase** of the tree-based algorithm.

The time required to compute this implicit representation of final set of IPs for every test is same as the time required for computing NPs, i.e., $O(N)$, where N is the number of lines in the circuit. However, the price paid for this implicit representation of final set of IPs is in the expensive step of coverage computation. This forms the **second phase** of the tree-based algorithm.

The second phase has **two parts**. In the **first part**, the number of undetected BFs between nodes within each NP, attached to a leaf node in S_3, is computed. This is given by 1 (from E8) + 1 (from E9) + 1 (from E10) + 0 (from E11) + 0 (from E12) = 3.

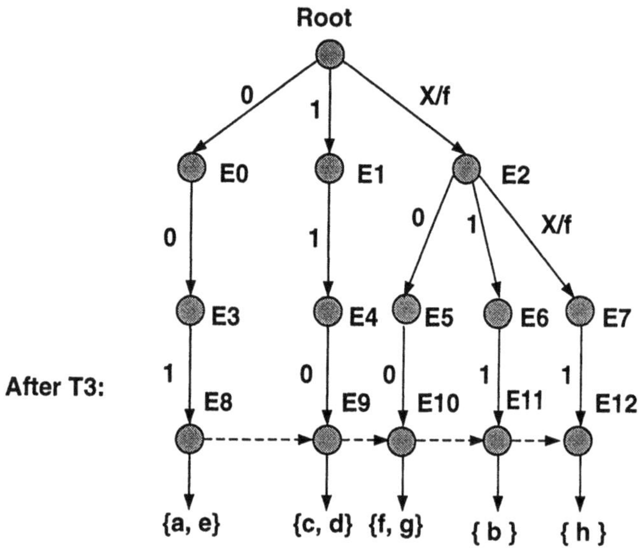

Figure 6.12 Tree after test T_3

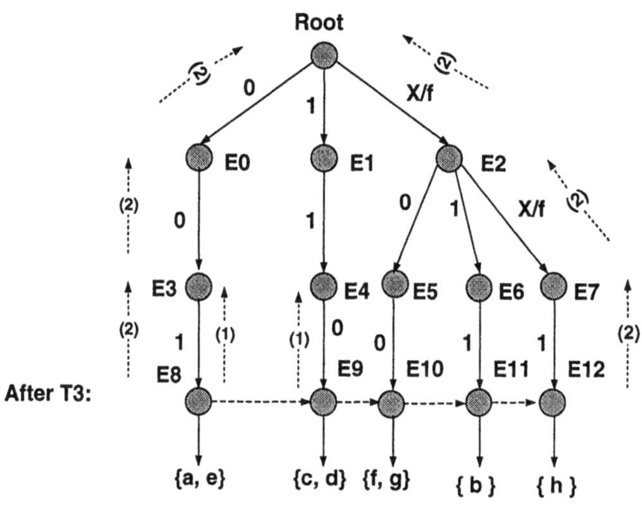

Figure 6.13 Extracting coverage information from the tree

In the **second part**, the number of undetected BFs from the implicit representation of the final set of IPs is computed. Its only here, and not after S_0, S_1 and S_2, that the set of IPs represented by S_3 is enumerated. This is illustrated using S_3 of Figure 6.13. Start with the leftmost leaf node E_i of the tree and visit each leaf node E_j to its right. The task is to determine the number of BFs involving nodes in E_i that are undetected. This step is illustrated, using arrows labeled (1), in Figure 6.13 for $E_i = E8$ and $E_j = E9$. Here since $P8[1] \neq P9[1]$, $P8[1] \neq X/f$ and $P9[1] \neq X/f$, $\langle E8, E9\rangle$ is not an IP. Next E_j is set to $E10$. On traversing up edges $P8[1]$ and $P10[1]$, we conclude that $\langle E8, E10\rangle$ is not an IP. Next, E_j is set to $E11$. On traversing edges $P8[2], P11[2]$ we conclude that $\langle E8, E11\rangle$ is not an IP. Next, E_j is set to $E12$. The traversal of the corresponding paths is illustrated in Figure 6.13 using arrows marked (2). It can be verified that $\langle E8, E12\rangle$ is an IP. Number of undetected BFs for this IP is 2. The reader can verify that the total number of undetected BFs is 6.

In the second phase of computation, all pairs of leaf nodes are visited. For each pair we traverse up the tree. Let K be the number of leaf nodes and L be the number of test vectors in the given test set. Thus the maximum number of tree edges traversed from any leaf to the *Root* node is L. Therefore, time taken to enumerate the IPs is $O(K^2L)$. We have assumed that for every pair of leaf nodes all edges up to the *Root* are traversed. In practice this almost never occurs. The maximum number of leaf nodes in the tree after K tests is N, the number of nodes in the circuit. Like the list-based algorithm, the worst case time complexity of the tree-based algorithm is also $O(N^2L)$. However in practice it is found that for several test sets and fault lists the tree-based algorithm performs better. This suggests that the average case performance of the tree-based algorithm is better.

For each test vector tree nodes are inserted corresponding to the 0, 1 and f subsets, only if they are nonempty. The maximum number of tree nodes inserted after processing each test vector is N. Hence, the space required by the tree-based algorithm is $O(NL)$. Compare this with the list based algorithm, where the space required was $O(N^2)$. Usually, $L\langle\langle N$ for I_{DDQ} test sets. Therefore, the space required by the tree-based algorithm is significantly less than the space required by the list-based algorithm.

6.2.4 Experimental Results

The algorithms of Sections 6.2.2 and 6.2.3 were implemented and experimental results are tabulated in Tables 6.4 through 6.15.

Table 6.4 BF simulation using GENTEST testset (all-BFs)

circuit	No of Vec	Total Faults	Cov	CPU s (Tree)	CPU s (List)	Ratio
s298	162	47278	98.4	2.86	6.91	2.41
s208.1	122	27261	20.4	0.95	2.28	2.41
s344	91	59685	99.6	1.56	3.48	2.22
s382	2463	66430	99.4	35.48	94.18	2.65
s400	471	72771	99.4	22.83	62.21	2.73
s444	1881	81003	99.3	33.58	87.71	2.62
s420.1	182	121771	18.0	3.95	9.60	2.43
s526	754	152628	98.7	35.96	95.11	2.64
s641	133	241860	98.5	4.61	14.53	3.15
s713	107	285390	98.0	5.06	13.06	2.57
s820	411	419070	99.3	38.41	100.35	2.61
s832	377	435711	99.3	37.41	98.65	2.63
s953	16	380628	27.0	1.86	4.00	2.14
s1238	349	869221	99.7	49.75	161.35	3.2
s1423	36	1261666	93.8	8.66	12.85	1.49
s1494	469	1929630	99.7	103.95	335.65	3.23
s1488	590	1902225	99.7	121.83	424.95	3.48
s5378	408	10878780	95.7	587.85	1134.68	1.93
s35932	86	620065720	99.4	5885.53	2425.00	0.41
Average Speedup ratio : 2.47						

Table 6.5 BF simulation using HITEC testset (all BFs)

circuit	No of Vecs	Total Faults	Cov	CPU s (Tree)	CPU s (List)	Ratio
s298	259	47278	98.4	5.16	11.61	2.24
s344	108	59685	99.6	1.56	3.73	2.38
s400	2069	72771	98.6	35.98	143.33	3.98
s420.1	175	121771	18.0	4.03	9.88	2.45
s526	192	152628	88.2	7.85	16.15	2.05
s641	211	241860	98.6	7.03	20.41	2.90
s713	175	285390	98.1	7.08	19.35	2.73
s820	968	419070	99.3	83.96	216.88	2.58
s832	967	435711	99.3	102.25	227.00	2.22
s953	14	380628	26.4	1.46	3.18	2.17
s1238	478	869221	99.7	73.20	239.71	3.27
s1423	88	1261666	94.7	14.81	29.38	1.98
s1494	1285	1929630	99.7	1965.00	794.71	0.40
s1488	1192	1902225	99.7	292.85	712.93	2.43
s5378	900	10878780	92.8	1454.63	2417.20	1.66
s35932	381	620065720	99.2	3724.03	8779.78	2.35
Average Speedup ratio : 2.36						

BF coverage was computed for three types of fault lists: all BFs, metal BFs and input fault list. An input fault list simulates extracted BFs. In these experiments, input fault list is a list of two-line BFs involving a node and $k = 15$ of its "topologically close" neighbors. Here we refer to topology as one available from the gate-level circuit description. For sequential circuits stuck-at test sets computed using GENTEST [71] and HITEC [224] were used. In Tables 6.4 through 6.9 "CPU s" is CPU seconds on a SUN SPARC 2 and circuits are the ISCAS 89 circuits [32]. Column "Ratio" is the time required by list-based scheme divided by the time required by the tree-based scheme.

For **all BFs** refer to Tables 6.4 and 6.5. The average speedup obtained using the tree-based algorithm is 2.4. There are only two cases where the tree-based algorithm performs badly compared to the list-based algorithm. In Table 6.4, for s35932 it was observed that the number of IPs continued to grow and hence there was no advantage of using the tree-based algorithm. In Table 6.5, for s1494 it is not clear why the tree-based algorithm performed badly. We conjecture that the number of IPs is small, unlike the other cases.

Table 6.6 BF simulation using GENTEST testset (input fault list)

circuit	No of Vecs	Total Faults	Cov	CPU s (Tree)	CPU s (List)	Ratio
s298	162	4312	97.8	1.983	5.600	2.82
s344	91	4844	98.9	1.367	3.633	2.65
s382	2463	5110	98.5	29.383	42.517	1.45
s1238	349	18467	99.5	18.167	25.350	1.40
s1423	36	22246	88.7	4.450	18.560	4.17
s1494	469	27510	99.3	31.350	42.083	1.34
s1488	590	27314	98.3	43.683	62.733	1.43
s5378	408	65310	90.8	82.200	332.167	4.04
s35932	86	493025	96.7	162.233	502.300	3.10
Average Speedup ratio : 2.5						

For **input fault list**, from Tables 6.6 and 6.7, note that the tree-based algorithm performed consistently better than the list-based algorithm. The average speedup ratio was around 2.4. Since the input fault list was in pairwise format $\langle x, y \rangle$, where x and y are nodes in the circuit, the list based algorithm reduced to explicitly processing each individual fault for every test in the given test set. This is equivalent to the fault simulation algorithms for extracted BFs discussed in Section 6.1. This implies that the tree-based algorithm may also be useful for extracted BFs. Since, these fault simulation algorithms are compiled code algorithms, it will be interesting to see if performance improvement can be obtained if features of parallel pattern, parallel fault and concurrent fault simulators are included.

From Tables 6.8 and 6.9 we observe that for **metal BFs**, the speedup was small for the following reason. Since only metal nodes were involved the number of nodes that can have a value other than 0 or 1 is greatly reduced. This leads to a considerable reduction in the number of IPs. Hence sometimes, the overhead for maintaining the tree outweighs the time required to explicitly store and process the few IPs that occur. For metal-BFs the two fault simulation algorithms will have comparable performance.

These algorithms can also be used for combinational circuits. However, because the number of nodes with X or f will be small, the number of IPs will be small. Therefore, the two algorithms are about the same with the list-based algorithm performing marginally better. There is another interesting feature of the list-

Table 6.7 BF simulation using HITEC testset (input fault list)

circuit	No of Vecs	Total Faults	Cov	CPU s (Tree)	CPU s (List)	Ratio
s298	259	4313	97.8	3.267	6.98	2.14
s344	108	4844	98.9	1.467	4.01	2.73
s1238	478	18467	99.4	23.583	35.73	1.52
s1423	88	22246	89.8	7.833	29.03	3.71
s1494	1285	27511	99.4	84.367	94.10	1.12
s1488	1192	27315	98.3	84.433	119.45	1.42
s5378	900	65310	87.0	269.300	774.16	2.88
s35932	381	493025	94.7	419.850	1542.43	3.67
Average Speedup ratio : 2.4						

Table 6.8 BF simulation using GENTEST testset (metal nodes only)

circuit	No of Vecs	Total Faults	Cov	CPU s (Tree)	CPU s (List)	Ratio
s298	162	9180	99.6	0.71	0.91	1.30
s344	91	16836	99.7	0.56	0.56	1.00
s382	2463	16471	99.8	13.91	14.93	1.10
s1238	349	145530	100	9.31	5.51	0.60
s1423	36	279378	97.2	1.46	2.13	1.44
s1494	469	218130	100	14.61	10.83	0.74
s1488	590	222111	100	17.00	13.63	0.80
s5378	408	4477528	98.4	78.08	63.93	0.80
s35932	86	158909878	99.7	1793.21	53.41	0.03
Average Speedup ratio : 0.86						

based algorithm that makes it useful. This will be discussed in the context of computing I_{DDQ} Tests.

We next compare the implicit techniques for metal BFs and all BFs with the explicit enumeration techniques such as those used for extracted BFs. In Table 6.10 we compare the performance of fault simulation for extracted BFs and metal BFs for sequential circuits. The test set used is computed using GENTEST [71]. Only results for ISCAS89 [32] circuits with more than 1000 gates

Table 6.9 BF simulation using HITEC testset (metal nodes only)

circuit	No of Vecs	Total Faults	Coverage	CPU s (Tree)	CPU s (List)	Ratio
s298	259	9180	99.6	1.28	1.53	1.20
s344	108	16836	99.7	0.68	0.73	1.10
s1238	478	145530	100	13.65	7.50	0.55
s1423	88	279378	97.5	2.61	2.75	1.05
s1494	1285	218130	100	34.40	31.03	0.90
s1488	1192	222111	100	30.61	28.18	0.92
s5378	900	4477528	97.0	123.18	164.85	1.34
s35932	381	158909878	99.6	511.80	249.81	0.48
Average Speedup ratio : 0.94						

for which GENTEST test set was available are presented. The times are CPU seconds on a SUN 4/670 workstation. The number of faults in the two sets is given in column *Faults* under headings *Extracted BFs* and *metal-BFs*. Column *Faults* under heading *Ratio* gives the ratio of the size of metal BFs and extracted BFs. Columns labeled *CPU s* is similarly defined.

Close examination of columns *Faults* and *CPU s* under *Ratio* suggests that fault simulation of metal BFs, in spite of having a considerably larger number of faults, actually requires less time than the simulation of extracted BFs. The column *Adj Time* is used to account for the possibility that some improvements to the explicit enumeration approach such as parallel pattern and concurrent fault simulation cannot be applied to implicit techniques. So, a multiplicative factor of 4 is chosen. After making these adjustments, there is still only a marginal difference in the running times. Table 6.11 contain similar data for extracted BFs and all BFs. Examining columns *Faults* and *CPU s* under *Ratio*, one notices that conclusions similar to those drawn earlier for metal-BFs can be made.

Tables 6.12 and 6.13 compare the performance of fault simulation of extracted BFs with metal-BFs and all BFs, respectively, in combinational circuits. The test set used was generated using the test generator available with Octtools [163].

From above tables we can conclude that using implicit techniques, a larger number of faults can be targeted with little or no increase in computational overhead. We can also conclude that with appropriate techniques it is possi-

Table 6.10 Fault simulation results for sequential circuits (metal BF)

	# of	Extracted-BFs		metal-BFs		Ratio		
	Tests	Faults	CPU s	Faults	CPU s	Faults	CPU s	Adj CPU s
S1238	349	18467	18.17	145530	5.52	7.88	0.30	1.22
S1423	36	22246	4.45	279378	2.13	12.56	0.48	1.91
S1494	469	27510	31.35	218130	10.83	7.93	0.35	1.38
S1488	590	27314	43.68	222111	13.63	8.13	0.31	1.25
S5378	408	65310	82.20	4477528	63.93	68.56	0.78	3.11
S35932	86	493025	162.23	158909878	53.42	322.32	0.33	1.32

Table 6.11 Fault simulation results for sequential circuits (all BF)

	# of	Extracted-BFs		all BFs		Ratio		
	Tests	Faults	CPU s	Faults	CPU s	Faults	CPU s	Adj CPU s
S1238	349	18467	18.17	869221	49.75	47.07	2.74	10.95
S1423	36	22246	4.45	1261666	8.67	56.71	19.48	7.79
S1494	469	27510	31.35	1929630	103.95	70.14	3.32	13.26
S1488	590	27314	43.68	1902225	121.83	69.64	2.79	11.16
S5378	408	65310	82.20	10878780	587.85	166.57	7.15	28.61
S35932	86	493025	162.23	620065720	5885.53	1257.68	36.28	145.12

Table 6.12 Fault simulation results for combinational circuits (metal-BFs)

	# of	Extracted BFs		metal BFs		Ratio		
	Tests	Faults	CPU s	Faults	CPU s	Faults	CPU s	Adj CPU s
C1355	99	16282	6.96	171991	2.03	10.56	0.29	1.17
C1908	146	22316	11.38	416328	5.20	18.66	0.46	1.83
C2670	310	39130	36.33	1127251	19.26	28.81	0.53	2.12
C3540	227	50106	37.20	1476621	14.96	29.47	0.40	1.61
C5315	370	76944	90.75	3086370	34.68	40.11	0.38	1.53
C6288	83	71232	23.57	2995128	7.18	42.05	0.30	1.22
C7552	278	103195	88.87	6913621	40.83	67.00	0.46	1.84

Table 6.13 Fault simulation results for combinational circuits (all BFs)

	# of	Extracted-BFs		all BFs		Ratio		
	Tests	Faults	CPU s	Faults	CPU s	Faults	CPU s	Adj CPU s
C1355	99	16282	6.96	675703	26.62	41.50	3.82	15.30
C1908	146	22316	11.38	1269621	22.38	56.89	1.97	7.87
C2670	310	39130	36.33	3904615	136.48	99.79	3.76	15.03
C3540	227	50106	37.20	6402831	98.60	127.79	2.65	10.60
C5315	370	76944	90.75	15100260	863.03	196.25	9.51	38.04
C6288	83	71232	23.57	12941328	112.50	181.68	4.77	19.09
C7552	278	103195	88.87	27162135	435.22	263.21	4.90	19.59

F_{BG}	F_{BD}	F_{BS}	F_{SD}	F_{GD}	F_{GS}

Figure 6.14 6-bit representation of six leakage faults in a transistor

ble to have fault simulation algorithms requiring run times that grow slower compared to increase in number of faults. In addition, simulating metal-BFs is computationally not very time consuming.

6.3 SIMULATION OF LEAKAGE FAULTS

Leakage faults are simulated using a hierarchical scheme to store detection status of targeted leakage faults [198]. This method has two stages. In the preprocessing stage, all logic components are characterized (**Fault Table Generation**). This is a one time process and is done for all cells in the given cell library. In the next stage, during evaluation of an I_{DDQ} test set, for each test vector, full circuit gate level logic simulation is used to compute the I/O values for each component instantiation. From the I/O values computed and the fault tables, detection status for all targeted faults is computed. Although we explain this method with leakage faults as example, it can be used for simulation of other fault models.

In the preprocessing phase, for each component type, the transistor terminal node values are computed for all possible input-output combinations. From the transistor terminal node values detection status for each leakage fault can be determined. For example, if the logic values at the terminals of an nMOS transistor are (G,S,D,B) = $(1, 0, 0, 0)$, then leakage faults f_{GS}, f_{GD} and f_{GB} are detected. This information on each transistor, for each input combination is stored in a table-like data structure. Such a table, where rows correspond to I/O combinations and columns correspond to transistors in the component, the entries can be some encoding of the detection status of all leakage faults involving the particular column transistor. An octal encoding scheme is used where the six bits shown in Figure 6.3 correspond to the six leakage faults. Each bit is set to 1 if the corresponding leakage fault is detected and a 0 otherwise. So, using this encoding scheme, each table entry is a two digit octal number. As an example, for the two-input NAND gate in Figure 6.14, the fault table is given in Table 6.14. In general, for each component with k inputs and m transistors, a table of size $m2^k$ is pre-computed and stored.

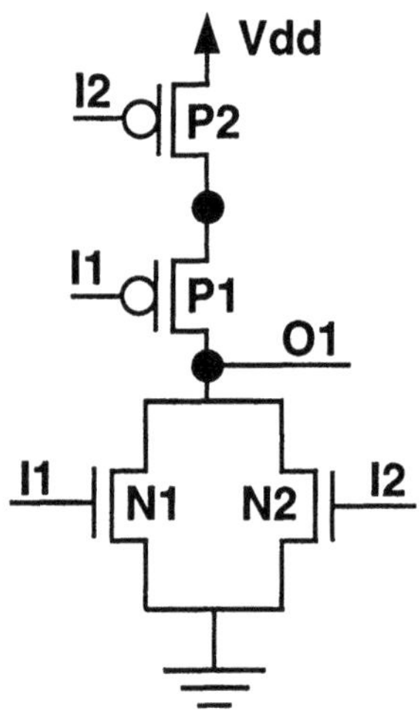

Figure 6.15 Example to illustrate leakage fault simulation

Table 6.14 Leakage fault table

Row index	N1	N2	P1	P2
1	0	0	0	0
2	15	15	43	43
3	0	43	70	26
4	0	0	0	0
5	43	0	26	43
6	0	0	0	0
7	43	43	22	0
8	0	0	0	0

During the phase of evaluating a test set for leakage faults, logic simulation is done for each test vector to compute the I/O values for all components. These I/O values correspond to the row index of the fault table. So, using table lookup the indexed row gives the list of leakage faults detected by the given test vectors. For example, for the circuit in Figure 6.14, the test input $I1 = 0, I2 = 0$ detects the leakage faults given in row 0 of the fault table in Table 6.14.

6.4 SIMULATION OF WEAK FAULTS

For a given test set, coverage of weak faults is computed using a hierarchical scheme similar to the scheme used for leakage fault simulation [198]. During pre-processing (fault table generation) for each component type, nodes with weak faults are classified as: (i) **internal fault node**, if there is no path to component output; (ii) **boundary fault node**, if there is at least one path to component output and this node does not drive any other transistor gate in the same component; and (iii) **internal boundary fault node**, if there is at least one path to component output and this node drives at least one transistor gate in the same component. For example, in Figure 5.12(b) node A is of type (i), node $O2$ is of type (ii) and node $O1$ is of type (iii).

Table 6.15 Weak fault detection table

Input id	A		$O1$	
	Weak-0	Weak-1	Weak-0	Weak-1
0	0	0	1	0
1	0	0	0	0
2	0	0	0	1
3	0	0	0	0

Based on this classification of nodes in the components, three tables are generated during the fault characterization phase. First table, called **fault detection table** lists (i) and (ii) type faults detected for all possible I/O values for a given component. The second table, called **fault sensitization table** lists all type (ii) and (iii) that are sensitized but not detected for all I/O values for a given component. The third table, called **fault propagation table** gives I/O conditions for which a sensitized weak fault at the component input will be propagated. Using the circuit in Figure 5.12(b), we illustrate the generation

Table 6.16 Weak fault sensitization table

Input id	$O1$		$O2$	
	Weak-0	Weak-1	Weak-0	Weak-1
0	1	0	0	1
1	0	0	0	0
2	0	1	1	0
3	0	0	0	0

of these three types of fault tables. The generation of weak **fault detection table** is illustrated in Table 6.15, for weak-0 and weak-1 faults at nodes A and $O1$. Weak faults at A and $O1$ do not have to be propagated. For weak faults

Table 6.17 Weak fault propagation table

Input id	$O1$		$O2$	
	Weak-0	Weak-1	Weak-0	Weak-1
0	1	1	0	0
1	0	0	0	0
2	1	1	0	0
3	0	0	0	0

at $O1$ and $O2$ of the circuit, fault sensitization can be done using the fault sensitization table given in Table 6.16. Weak faults propagated to the input $I1$ or $I2$ of the circuit can be detected as given by Table 6.17.

6.5 SUMMARY

In summary we note that if extracted BFs are used as the fault model then modifications of known techniques for fault simulation of stuck-at faults can be used. For metal-BFs and all-BFs new techniques have been developed. Experimental results show that fault simulation of metal-BFs and all-BFs is computationally feasible. Leakage faults are a special case of extracted bridges. Fault simulation algorithms for extracted as well as all-BFs can be used to simulate leakage faults. However, better methods tailored for leakage faults have been developed. Note that all these fault models target subsets of bridges. Fault simulation algorithms for opens have not been studied. Fault simulation algorithms for weak-faults target a small subset of opens. More research is needed to understand the fault simulation problem for opens.

7

SELECTING I_{DDQ} TESTS

Issues in computing I_{DDQ} test sets are discussed. I_{DDQ} test sets are computed using different fault models. They also differ in the source of the tests. Based on this, I_{DDQ} test generation is classified as either: (i) **direct**; or (ii) **indirect**.

In *indirect methods*, a subset S of test vectors is selected from a pre-existing test set T, such as stuck-at or functional test sets. It is assumed that during testing the complete test set T is applied. However, I_{DDQ} measurements are made only after application of tests in S. The selected vectors identify the I_{DDQ} **measurement points**. Indirect methods are further subdivided as **every vector** and **selective**. In the *every vector indirect method* the subset S is the set T itself. That is, a test set T generated for logic or functional testing is used as the I_{DDQ} test set. It is motivated by some of the good properties of pseudo stuck-at test sets that we discussed in Chapter 5. However, we have also seen in Chapter 3 that I_{DDQ} measurement is a slow process. In addition, it is possible to get a good coverage with considerably fewer measurements. These motivate the *selective indirect method* where a small subset S of T is selected as the I_{DDQ} test set.

In a *direct method*, an I_{DDQ} test set is computed by using an ATPG technique where defects are specifically targeted. No other source of test vectors is used to compute I_{DDQ} test sets. Tests so generated are also referred to as **supplemental** [301]. Tests obtained using indirect methods are quite effective in production test environment [198, 204, 205, 238]. However, I_{DDQ} tests computed using direct methods are better in that they have a higher fault coverage with fewer I_{DDQ} measurements. Another advantage of direct methods is the ability to identify undetectable faults.

In Chapter 8 we discuss direct methods for computing I_{DDQ} test sets. Here we discuss indirect methods. The basic problem in indirect methods is how to compute a minimum cardinality subset S of the given test set T. Details of different approaches to select an I_{DDQ} subset are given. Prior to that, we present experimental results for I_{DDQ} measurement on every vector.

7.1 EVERY VECTOR METHOD

Functional tests and stuck-at tests for logic testing have been used as I_{DDQ} test sets. High fault coverages are reported for metal BFs and intra-transistor faults [133]. Metal BFs were discussed in Chapter 5. Intra-transistor faults refer to the BF between the three terminals of a transistor. Thus, for each transistor we have $\langle source, drain \rangle$, $\langle source, gate \rangle$ and $\langle drain, gate \rangle$ faults. This set of faults is a subset of leakage faults discussed in Chapter 5. We have already seen in Chapter 5 that a complete stuck-at test set detects all leakage and hence all intra-gate faults. Even if the stuck-at test set is not complete, which is usually the case, it has been observed that logic tests still have good fault coverage for intra-transistor faults [133]. In the results given in Table 7.1, "TF" is the total number of faults targeted and "FC" is the fault coverage. Fault coverages in excess of 99.0% were obtained for all circuits. It is interesting to note that the stuck-at test set also detected a large percentage of metal BFs.

Table 7.1 Intra-transistor and gate-level BF coverage using stuck-at tests

Circuit	Number of	FC	Intra-transistor		Metal-BFs	
	Vectors	(Stuck-at)	TF	FC	TF	FC
c3540	265	95.89	22512	99.74	1476621	99.85
c5315	213	98.79	33786	99.98	3086370	99.95
c6288	60	99.42	30336	99.72	2995128	99.99
c7522	338	98.10	46188	99.92	6913621	99.98
s1238	353	94.69	8694	99.86	145530	99.99

The above results indicate that the method of using the entire logic test set, computed using the stuck-at fault model, can be very effective in detecting intra-transistor faults and a large percentage of metal BFs. The drawback is that a large number of I_{DDQ} measurements have to be made. This makes the method impractical for production testing. Υ

7.2 I_{DDQ} TEST SELECTION

The basic I_{DDQ} selection problem is defined as follows. Given a set of input vectors $T = \{T_1, \ldots, T_n\}$ let X be the percentage of target BFs detected when I_{DDQ} measurement is made after application of every vector in T. The objective is to select a subset S of vectors from T, of minimum cardinality, satisfying the following property. If I_{DDQ} measurement is made after application of only the vectors in S then the percentage of target BFs detected is also X [199]. Recall that every vector in the set T is applied. However, I_{DDQ} measurement is made only after application of vectors in S. The smaller the cardinality of S the shorter the test time. There are two other related problems.

- Typically, T is a stuck-at test set or a set of functional test vectors. If T is itself an I_{DDQ} test set then solving the subset selection problem helps in **compressing the I_{DDQ} test set.**
- In many IC production facilities, there is a bound, say K, on the number of I_{DDQ} measurements that can be made. Thus, a selection of the best K vectors to maximize fault coverage, when I_{DDQ} measurements are made only after applying the K selected vectors, is important.

We discuss solutions to the selection problem with no bound on the size of the selected subset. These solutions are easy to modify when the number of I_{DDQ} measurements is bounded.

Solutions to the selection problem are modeled as solutions to **set cover** [146], a well known combinatorial optimization problem. Several heuristics have been studied for set cover [53, 64, 126, 146, 180, 341]. We present these heuristics and show that there is a difficulty in using these heuristics for our problem. The primary source of this difficulty lies in the large number of target BFs.

7.2.1 Set Cover and Selecting Measurement Points

To define **set cover** let $F = \{S_1, S_2, \ldots, S_k\}$ be a collection of finite sets such that $U = \bigcup_{i=1}^{k} S_i$. Then $F' \subset F$ covers F if and only if $\bigcup_{s \in F'} S = U$. The optimization problem is to compute a cover F' of minimum cardinality.

Example 7.2.1. Assume that we have four sets $S_1 = \{3\}$, $S_2 = \{2\}$, $S_3 = \{2,3\}$, $S_4 = \{1,3\}$. The **universal set** $U = \bigcup_{i=1}^{4} S_i = \{1,2,3\}$. A **cover** for this collection of sets is a subset of this collection such that every element in the universal set is in at least one set in the *cover*. For our example there are 5 covers: $\{S_1,S_2,S_3,S_4\}$, $\{S_1,S_2,S_4\}$, $\{S_2,S_3,S_4\}$, $\{S_2,S_4\}$ and $\{S_3,S_4\}$. The minimum cardinality covers are $\{S_2,S_4\}$ and $\{S_3,S_4\}$. Note that the solution need not be unique.

To understand the **relationship between set cover and subset selection** assume that a test set $T = \{T_1,\ldots,T_N\}$, from which we select measurement points, is given. An instance of *set cover* is created as follows. For each i, let S_i be the set of faults detected by T_i. Then, $U = \bigcup_{i=1}^{N} S_i$ is the set of all faults detected by the test set. Thus, the S_i's and U define an instance of set cover. Let $\{S_{j1},\ldots,S_{jk}\}$ be a solution to this instance of set cover. This implies that the I_{DDQ} measurement points are $T_{j1},\ldots,T_{jk}$.

Example 7.2.2. Let the instance of the selection problem be given by the test set $\{T_1,T_2,T_3,T_4\}$ and the set of target faults $\{1,2,3\}$. Let T_1,T_2,T_3 and T_4 detect, respectively, faults $\{3\}$, $\{2\}$, $\{2,3\}$ and $\{1,3\}$. The instance of *set cover* that we get is same as in Example 7.2.1. For that instance of set cover, assume that we get the solution $\{S_2,S_4\}$. We deduce that I_{DDQ} measurements be made only after applying tests T_2 and T_4.

7.2.2 Heuristics for Set Cover

QUIETEST Heuristic [53, 197]:

From the given collection of sets, traverse the sets in a predetermined order. If, S_i is the set being visited and S_i covers an element not covered by any of the sets in the current solution, then we include S_i in the solution. The selection is complete when all elements of the universal set are covered. The quality of the solution depends on the order in which the list of sets is traversed.

Example 7.2.3: Assume that the sets of Example 7.2.1 are traversed in the order $\langle S_1,S_2,S_3,S_4\rangle$. We initially pick S_1 thus covering element $\{3\}$. We next pick S_2 and the new set of elements covered is $\{2,3\}$. We do not pick S_3 since both 2 and 3 have already been covered. We then pick S_4 thus covering $\{1\}$. Since all elements in the universal set have been covered the selection process terminates giving us the solution $\{S_1,S_2,S_4\}$.

COVER Heuristic [146]:

We start by selecting the set with the largest number of elements. Next, for each set we determine the number of elements, not covered by the sets in the current solution. We add to the solution the set that covers the maximum number of elements not covered. The process terminates when all elements are covered.

Example 7.2.4: Consider the instance of *set cover* given in Example 7.2.1. We first select S_3 which covers $\{2,3\}$. Next we remove the covered elements to get the updated sets as $S_1 = S_2 = \Phi, S_4 = \{1\}$. Next the heuristic picks S_4. All elements are now covered and the solution is $\{S_3, S_4\}$. Note that this solution is smaller than the solution obtained by the first heuristic.

W_Cover Heuristic [180]:

The idea here is based on the notion of weighted cover. For each element x in U, the set of all elements, let weight $c(x) = \frac{1}{N}$, where N is the number of sets that cover x. For each set S_i, its weight is the sum of the weight of its elements.

Start by picking the set with maximum weight. Next, for each set determine the elements not covered by the current solution. Update the weights of each set and select the set with the maximum weight. Repeat the process until all elements are covered.

Example 7.2.5: Consider the instance of *set cover* given in Example 7.2.1. First compute weights of each element in U. Hence, $c(1) = 1$, $c(2) = \frac{1}{2}$ and $c(3) = \frac{1}{3}$. Next, weight of each set S_1, S_2, S_3 and S_4 are $\frac{1}{3}$, $\frac{1}{2}$, $\frac{5}{6}$ and $\frac{4}{3}$ respectively. Since, S_4 has maximum weight, it is selected. Next, weights for each of the remaining sets are computed. This gives the weight for S_1 as 0 and for S_2 and S_3 as $\frac{1}{2}$. We could select either S_2 or S_3. Suppose, we select S_2. On picking S_2, all the elements are covered. Thus the final solution is $\{S_2, S_4\}$. Note that this solution is better than the solution obtained by the first heuristic. However, the quality of this solution is the same as the second heuristic.

W_ECover Heuristic [180]:

As in W_Cover, weights for the sets are first computed. Then, the elements of the universal set are considered one at a time. Let, during the current iteration, e be the uncovered element being targeted. Select among all sets containing e the one with the maximum weight. Next, delete all covered elements, update

Table 7.2 Classical data structure for set cover heuristic

	1	2	3	4	5	6
S_1	X				X	X
S_2		X	X	X	X	
S_3		X		X		X

the weights of the sets and select the next uncovered element. This is repeated until all elements are covered.

Example 7.2.6: For Example 7.2.1, assume the weights of the sets to be the number of elements in it. So, the weights of S_1, S_2, S_3 and S_4 are 1, 1, 2 and 2, respectively. Starting with element 1, the set with maximum weight that covers 1 is S_4. Hence, we select S_4. Next, delete all covered elements from each of the sets. Now, the only uncovered element is 2. The set with maximum weight that covers 2 is either S_2 or S_3. We pick S_2, hence the solution we obtain is $\{S_2, S_4\}$. This solution is better than the solution obtained by the first heuristic.

Fault Tables:

Let $U = \{1, 2, 3, 4, 5, 6\}$ be the universal set and $S_1 = \{1, 5, 6\}, S_2 = \{2, 3, 4, 5\}$ and $S_3 = \{2, 4, 6\}$ be the collection of sets. The standard representation used for this problem is a two-dimensional array. There is one column for each member of the universal set and one row for each set. An X in row i and column j implies that element j is in set S_i. For our example the data structure is shown in Table 7.2. It is henceforth referred to as **fault table**.

It is the representation used in all known implementation of these heuristics [53, 146, 180, 198, 341]. We refer to methods using this representation as **fault table based** methods. To use such a two dimensional structure the faults have to be explicitly enumerated. Consequently, if the number of targeted faults is large, the number of columns will be large too. Such an approach is therefore ruled out. We will see in Section 7.5 how we can get around this.

7.3 SELECTING I_{DDQ} TESTS FOR EXTRACTED BFS

We discuss CURRENT, a system for selecting I_{DDQ} tests for extracted faults in full scan circuits [180]. It was proposed as a method for compressing an I_{DDQ} test set computed using a direct method. The heuristic used is based on **W_ECOVER** discussed in Section 7.2.2, which is implemented using a fault-table of size $N_p \times N_f$. Here N_p is the number of patterns and N_f is the number of faults. Each entry $t[i, j]$ is set to 1 if test vector T_i detects fault j and to 0 otherwise.

The idea is to pick a fault j, then select the test input i with maximum weight that detects j. Weights for all unselected tests are recomputed and the process is repeated. In each stage of the algorithm, the weight of a test vector i is the number of additional faults detected by it. As a pre-processing step, all tests that uniquely detect a fault are selected and the table is appropriately updated. The procedure is summarized below:

1. Initialize the fault table using fault simulation.
2. For each row, compute weight of test T_i, the number of faults detected by T_i (row addition).
3. Pick an undetected fault f.
4. Select the test input T_i with maximum weight that detects f.
5. Delete row i and all columns corresponding to faults detected by test T_i.
6. If all vectors have been inspected, or maximum fault coverage attained, terminate. Else, go to 2.

The method is illustrated with the circuit of Figure 7.1(a) and the test set of Table 7.3. Figure 7.1(b) is the switch-level representation of the circuit in Figure 7.1(a). Let L, the list of extracted faults, be $\{\langle a,b\rangle, \langle a,c\rangle, \langle a,d\rangle, \langle a,f\rangle, \langle b,c\rangle, \langle b,d\rangle, \langle c,e\rangle, \langle d,e\rangle\}$. The initialized fault table is given in Table 7.4. Column "Weight" gives the number of faults detected by the corresponding test T_i. For this example, total faults in L is 8.

For the first iteration pick the undetected fault $\langle a,b\rangle$. Tests with maximum weight that detect $\langle a,b\rangle$ are T_1 and T_2. Hence, we can select either of these

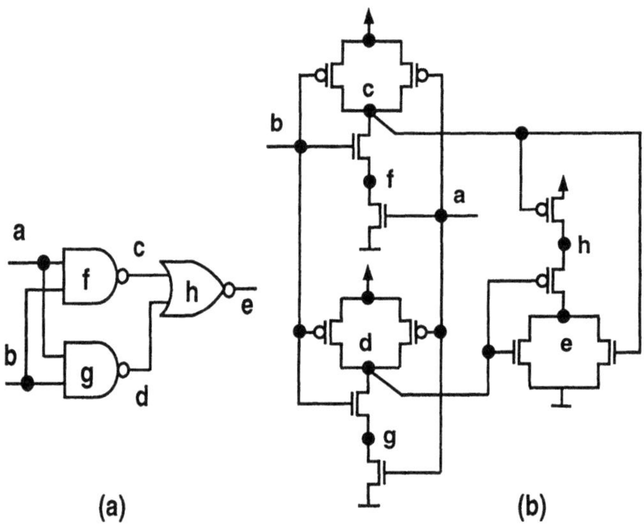

Figure 7.1 Example illustrating selection algorithms for combinational circuits

Table 7.3 Test set used to illustrate simulation of extracted faults

	a	b	c	d	e	f	g	h
T_1	1	0	1	1	0	0	0	f
T_2	0	1	1	1	0	1	1	f
T_3	1	1	0	0	1	0	0	1
T_4	0	0	1	1	0	f	f	f

Table 7.4 Illustration of I_{DDQ} selection for extracted BFs (xy refers to BF between x and y)

Tests	Weight	*ab*	*ac*	*ad*	*af*	*bc*	*bd*	*ce*	*de*
T_1	6	1	0	0	1	1	1	1	1
T_2	6	1	1	1	1	0	0	1	1
T_3	7	0	1	1	1	1	1	1	1
T_4	6	0	1	1	0	1	1	1	1

Table 7.5 Test compaction by the CURRENT system

circuit	Faults	FC	Tests	Subset	CPU (h:m:s)
c3540	66214	99.72	64	50	0:0:22
c5315	101478	99.89	64	49	0:0:33
c6288	110176	99.88	38	25	0:0:22
c7552	138068	99.87	96	70	0:0:52
s13207	232269	99.92	236	206	0:4:28
s15850	279102	99.90	198	172	0:4:29
s38417	667346	99.96	310	250	0:15:08
s38584	716616	99.52	424	344	0:26:35
s100K	2432130	99.80	1716	868	4:43:27

tests. We pick T_1 and delete all columns corresponding to faults detected by it. The remaining columns correspond to faults $\langle a, c\rangle$ and $\langle a, d\rangle$.

In the second iteration, pick the undetected fault $\langle a, c\rangle$. All three tests $T_2 - T_4$ detect it and they all have weight 2. Hence, we can pick any one of the 3 tests, say T_2. On deleting columns corresponding to faults covered by T_2, say $\langle a, c\rangle$ and $\langle a, d\rangle$, we note that all faults are covered by the solution computed so far. Hence, the algorithm terminates. The selected subset is $\{T_1, T_2\}$.

Experimental results obtained by this approach are given in Table 7.5. Results for a 100k gate design are also given. Column "Faults" gives the total number of faults targeted, "FC" is the fault coverage, "Tests" is the size of initial test set and "Subset" is the size of the selected subset. Column "CPU" gives the run time in hours:minutes:seconds on an HP 9000/735 workstation. For the largest ISCAS benchmark circuit the time required is less than 30 minutes. An average test set reduction of 42% was obtained. For the largest design s100k, the memory requirement was about 500MB.

We conclude this section by noting that if the number of target faults is large, the size of the fault table is large, leading to a high memory requirement.

7.4 SELECTING I_{DDQ} TESTS FOR LEAKAGE AND WEAK FAULTS

Two procedures to select I_{DDQ} tests for leakage faults [199, 341] and a procedure for selecting I_{DDQ} tests for weak faults [199] are discussed. Note that selection algorithms from the previous section can also be used for leakage faults. Heuristics discussed in this section can also be used for extracted faults.

The selection algorithm in [341] is valid only for a full-scan circuit. It uses a fault-table of size $N_p \times N_f$, where N_p is number of patterns and N_f is number of faults. As a pre-processing step all tests that uniquely detect a fault are selected and the fault table is appropriately updated. This reduces the number of rows and columns. A fault collapsing procedure is used to reduce the number of target leakage faults. This reduces the number of columns.

The basic idea is to select patterns detecting those faults that are detected by the smallest number of patterns in the test set. It uses the **W_COVER heuristic**. In the fault table, entry $t_{i,j} = 1$ if test T_i detects fault j, else $t_{i,j} = 0$. A weight c_j is associated with each fault where c_j is the number of equivalent faults represented by fault j. This weight is used during selection in Step 4, where instead of using only $t_{i,j}$, $t_{i,j} \times c_j$ is used.

1. Initialize fault table using fault simulation.
2. For each fault, compute the number of patterns (d_j) that detect it (column addition).
3. For each test input, compute a weight $w_i = \sum_{j=1}^{N_f} \frac{c_j \times t_{i,j}}{d_j}$ (row addition).
4. Select test T_i with maximum w_i.
5. Update fault table by removing all faults detected by the selected test and go to Step 3.

This algorithm is illustrated in Table 7.6. The leakage faults considered are $f1 - f8$. Set $\{T_1, \ldots, T_4\}$ is the initial test set. Table 7.6 is the fault table computed in Step 1 of the algorithm. Value of d_j, for each fault fj, computed in step 2 is given in row "d_j" of the fault table.

Weights w_i for test T_i are: $w_1 = 2$, $w_2 = 2$, $w_3 = \frac{13}{6}$ and $w_4 = \frac{11}{6}$. Since T_3 has maximum weight it is selected. On deleting columns corresponding to faults

Table 7.6 Fault table used to illustrate I_{DDQ} selection for leakage faults

Tests	$f1$	$f2$	$f3$	$f4$	$f5$	$f6$	$f7$	$f8$
T_1	1	0	0	1	1	1	1	1
T_2	1	1	1	1	0	0	1	1
T_3	0	1	1	1	1	1	1	1
T_4	0	1	1	0	1	1	1	1
d_j	2	3	3	3	3	3	4	4

Table 7.7 Selecting tests for leakage and weak faults

	Leakage faults			Weak faults		
Circuit	Faults	FC %	% of tests Selected	Faults	FC %	% of tests Selected
circuit 1	39295	94.84	0.50	1923	85.3	0.35
circuit 2	220571	90.50	0.99	1497	87.64	0.21

detected by T_3, the remaining column is $f1$. In the second iteration, we can select either T_1 or T_2 since they have the same weight. We select T_2 and stop. The selected subset is $\{T_3, T_2\}$.

QUIETEST [199] selects I_{DDQ} vectors to detect all targeted leakage and weak faults. Its heuristic is implemented using fault tables. The full circuit is simulated at the gate level using production test patterns. Logic levels at the inputs and outputs of all components are captured. The I_{DDQ} test vector selection is done using component fault tables for leakage and weak faults. If required more than one fault tables are used to infer detection status of weak faults. If a vector catches even one new fault, it is selected for I_{DDQ} measurement.

QUIETEST was used to select I_{DDQ} vectors for detecting leakage and weak faults in two circuits. The results are shown in Table 7.7. For each circuit QUIETEST selected less than 1% of the total test vectors. For circuit 1(circuit 2), 80.2%(95%) of the I_{DDQ} vectors selected for testing weak faults were common with the vectors selected for leakage faults.

Results on selection from [341], targeting leakage faults, are presented in Table 7.8. The circuits are the ISCAS85 combinational benchmark circuits and the stuck-at test set was generated using FAN [98]. Column "Faults" give the

Table 7.8 Test selection results

Circuit	Faults	FC	NT	QUIETEST	COVER	W_COVER	CPU s
c2670	27488	98.5	150	50	37	30	38.8
c3540	39672	98.7	250	81	48	41	43.0
c5315	60060	99.1	191	69	45	38	141.6
c6288	60672	99.5	48	35	28	26	163.9
c7552	79560	98.5	274	95	58	49	287.2

Table 7.9 Results for selecting 20 tests

Circuit	Faults	NT	FC(COVER)	FC(W_COVER)
c2670	27488	150	97.9	97.7
c3540	39672	250	96.8	96.3
c5315	60060	191	98.5	97.5
c6288	60672	48	99.2	99.3
c7552	79560	274	97.6	96.5

number of target faults, "NT" is the initial test set size and "CPU s" is the run time in CPU seconds. To compare the quality of this solution (**W_COVER**) with other methods, the subset sizes obtained using *QUIETEST heuristic* and *COVER heuristic* are also included. We note that W_COVER results in the selection of fewer tests.

In Table 7.9, experimental results obtained for the $K - I_{DDQ}$ selection problem, where K is the bound on the number of tests, are given for $K = 20$. For all circuits, *COVER heuristic* gives better coverage than *W_COVER heuristic.*

Note that, although heuristics like W_COVER, COVER, etc., are better for leakage faults they can only be used for full-scan circuits. For sequential circuits, QUIETEST is the only known heuristic. Techniques to implement other heuristics for sequential circuits need to be investigated. The main difficulty is in simulating, without fault dropping, leakage and weak faults in sequential circuits.

7.5 SELECTING I_{DDQ} TESTS FOR METAL BF AND ALL BF

The fault table based methods discussed in the previous subsections require that faults be explicitly enumerated. The number of columns required is proportional to the number of faults. Such methods cannot be used when the fault list is very large, as is the case for all BFs and metal BFs. Here we discuss methods that enumerate the target faults implicitly and can be used for both the metal BF and all BF fault models.

We next discuss how we adapt the QUIETEST heuristic for combinational and sequential circuits and the COVER heuristic for combinational circuits. The technique presented here [308] is an improvement over the one presented in [55].

7.5.1 QUIETEST heuristic for BFs

A procedure to select I_{DDQ} tests, targeting metal BF and all BF, using **QUIETEST heuristic** is given below:

1. Initialize F to the target set of faults.
2. For each test $T_i \in T$, do steps 3, 4, 5 and 6.
3. Perform logic simulation using T_i.
4. (For sequential circuits) If a fault $f \in F$ is a PDF w. r. t. T_i, select T_i.
5. Compute the new list of undetected BFs F'.
6. If $|F'| < |F|$, select T_i. Let $F = F'$.

The method can be used both for combinational and sequential circuits. We illustrate the method only for sequential circuits. First we explain steps 4 and 5. During simulation of test T_i we make the following assumption: initial state is unknown; fault-free next state is used; and if a fault f is a PDF w.r.t. T_i then it is assumed not detected by T_i. These assumptions were discussed in more details in Chapter 6. In the adaptation of the **QUIETEST heuristic** for sequential circuits, test T_i is selected if: (i) T_i detects an additional fault; or (ii) there exists a fault $f \in F$ such that f is a PDF w.r.t. T_i. Condition (i) is based on the heuristic itself. The need for condition (ii) is discussed below. For combinational circuits only condition (i) is required.

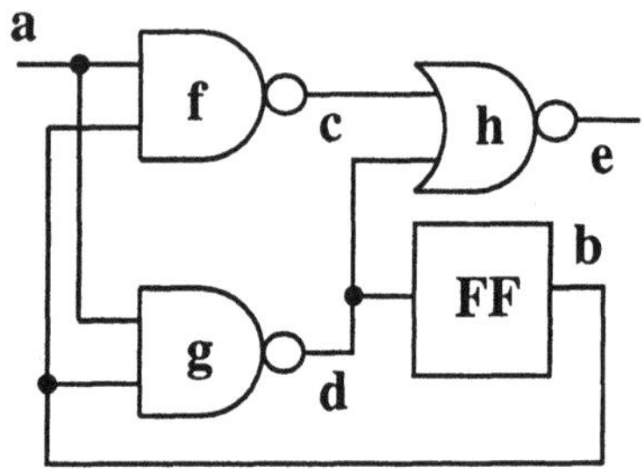

Figure 7.2 Circuit used in the sequential **QUIETEST** example

Consider a PDF $\langle a, b\rangle$ where $T_i(a) = X$ and $T_i(b) = 0$. During testing, if $T_i(a) = 0$ then the PDF is not detected and we can use the fault-free next state. If $T_i(a) = 1$ and T_i was not selected then the next state obtained can be different from the fault-free next state assumed during evaluation of the test set. To avoid any error due to this test T_i is always selected.

Implementation of the above idea requires careful consideration of the representation of the undetected set of BFs. The representation of the fault list for metal BF and all BF using NPs and IPs, as discussed in Chapter 6, is used. Fault dropping is done as in the list-based fault simulation algorithm of Chapter 6. We use the circuit of Figure 7.2 and the test set given in Table 7.10 to illustrate the selection procedure:

- Initially, $NP = E0$ and $IP = \Phi$, where $E0 = \{a, b, c, d, e, f, g, h\}$ and the number of undetected faults F=28.

- After T_1 is evaluated, $NP = \{E0, E1, E2\}$ and $IP = \{\langle E0, E2\rangle, \langle E1, E2\rangle\}$, where $E0 = \{a, e\}, E1 = \{c, d\}, E2 = \{b, f, g, h\}$ and $F = 24$. Since $\langle a, b\rangle$ is a PDF w.r.t. T_1, T_1 is selected. The partial solution is $\{T_1\}$.

- After T_2 is evaluated, $NP = \{E0, E1, E2, E3, E4\}$ and $IP = \{\langle E0, E2\rangle, \langle E0, E4\rangle, \langle E1, E3\rangle, \langle E1, E4\rangle, \langle E2, E4\rangle, \langle E3, E4\rangle\}$, where $E0 = \{a, e\}, E1 = \{c, d\}, E2 = \{f, g\}$, $E3 = \{b\}$ and $E4 = \{h\}$. Since T_2 detects 8 additional faults, T_2 is selected. The partial solution is $\{T_1, T_2\}$.

- After T_3 is evaluated, $NP = \{E0, E1, E2, E3, E4\}$ and $IP = \{\langle E0, E4\rangle, \langle E3, E4\rangle\}$, where $E0 = \{a, e\}, E1 = \{c, d\}, E2 = \{f, g\}$, $E3 = \{b\}$ and $E4 = \{h\}$. Since T_2 detects 8 additional faults, T_3 is selected. The partial solution is $\{T_1, T_2, T_3\}$.

Table 7.10 An example test set

	a	b	c	d	e	f	g	h
T_1	0	X	1	1	0	X	X	X
T_2	0	1	1	1	0	0	0	f
T_3	1	1	0	0	1	0	0	1
T_4	0	0	1	1	0	f	f	f

- Finally, after T_4 is evaluated, $NP = \{E0, E1, E2, E3, E4\}$ and $IP = \{\langle E0, E4\rangle, \langle E3, E4\rangle\}$, where $E0 = \{a, e\}, E1 = \{c, d\}, E2 = \{f, g\}$, $E3 = \{b\}$ and $E4 = \{h\}$. Since T_4 detects 0 additional faults, T_3 is not selected. Hence the final set of selected tests is $\{T_1, T_2, T_3\}$.

7.5.2 COVER heuristic for BFs

The **COVER** heuristic can be adapted for selecting I_{DDQ} tests for BFs in combinational circuits. A synopsis of the selection algorithm follows. From the given test set T, first select test T_i that detects the maximum number of as yet undetected faults. Next delete T_i from T. Repeat the above process of selecting the next vector until all faults are covered.

An efficient implementation of this idea requires careful consideration of several issues: (i) an efficient representation for the set of undetected faults; (ii) an efficient fault dropping method; and (iii) an efficient solution to the **incremental coverage computation** problem defined below. Let $\{T_1, \ldots, T_N\}$ be a test set and F be a set of faults. The problem of computing the number of faults from F detected by each of $T_1, \ldots, T_N$ is known as incremental coverage computation.

The need to efficiently represent the set of undetected faults and drop the detected faults should be clear. NPs and IPs, as discussed for the list-based simulation algorithm of Chapter 6, are used to represent the set of undetected faults. The fault dropping algorithm of the list-based algorithm is used. The need for an efficient solution to incremental coverage computation is illustrated below using the circuit of Figure 7.1(a), whose switch level implementation is shown in Figure 7.1(b), and the test set $T = \{T_1, T_2, T_3, T_4\}$ of Table 7.3.

Let L be the list of undetected faults, IF_i be the fault list after dropping faults from L that are detected by test T_i and F be the incremental coverage of T_i. IF_is are referred to as the **incremental fault lists**. Initially, L is represented by $NP = \{E0\}$ and $IP = \Phi$, where $E0 = \{a, b, c, d, e, f, g, h\}$. We proceed as follows:

Iteration 1.

- IF_1 is represented by $NP = \{E1, E2, E3\}$ and $IP = \{\langle E1, E3\rangle, \langle E2, E3\rangle\}$, where $E1 = \{b, e, f, g\}, E2 = \{a, c, d\}$ and $E3 = \{h\}$. The incremental coverage of T_1 is 12.
- IF_2 is represented by $NP = \{E4, E5, E6\}$ and $IP = \{\langle E4, E6\rangle, \langle E5, E6\rangle\}$, where $E4 = \{a, e\}, E5 = \{b, c, d\}$ and $E6 = \{f, g, h\}$. The incremental coverage of T_2 is 6.
- IF_3 is represented by $NP = \{E7, E8\}$ and $IP = \Phi$, where $E7 = \{c, d, f, g\}$ and $E8 = \{a, b, e, h\}$. The incremental coverage of T_3 is 16.
- IF_4 is represented by $NP = \{E9, E10, E11\}$ and $IP = \{\langle E9, E11\rangle, \langle E10, E11\rangle\}$, where $E9 = \{a, b, e\}, E10 = \{c, d\}$ and $E11 = \{f, g, h\}$. The incremental coverage of T_4 is 6.

  Since T_3 has the highest incremental coverage, we select T_3. L is updated and represented by $NP = \{E7, E8\}$ and $IP = \Phi$, where $E7 = \{c, d, f, g\}$ and $E8 = \{a, b, e, h\}$. This completes one iteration.

Iteration 2.

- IF_1 is represented by $NP = \{E12, E13, E14, E15, E16\}$ and $IP = \{\langle E14, E16\rangle, \langle E15, E16\rangle\}$, where $E12 = \{f, g\}, E13 = \{c, d\}, E14 = \{b, e\}, E15 = \{a\}$ and $E16 = \{h\}$. The incremental coverage of T_1 is 6.
- IF_2 is represented by $NP = \{E17, E18, E19, E20, E21\}$ and $IP = \{\langle E19, E20\rangle, \langle E20, E21\rangle\}$, where $E17 = \{c, d\}$, $E18 = \{f, g\}$, $E19 = \{a, e\}$, $E20 = \{h\}$ and $E21 = \{b\}$. The incremental coverage of T_2 is 6.
- IF_4 is represented by $NP = \{E22, E23\}$ and $IP = \Phi$, where $E22 = \{c, d, f, g\}$ and $E23 = \{a, b, e, h\}$. The incremental coverage of T_3 is 0.

  Since T_1 is one of the vectors with highest incremental coverage, T_1 is selected. Also, since T_4 has no incremental coverage, it is dropped from further consideration. L is updated and the new list of undetected faults is represented by $NP = \{E12, E13, E14, E15, E16\}$ and

$IP = \{\langle E14, E16\rangle, \langle E15, E16\rangle\}$, where $E12 = \{f, g\}$, $E13 = \{c, d\}$, $E14 = \{b, e\}$, $E15 = \{a\}$ and $E16 = \{h\}$.

Iteration 3.

- IF_2 is represented by $NP = \{E24, E25, E26, E27, E28, E29\}$ and $IP = \{\langle E26, E29\rangle, \langle E27, E29\rangle, \langle E28, E29\rangle\}$, where $E24 = \{f, g\}$, $E25 = \{c, d\}$, $E26 = \{b\}$, $E27 = \{e\}$ $E28 = \{a\}$ and $E29 = \{h\}$. The incremental coverage of T_2 is 1.

  Since T_2 has incremental coverage greater than 0, we select T_2.

At this point, there are no more tests to be evaluated. Therefore, the selection process terminates. The selected set of tests is $\{T_1, T_2, T_3\}$.

To summarize, each iteration is divided into two phases: (i) *evaluation*; and (ii) *update*. In the *evaluation* phase, incremental coverage of each unselected test T_i is computed. In the *update* phase, the test with the highest incremental coverage is selected. The fault list is updated to reflect the new set of undetected faults. The evaluation phase is very expensive and requires repeated computation of incremental coverage.

In the above example the incremental fault lists are computed from which the incremental fault coverage is derived. This is done for each unselected test in each iteration. The incremental fault lists so computed are discarded. This is a terrible waste of time and space. So, if the list-based fault simulation is used the resulting system will be very inefficient. A major source of efficiency in the algorithm given here is the way it computes incremental coverage [308]. This is explained next.

Given a list L of undetected faults, represented using NPs and IPs, the incremental coverage of T_j is computed as follows using logic simulation. The required logic simulation is done only once and the results are stored. Recall that in the list-based algorithm there are three parts in the computation of IF_j from L. We refer to them as the first, second and third parts.

- For each NP E_i in L, using logic simulation of T_j only, r_i^0, r_i^1 and r_i^2, respectively, the cardinality of the 0, 1 and X/f subsets of E_i w.r.t. T_j, are computed. For example, for the NPs in L at the start of iteration 2 for T_2 we have $r_7^0 = 2, r_7^1 = 2, r_8^0 = 2, r_8^1 = 1$ and $r_8^2 = 1$.

- In the first part the splitting of NPs $E_i \in L$, by test T_j, results in NPs in IF_j. The number of faults represented by these NPs is given by $\sum_{k=0}^{2} \frac{r_i^k \times (r_i^k - 1)}{2}$.
For example, for the NPs in L at the start of iteration 2 for T_2 the contribution of the NPs resulting from the splitting by T_2 of the NPs in L is calculated as follows:

$$\sum_{k=0}^{2} \frac{r_7^k \times (r_7^k - 1)}{2} + \sum_{k=0}^{2} \frac{r_8^k \times (r_8^k - 1)}{2} = 1 + 1 + 0 + 1 + 0 + 0 = 3$$

- In the second part, the splitting of each NP in L creates some IPs in IF_j, according to Rule 1 of Figure 6.9(a). The contribution of the new IPs obtained by the splitting of NP E_i by T_j is computed as $r_i^0 \times r_i^2 + r_i^1 \times r_i^2$.
For example, for the NPs in L at the start of iteration 2 for T_2 the contribution of the IPs resulting from the splitting by T_2 of the NPs in L is calculated as follows:

$$r_7^0 \times r_7^2 + r_7^1 \times r_7^2 + r_8^0 \times r_8^2 + r_8^1 \times r_8^2 = 0 + 0 + 2 + 2 = 4.$$

- In the third part, the new IPs in IF_j are created from IPs in L using Rule 2 of Figure 6.9(b). The contribution of the new IPs so obtained from the IP $\langle E_h, E_k \rangle$ is computed as follows:

$$r_h^0 \times r_k^0 + r_h^1 \times r_k^1 + \quad r_h^2 \times r_k^2 \quad + r_h^0 \times r_k^2 + r_h^2 \times r_k^0.$$

For example, for the IP $\langle E14, E16 \rangle$ in L at the start of iteration 3 for T_2, the contribution of the new IPs in IF_2 is calculated using $r_{14}^1 = 1, r_{14}^0 = 1, r_{14}^2 = 0, r_{15}^1 = 0, r_{15}^0 = 0$ and $r_{15}^2 = 1$. The total contribution is: $0 + 0 + 0 + 1 + 0 = 1$.

Thus, to compute incremental coverage, only the size of each new partition is required. Note that we compute and store the new list of NPs and IPs only in the *update* phase. This helps in saving memory and time required to compute incremental coverage. This algorithm, henceforth referred to as **FULL_SELECT** [308], is summarized below. We use the following notations. T is the set of unselected test vectors, S is the set of selected test vectors, NP is the list of NPs after simulating vectors in S, IP is the list of IPs after simulating vectors in S, X is the set of all or metal nodes in the circuit and F is the number of undetected BFs. The algorithm follows:

1. $S = \Phi, NP = X$ and $IP = \Phi$. Perform logic simulation and store the results in a table.
2. Repeat Steps 3 through 5 until $T = \Phi$.
3. *Evaluate:* Compute incremental coverage of each test $T_i \in T$ by calculating the size of the new NPs if T_i is selected. Compute the incremental coverage of T_i.
4. *Update:* Delete all tests from T with incremental coverage = 0.
5. *Update:* Let T_j be test with maximum incremental coverage. $T = T - \{T_j\}$ and $S = S \cup \{T_j\}$. Update NP and IP.

To use FULL_SELECT for the K-I_{DDQ} selection problem, modify Step 2 as follows:

2. Repeat Steps 3 through 5 until $T = \Phi$ or $\mid S \mid = K$.

7.5.3 Experimental Results

We present results obtained by the **QUIETEST** and **COVER** heuristics. For experiments on algorithms for combinational circuits, we used ISCAS85 [34] and combinational parts of ISCAS89 [32] benchmark circuits. For experiments on algorithms for sequential circuits, we use ISCAS89 [32] benchmark circuits. All times shown are in CPU seconds for a SUN 4/670 workstation.

Results for sequential circuits using the **QUIETEST** heuristic are given in Tables 7.11 and 7.12. In Table 7.11, an initial stuck-at test set is used. This stuck-at test set was generated using GENTEST [71]. The subset selected is quite small compared to the original test set. For example, for s1488 the selected test set is only 16% of the original test set. In Table 7.12, the initial test set is a set of randomly generated vectors. The size of this initial test set is approximately the same as the size of the stuck-at test set used in Table 7.11. Here again, we get large reduction in the number of I_{DDQ} measurements.

In Tables 7.13 and 7.14, an improved implementation of the **COVER** heuristic as discussed here [308] is compared with the results reported in [55]. The difference between the two approaches is in: (i) the representation of the list of undetected faults; and (ii) the solution to the problem of computing incremental

Table 7.11 **QUIETEST** heuristic for sequential circuits with an initial stuck-at test set

	s1238	s1423	s1488	s1494	s27	s298	s344
Test Set Size	349	36	590	469	15	162	91
Subset Size	272	35	93	82	9	52	40
Coverage	98.98	89.11	95.61	96.20	91.80	97.82	97.74
CPU s	130.72	21.05	449.45	356.58	0.02	5.28	3.27
	s382	s444	s526	s641	s713	s820	s832
Test Set Size	2463	1881	754	133	107	411	377
Subset Size	130	176	47	81	77	141	130
Coverage	99.33	99.18	73.72	97.92	93.74	93.69	91.24
CPU s	56.78	65.72	136.93	9.80	10.78	183.52	176.45

Table 7.12 **QUIETEST** heuristic for sequential circuits with an initial test set of randomly generated vectors

	s1238	s1423	s1488	s1494	s27	s298	s344
Test Set Size	350	35	590	470	15	160	90
Subset Size	256	34	91	87	7	46	47
Coverage	98.76	85.16	96.31	96.30	92.86	94.25	98.77
CPU s	134.63	19.27	421.00	339.22	0.02	7.23	3.08
	s382	s444	s526	s641	s713	s820	s832
Test Set Size	2400	1800	750	130	100	400	370
Subset Size	52	43	51	73	69	150	146
Coverage	77.07	75.54	75.36	95.31	94.48	91.04	90.73
CPU s	176.62	157.33	148.97	9.02	8.02	203.30	188.33

Table 7.13 Comparison with a previous approach (stuck-at test set)

				A [55]	B	
Circuit	Tests	Subset size	Coverage	CPU s	CPU s	Speedup
c880	98	61	99.85	72.88	42.1	1.73
c1355	99	96	99.76	190.55	115.92	1.64
c1908	146	126	99.66	671.85	356.78	1.88
c2670	310	246	99.78	4181.08	1678.53	2.49
c3540	227	180	99.85	2850.45	1608.70	1.77
c5315	370	319	99.91	13269.90	6787.10	1.95
c6288	83	70	98.64	592.50	310.30	1.90
c7552	278	235	99.92	11822.63	6897.40	1.71
Average Speedup = 1.88						

Table 7.14 Comparison with a previous approach (random test set)

				A [55]	B
Circuit	Tests	Subset size	Coverage	CPU s	CPU s
c6288	200	38	99.98	602.78	434.18
c7552	200	153	99.17	NA	5757.95
s9234	280	273	99.85	NA	30096.44
s13207	900	418	99.16	NA	136327.92
s15850	1100	496	99.81	NA	236080.67
s38417	480	477	99.98	NA	521170.96

coverage. In Table 7.13, the test set used is stuck-at test set computed using the ATPG program available with Octtools [163]. Column "A" gives time taken by [55] and column "B" gives time taken by the improved implementation discussed here. Column "Speedup" gives the ratio of A and B. The average speedup obtained is 1.88[1]. Table 7.14 gives comparative result for a random test set. I_{DDQ} subsets could not be computed for larger circuits using [55] because of the excessive memory requirement. On the other hand the improved method discussed here computed the subsets for all benchmark circuits.

[1]Note that since both solutions use the same heuristic, the size of the subset obtained by the two should be the same. However, they use slightly different circuit representations, leading to slight difference in fault coverages and subsets selected.

Table 7.15 Comparison of **COVER** scheme with **QUIETEST** heuristic

		COVER		QUIETEST			
Circuit	TS	S	CPU s	S	CPU s	FC	Improv
c2670	105	73	227.83	96	65.26	99.82	23
c3540	153	134	1019.86	148	187.61	99.84	14
c5315	113	93	1001.11	109	192.53	99.91	16
c7552	210	169	5018.03	202	726.00	99.87	33
s1196	144	123	345.38	140	68.70	99.78	17
s13207	469	326	28256.13	411	4180.73	99.95	85
s15850	440	374	42830.80	419	4003.28	99.92	45
s35932	70	69	7780.70	70	2630.55	99.41	1

In Table 7.15, we compare the COVER and QUIETEST heuristics for the I_{DDQ} selection problem of combinational circuits. The test set used is the stuck-at test set obtained using ATALANTA [168]. Column "TS" gives the size of the test set, "FC" is BF fault coverage, S is the size of I_{DDQ} subset, and "Improv" is the improvement obtained using **COVER**. "Improv" is calculated as the difference in the size of the I_{DDQ} subset obtained. Observe that **QUIETEST** uses considerably smaller CPU time. However, the size of the subset computed by **COVER** is considerably smaller, sometimes by as much as 15.5% of the given test set.

In Table 7.16 results for the I_{DDQ} test set size bounded by 20 are presented. The test sets used are the stuck-at test sets obtained using ATALANTA [168]. Results obtained using **QUIETEST** and **COVER** are compared. Column "TestSet" gives the size of the initial test set, "FC" gives fault coverage, and "Improv" gives the difference in fault coverage of the sets selected using **COVER** and **QUIETEST**. In all cases, the set selected by **COVER** has better fault coverage than the set selected by **QUIETEST**. For more than half of the circuits the improvement in coverage is over 0.5%. This improvement translates to a large number of BFs because the total number BFs targeted is very large. For example, in circuit s13207, the improvement is 1.16%. This translates to about 1059949 more BFs detected by the vectors selected by **COVER**. On the other hand, the running time for **QUIETEST** is consistently less than that for **COVER**.

Finally, we make an interesting comparison. We have already seen in Chapter 5 that it is sometimes advisable to target metal BFs. One reason that extracted

Table 7.16 K-I_{DDQ} Selection: Comparison of **COVER** heuristic with **QUIETEST** heuristic

20 vectors						
		COVER		QUIETEST		
Circuit	TestSet	FC (%)	CPU s	FC (%)	CPU s	Improv
c432	47	98.64	6.8	98.22	2.68	0.42
c499	53	98.33	56.17	97.08	13.33	1.25
c880	53	99.79	13.47	99.45	4.76	0.34
c1355	85	98.55	41.01	97.86	7.81	0.69
c1908	120	98.23	86.70	97.1	9.83	1.13
c2670	105	99.72	111.08	99.07	16.13	0.65
c3540	153	99.46	375.53	98.66	34.23	0.80
c5315	113	99.74	512.16	99.26	72.26	0.48
c6288	28	99.95	28.96	99.93	16.85	0.02
c7552	210	99.47	1589.71	98.86	94.66	0.61
s1196	144	99.05	151.11	98.09	15.15	0.96
s1238	149	98.81	193.51	96.42	17.16	2.39
s1488	127	99.17	196.33	97.69	21.03	1.48
s1494	119	99.12	204.53	98.4	21.6	0.72
s13207	469	99.11	6616.0	97.95	193.88	1.16
s15850	440	99.42	7475.2	98.57	235.28	0.85
s35932	70	99.39	978.38	99.36	229.88	0.03

faults are used in I_{DDQ} tool development is to avoid targeting a very large number of faults. The concern is that, if all BFs or metal BFs are targeted the tools would require inordinate amount of time. What is interesting is that, although the number of metal BFs is very large, compared to the number of extracted BFs, the resource requirement does not scale up as rapidly. This is supported by experimental results presented in Table 7.17.

In Table 7.17 results obtained for extracted BFs [180] are compared with results obtained for metal BFs using the more expensive of the two heuristics, viz., the **COVER** heuristic. The column "Faults" refers to the number of target BFs and the column "Ratio" gives the following ratios. The column labeled "Faults" under "Ratio" is the ratio of the number of target BFs. The column labeled "Tests" under "Ratio" is the ratio of the size of the initial test sets. Finally, the column labeled "Adj Time" under "Ratio" gives the value of $\frac{Time(\textbf{COVER})}{Time(\text{[180]})}/4.47$.

The factor of 4.47 was used to account for the difference between the speeds of the two machines used. Experiments on **COVER** were done on a SUN4/670 workstation and experiments in [180] were done on a HP9000/735 workstation. The performance ratings for these two machines, for various measures like MIPS, etc., are shown in Table 7.18. The average ratio of these ratings is 4.47.

Table 7.17 tabulates the comparisons for metal BF. Note that the run time for metal BF is marginally larger than the run time for extracted BF even though the ratio of the number of faults is very large. Thus, just as in the case of fault simulation, we see that targeting metal BF is computationally feasible.

7.6 SUMMARY

In summary, we note that I_{DDQ} test selection algorithms have been discussed for a variety of fault models. All fault models, except weak faults, target some subset of bridging faults. No work exists for opens like open gate, open source, etc. that are not targeted by the weak fault model. The algorithms that have been studied are based on heuristics for set cover. If the number of faults is small, as is the case for extracted BFs and leakage faults, the classical implementation technique for these heuristics can be used. This implementation technique is not suitable for metal BFs and all BFs. New techniques to handle the large number of metal BFs and all BFs have also been discussed.

Table 7.17 Comparison of Metal-BFs with extracted BFs. All times in CPU minutes

	Metal BFs				Extracted BFs [180]				Ratio		
Circuit	Faults	Tests	Subset	Time	Faults	Tests	Subset	Time	Tests	Faults	Adj. Time
C5315	14121955	102	37	5	101478	87	49	0.5	1.1	139	2.2
C6288	19766328	57	29	2	110176	50	25	0.3	1.1	179	1.5
C7552	28512576	599	46	64	138068	131	70	1.0	4.5	206	14
S13207	87205821	884	224	970	232269	383	206	4.5	2.3	375	48
S15850	125603325	1091	163	1065	279102	317	172	4.5	3.4	450	53
S38417	737913736	1133	198	3113	667346	480	250	15.0	2.3	1105	46
S5378	14458753	392	93	57	91546	206	113	1.0	1.9	157	12.8
S9234	42628761	610	137	282	163749	280	127	2.1	2.1	260	30.0

Table 7.18 Performance ratings of workstations

	MIPS	SPECint92	SPECfp92	SPECrateint92	SPECratefp92
SUN 4/670	28.50	23.90	21.80	567.00	517.00
SUN 5/70	105.40	59.60	46.80	1414.00	1110.00
HP9000/735				1832.00	2950.00

8

COMPUTING I_{DDQ} TESTS

In this chapter direct methods to compute I_{DDQ} tests are discussed. The advantage of using direct methods are: the resulting test set is smaller; the fault coverage is better; and undetectable faults can be identified. In addition to fault coverage, **fault efficiency** is another measure used to specify the quality of test sets. Fault efficiency is the percentage of all target faults that are either detected by the test set or proved to be undetectable by the direct method.

In Section 8.1 we discuss fault collapsing techniques. That is followed by test generation algorithms for combinational circuits. Three distinct classes of I_{DDQ} test generators, each with its own advantages, have been investigated. We discuss one representative, the one with the best known performance in its group, for each group in Sections 8.2 through 8.4. An analysis of the test generation results for combinational circuits is presented in Section 8.5. I_{DDQ} test generation for metal BFs and all BFs in sequential circuits is discussed in Section 8.6. As the readers will discern for themselves I_{DDQ} ATPG for combinational circuits has matured considerably. However, more research is required before the same can be said about I_{DDQ} ATPG for sequential circuits.

8.1 FAULT COLLAPSING

When a sequence of input vector is an I_{DDQ} test for f_1 if and only if it is also an I_{DDQ} test for f_2 then f_1 and f_2 are said to be I_{DDQ} equivalent. I_{DDQ} equivalent faults can be grouped into disjoint sets called **fault equivalence classes.** In order to speed up test generation time, ideally, the set of target

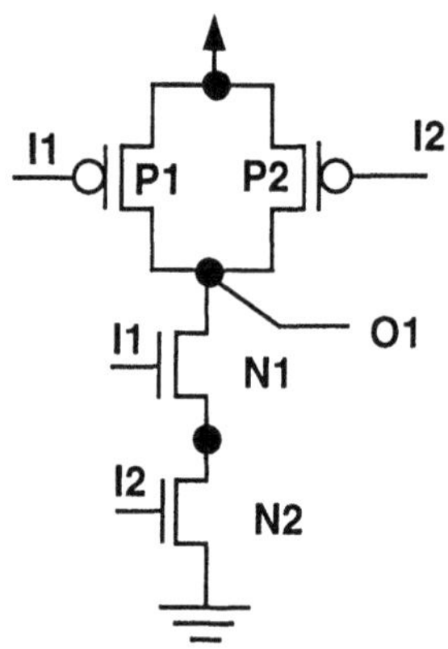

Figure 8.1 Illustration of leakage fault collapsing

faults should contain one fault per fault equivalence class. However, it is difficult to compute the fault equivalent classes. Thus, heuristics are used to include a small number of faults per fault equivalence class. This is known as **fault collapsing**. We discuss fault collapsing techniques for leakage and bridging faults in the next two subsections.

8.1.1 Leakage Faults

The method [341] described below uses **component fault tables** (**CFT**s). In CFT, there is one column for each leakage fault in the component and one row for each possible input combination to the component. Entry (i, j) in CFT is set to 1 (0) if leakage fault j is detected (not detected) by input combination i. The CFT for the NAND gate of Figure 8.1 is given in Table 8.1, where columns $i1$ and $i2$ represent the input combination. For each transistor, $p1, p2, n1, n2$ there are six columns. Column bg is for the leakage fault between the bulk (b) and gate (g). The other columns are similarly defined.

The columns in Table 8.1, corresponding to the leakage faults, are numbered 1 through 24. Columns 3, 9 and 21 correspond to leakage faults that are not detected by any input vector. Columns 1 and 6 are identical and the corresponding faults are collapsed into a single fault $f1$. Similarly, columns 2, 4, 8, 10 and 18 are collapsed into a single fault $f2$. It can be verified that, by collapsing all equivalent faults, we get the collapsed CFT (**CCFT**) of Table 8.2.

Table 8.1 Component fault table (CFT) for a NAND gate

		$p1$						$p2$					
$i1$	$i2$	bg	bs	bd	sd	gd	gs	bg	bs	bd	sd	gd	gs
		1	2	3	4	5	6	7	8	9	10	11	12
0	0	1	0	0	0	1	1	1	0	0	0	1	1
0	1	1	0	0	0	1	1	0	0	0	0	0	0
1	0	0	0	0	0	0	0	1	0	0	0	1	1
1	1	0	1	0	1	1	0	0	1	0	1	1	0
		$n1$						$n2$					
$i1$	$i2$	bg	bs	bd	sd	gd	gs	bg	bs	bd	sd	gd	gs
		13	14	15	16	17	18	19	20	21	22	23	24
0	0	0	1	0	0	1	0	0	0	0	0	0	0
0	1	0	1	0	1	1	0	1	0	0	0	1	1
1	0	1	1	1	0	0	0	0	1	0	1	1	0
1	1	1	0	0	0	1	1	1	0	0	0	1	1

Table 8.2 Collapsed CFT for a NAND gate

$i1$	$i2$	$f1$	$f2$	$f3$	$f4$	$f5$	$f6$	$f7$	$f8$	$f9$	$f10$	$f11$
0	0	1	0	1	1	1	0	1	0	0	0	0
0	1	1	0	1	0	0	0	1	0	1	1	1
1	0	0	0	0	1	1	1	1	1	0	0	1
1	1	0	1	1	0	1	1	0	0	0	1	1

Table 8.3 Fault collapsing results for leakage faults

Circuit	Number of Gates	Number of Faults (before)	Number of Faults (after)	Reduction (%)
c880	383	10188	3933	38.6
c1355	546	13080	5700	43.6
c1908	876	16788	6984	41.6
c2670	1193	27488	10258	37.3
c3540	1669	39672	14774	37.2
c5315	2307	60060	22880	38.1
c6288	2416	60672	26576	43.8
c7552	3512	79560	30480	38.3
Average				39.8

Note that this technique can identify two equivalent faults if and only if they belong to the same component. Leakage faults belonging to two different components can also be equivalent. No method to find such equivalent faults is known.

In Table 8.3, performance of the above leakage fault collapsing technique is presented [341]. ISCAS85 benchmark circuits [34] were used. An average reduction of 40% in the size of the fault list is reported.

8.1.2 Metal BFs and all BFs

The method described [309] uses the following idea. If nodes a and b are such that for any input vector T, $T(a) = T(b)$, then they are **identical nodes**. If a and b are identical nodes then for any X, BFs $\langle a, X\rangle$ and $\langle b, X\rangle$ are I_{DDQ} equivalent. Hence, $\langle a, X\rangle$ can replace $\langle b, X\rangle$. Care should be taken to avoid duplication. In addition, we infer that BF $\langle a, b\rangle$ is undetectable. The fault collapsing algorithm reduces to finding sets of identical nodes. Two heuristics for identifying identical nodes are discussed [309].

The **first heuristic (heuristic 1)** identifies trees containing buffers and/or inverters in the given circuit. Let the subcircuit T be a maximal tree of gates rooted at node G_0 containing nodes $\{G_1, G_2, \ldots, G_i, \ldots, G_n\}$, where each node G_i, $1 \leq i \leq n$, in T is either a buffer or an inverter. We have two cases.

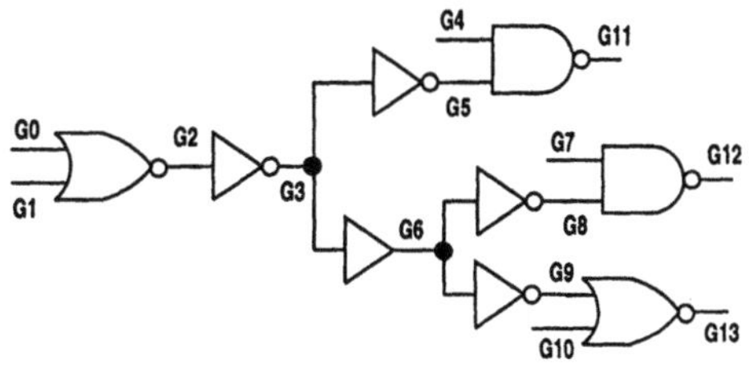

Figure 8.2 Illustration of heuristic 1 for fault collapsing

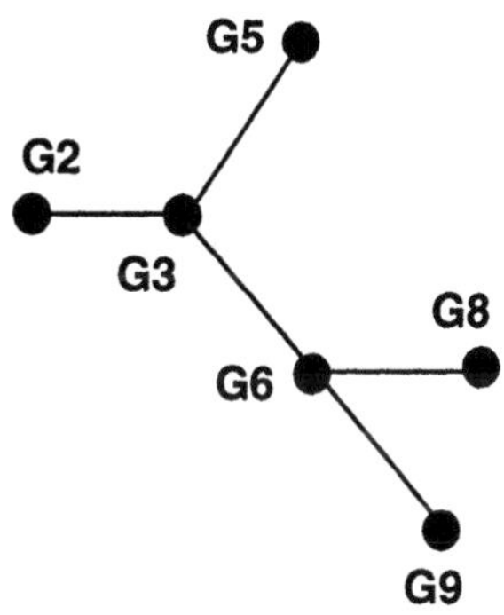

Figure 8.3 DFS tree T

- If there are no inverters in T, then $\{G_0, G_1, \ldots, G_n\}$ are identical nodes.
- There exists at least one inverter. Let S_E be the set of nodes in T such that for any $G_i \in S_E$, the path from G_0 to G_i contains an even number of inverters. Similarly, let S_O be the set of nodes in T such that for any $G_i \in S_O$, the path from G_0 to G_i contains an odd number of inverters. Then nodes in S_E, as well as those in S_O, are identical. In addition, BFs $\langle x, y\rangle$ such that $x \in S_E, y \in S_O$ are detected by every vector. Hence, they need not be targeted.

To collapse the largest number of nodes a maximal T is computed. It is computed using either breadth first search (BFS) or depth first search (DFS) starting from G_0.

The heuristic is illustrated with the example of Figure 8.2. There are 14 nodes. Therefore, before collapsing there are 91 BFs between these nodes. The DFS

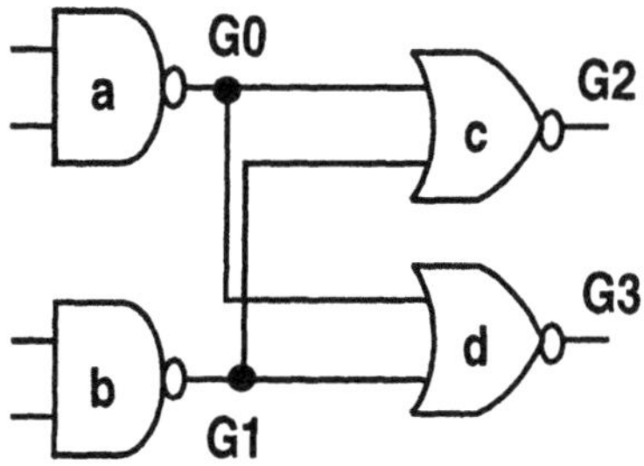

Figure 8.4 Illustrating fault collapsing using heuristic 2

tree rooted at $G2$ is shown in Figure 8.3. Nodes $G3$, $G5$, $G8$ and $G9$ are inverters, $S_E = \{G3, G6\}$, and $S_O = \{G2, G5, G8, G9\}$. Thus, the collapsed fault list contains only 45 faults. A restricted version of this heuristic was used previously [257]. If that is used the collapsed list will contain 78 faults.

To understand the second heuristic (**heuristic 2**), consider node G, which is a gate output. Let $Fanin(G)$ be the set of all nodes connected to G's inputs. Let the set of nodes $F = \{G_1, G_2, \ldots, G_i, \ldots G_k\}$ be such that $Fanin(G_1)= Fanin(G_2)\ldots = Fanin(G_i) = \ldots = Fanin(G_k)$ and each node in F implements the same function. Then $\{G_1, \ldots, G_k\}$ are identical nodes. In Figure 8.4, $Fanin(G2) = \{G0, G1\}$, $Fanin(G3) = \{G0, G1\}$ and both $G2$ and $G3$ are NOR gates. Therefore, $\{G2, G3\}$ are identical nodes.

The above heuristic can also be extended to include internal nodes [309]. In Figure 8.4 let nodes a, b, c and d be respectively internal nodes of $G0, G1, G2$ and $G3$. Since, $Fanin(G2) = Fanin(G3)$, c and d are identical nodes. The collapsed set of all BFs will contain BFs involving nodes $G0, G1, G3, a, b$ and d. The number of all BFs drops from 66 to 45.

In Table 8.4 the above fault collapsing technique for metal BFs is compared with the heuristic in [257]. Column "F1" gives the size of the collapsed fault list using the heuristic in [257], and "F2" gives the size of the collapsed fault list using the heuristic presented here. Two things should be noted. There is a drastic reduction after fault collapsing from the original number of faults. In addition, for six of the ten ISCAS85 circuits there is a drastic reduction in the number of collapsed faults when the above heuristic is used instead of [257].

Note that better heuristics to identify identical nodes or those that are complements of each other can be developed. However, care must be taken otherwise

Table 8.4 Comparison of fault collapsing results

Circuit	Faults (before)	F1	F2
c432	93,096	18,145	18,145
c499	124,251	22,155	22,155
c880	386,760	78,210	75,466
c1355	917,335	136,503	136,503
c1908	1,819,278	270,480	191,271
c2670	3,563,115	746,031	627,760
c3540	6,264,030	1,010,331	668,746
c5315	14,121,955	2,316,628	1,867,278
c6288	19,766,328	2,956,096	2,956,096
c7552	28,512,576	5,022,865	3,924,201

the fault collapsing technique may take too much time and the usefulness of fault collapsing might be lost. For example, if an ordered binary decision diagram (OBDD) based method is used to identify identical nodes [138] it will identify a large number of identical nodes. However, the method is very expensive. For example, to locate pairs of identical nodes using the OBDD based method for c7552 it takes 125 minutes on a Sun Sparc 2 workstation [138]. We will see later on that a complete I_{DDQ} test set for metal BFs can be computed in 25 minutes on the same workstation.

8.2 TC_ATPG FOR COMBINATIONAL CIRCUITS

The first class of I_{DDQ} ATPG systems that we discuss is characterized by its flexibility in that it can be used to compute I_{DDQ} tests for metal BFs, all BFs, leakage faults, and extracted faults. This is because this system uses implicit enumeration techniques for evaluating and compressing I_{DDQ} test sets discussed in Chapters 6 and 7. This class is represented by TC_ATPG [309]. RPRK_ATPG [257] has similarities with it. However, it can be used only for metal BFs. We will present experimental results which show that TC_ATPG is more efficient than RPRK_ATPG. Further analysis of TC_ATPG is presented in Section 8.5.

TC_ATPG consists of three phases: genetic algorithm (GA) phase, deterministic test generation phase and compression phase. Compression is done using

```
BEGIN
  Initialize_Population(Population_Size);
  WHILE (Stopping Criteria NOT satisfied)
   BEGIN
    Iterate Num_generation times
     BEGIN
      Reproduce(Population);
      Crossover(Population);
      Mutate(Population);
      Update_Best_Solution(Population);
     END
   Ti = Best_Vector;
   Compute F(Ti);
  END
END
```

Figure 8.5 Outline of the genetic algorithm phase

the selection algorithms discussed in Section 7.5. We next discuss the GA and deterministic phases.

8.2.1 Genetic Algorithm Phase

GA is an algorithmic paradigm which has been used for many optimization problems [107]. The primary reason for using GA is that no specific faults have to be targeted thereby making it attractive for large fault lists. ATPG is formulated as an optimization problem.

Vectors are computed one at a time. Assume that the set of vectors $\{T_1, ..., T_{i-1}\}$ have been computed. Let the list of undetected faults with respect to this set of vectors be $F(T_{i-1})$. Next an optimal input vector T_i such that it drops the maximum number of faults from $F(T_{i-1})$ is computed. Thus, at each step an optimization problem is solved. The selection process is outlined below followed by a clarification of the terms used.

To compute a test, the process starts with an initial population, whose size is Population_size, of randomly generated tests. This initial population is evolved for Num_generation generations. In our experiments we use Population_size =

25 and Num_generation = 10. Evolving the population through one generation consists of *reproduction*, *crossover* and *mutate* steps which are described next.

Fitness Function: The selection of the next vector T_i, to be included in the test set, is based on the fitness of the vector. The fitness function $FF(v)$, for vector v, is the number of BFs in $F(T_{i-1})$ detected by v.

Reproduction. A roulette wheel method is used to generate the initial population for the next generation. The wheel is divided into *Population_size* sectors and each sector in the population is assigned to one vector. The sectors are of unequal size. The sector assigned to vector v is proportional to $FF(v)$. The wheel is turned *Population_size* times. For each spin, if the wheel stops in the sector for vector z then z is reproduced in the next generation. A vector can be reproduced several times.

Crossover. It is the process of producing two new offsprings from two parents. If V_1 and V_2 are the two parent vectors, a bit position q is randomly selected for crossover. Two new vectors X_1 and X_2 are created as follows. Bit positions 1 through $q-1$ of $X_1(X_2)$ are the same as those of $V_1(V_2)$. The remaining bits of $V_1(V_2)$ are copied over into positions $q, q+1, ...$ of $X_2(X_1)$. In our experiments, pairs of input vectors are selected as follows. The first vector, is crossed with the second, the third with the fourth and so on.

Mutation: Each bit of every vector in the population is scanned and flipped with probability *MUT_Prob*. In our experiments we used *MUT_Prob* = 0.001.

Once we have created a new generation, the fitness function for all vectors are computed and the best solution computed so far is updated. Note that, evaluation of $FF(v)$ at any stage of the evolution is equivalent to the incremental coverage computation that we discussed in Chapter 7. After evolving through *Num_generation* generations the best vector generated during the process is included in the test set. This vector is now used to compute $F(T_i)$ from $F(T_{i-1})$ by dropping faults from it using the list based simulation algorithm of Chapter 6.

Stopping Criteria. We note that after computing a number of vectors, the best vector selected may have a fitness value of 0. If the fitness value of the best selected vector for five consecutive evolutions is 0 then test generation is terminated.

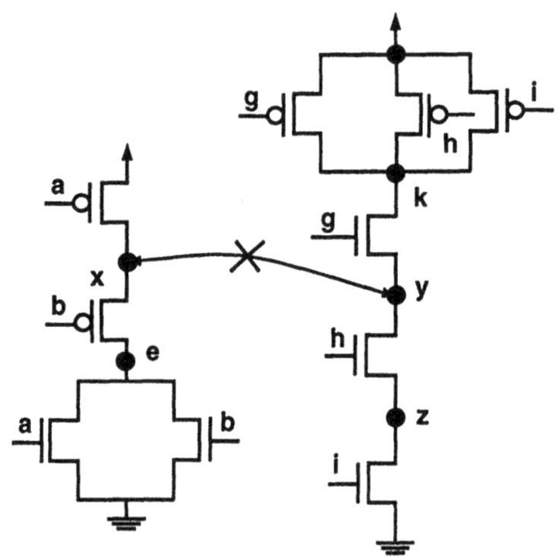

Figure 8.6 BFs involving internal nodes

8.2.2 Deterministic Phase

BFs not detected by the test set generated in the GA phase have to be detected or proved to be redundant. A stuck-at test generator is used with following modifications to the original circuit. For metal BFs an XOR gate suffices. For BFs where one of the nodes is an internal node of a gate the circuit gets to be a little more complicated.

Consider BF $\langle x, y \rangle$ in Figure 8.6. Two possible ways to detect this BF are: (i) $x = 0$ and $y = 1$ and (ii) $x = 1$ and $y = 0$. Condition (i) implies $a = 1, b = 0$, $g = 1$ and $h = 0$ or $i = 0$. The subcircuit shown in Figure 8.7(a) is such that $G1 = 1$ if (i) holds. Condition (ii) implies $a = 0$, $h = 1$ and $i = 1$. Subcircuit of Figure 8.7(b), with output $G2$, is such that $G2 = 1$ when (ii) holds. Using these two subcircuits a primary output of the circuit is created as shown in Figure 8.7(c). The stuck-at fault corresponding to BF $\langle x, y \rangle$ is G s-a-0.

In general, for each internal node i, we construct circuit C_1^i with output G_1^i such that $G_1^i(X) = 1$ if and only if X creates a conducting path from Vdd to node i. Similarly, we construct C_0^i with output G_0^i such that $G_0^i(X) = 1$ if and only if X creates a conducting path from node i to ground. For a BF $\langle x, y \rangle$, C_0^x, C_1^x, C_0^y and C_1^y are interconnected as shown in Figure 8.8. Then, G^{xy} s-a-0, where G^{xy} is a primary output, is the fault corresponding to the BF $\langle x, y \rangle$.

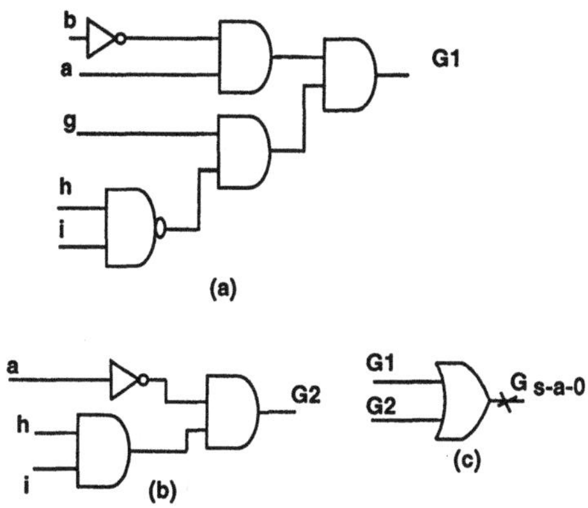

Figure 8.7 Modified circuit

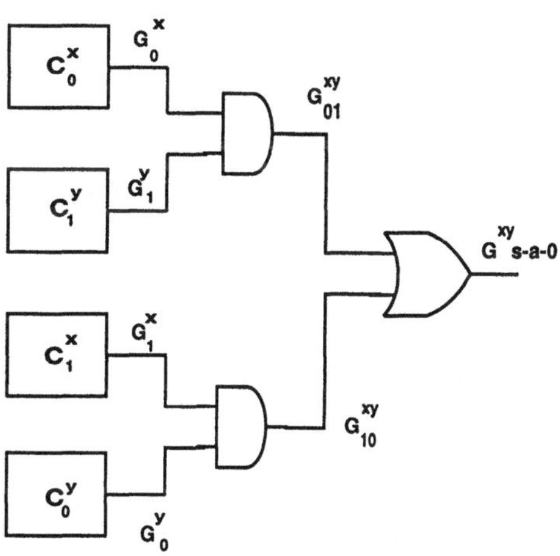

Figure 8.8 General circuit modification

8.2.3 Experimental Results

Performance results of TC_ATPG on a Sun Sparc 2 workstation for ISCAS85 [34] and combinational parts of ISCAS89 [32] benchmark circuits, are shown in columns under TC_ATPG of Table 8.5 for metal BFs and in Table 8.6 for all BFs. ATALANTA [169] was the stuck-at test generator used. The backtrack limit of the stuck-at fault test generator was set at 20. "FC" and "FE" are, respectively, the fault coverage and fault efficiency of the test sets computed. "Size" is the size of the test set. "CPU" is the total CPU time, in hours:minutes, for all three phases. The values have been rounded off upward to the nearest minute. More than half of the time was used by the third phase where test vectors are selected.

TC_ATPG differs from RPRK_ATPG [257] as follows. In the first phase, instead of using GA, RPRK_ATPG picks a "best" vector from a pool of 10,000 random vectors. This continues until 90% of the faults are detected. In the second phase, RPRK_ATPG uses a circuit transformation consisting of adder circuit blocks. Prior to the final phase of compressing the test set using reverse simulation, stuck-at test generation is used to compute a test set with 100% efficiency. RPRK_ATPG can be used only for metal-BFs.

Table 8.5 compares the performance of TC_ATPG with RPRK_ATPG. The run time of TC_ATPG is considerably shorter than that for RPRK_ATPG. RPRK_ATPG computes a smaller test set. The difference is small in most cases except for c7552. TC_ATPG can be used for larger circuits. The results presented here can be improved if the improved incremental simulation algorithm of Section 7.5 is used in the GA and the compression phases. These results are based on an earlier implementation of the incremental simulation algorithm.

8.3 CURRENT FOR EXTRACTED BFS

In this section we discuss ATPG for extracted BFs in combinational circuits [30, 91, 180]. In all these systems, faults are explicitly enumerated and stuck-at ATPG systems are modified to compute I_{DDQ} tests. They cannot therefore be used to compute tests for large fault lists like metal BFs or all BFs. We discuss CURRENT [180] which has the best performance among these systems. CURRENT computes I_{DDQ} tests in two phases:

Table 8.5 Performance of TC_ATPG for metal BFs and comparison with RPRK_ATPG

	TC_ATPG [257]					RPRK_ATPG [309]				
Circuit	Faults	Size	FC (%)	FE (%)	CPU (h:m)	Faults	Size	FC (%)	FE (%)	CPU (h:m)
c1355	136503	67	100.00	100.00	7:07	136503	68	100.00	100.00	0:01
c1908	270092	44	99.86	100.00	2:38	191271	44	99.77	99.98	0:06
c2670	745651	23	99.95	100.00	1:97	627760	29	99.90	99.99	0:08
c3540	1008864	52	99.85	100.00	5:88	668746	61	99.86	99.99	0:12
c432	18145	15	100.00	100.00	0:15	18145	19	99.96	100.00	0:01
c499	22155	32	100.00	100.00	0:15	22155	25	99.72	100.00	0:01
c5315	2316040	27	99.97	100.00	2:13	1867278	36	99.95	99.99	0:14
c6288	2955929	23	99.99	100.00	3:29	2956096	28	99.99	100.00	0:13
c7552	5021705	34	99.98	100.00	5:24	3924201	51	99.95	99.99	0:25
s1196	NA					126253	40	99.95	100.00	0:04
s1238	NA					142311	48	99.99	99.99	0:04
s1423	NA					179700	31	99.69	100.00	0:05
s13207	NA					8239770	243	99.94	100.00	8:21
s1494	NA					186355	39	99.97	100.00	0:04
s15850	NA					12819516	194	99.93	99.99	4:36
s298	5990	13	99.92	100.00	0:03	5565	13	99.62	100.00	0:01
s344	13691	13	99.97	100.00	0:04	13695	14	99.71	100.00	0:01
s35932	NA					114466015	32	99.73	100.00	1:18
s382	NA					11935	15	99.81	100.00	0:01
s38417	NA					94084903	257	99.98	99.99	18:48
s38584	NA					111116778	315	99.54	99.99	27:54
s386	NA					11781	21	99.87	100.00	0:01
s526	17939	17	99.91	100.00	0:10	16653	17	99.80	100.00	0:01
s5378	NA					1615503	96	99.93	100.00	0:28
s832	NA					41328	31	99.90	100.00	0:02
s9234	NA					3823995	162	99.85	99.99	2:25
s953	NA					95703	54	99.99	100.00	0:05

Table 8.6 Generation of tests for ALL-BFs

Circuit	Total Faults	Size	FC (%)	FE (%)	CPU (h:m)
c1355	603351	130	99.98	99.90	0:22
c1908	832695	190	99.72	99.93	1:12
c2670	2907666	64	99.82	99.98	0:49
c3540	4235505	146	99.86	99.99	2:10
c432	106030	39	98.88	99.96	0:04
c499	494515	170	99.72	99.97	0:50
c5315	11681361	93	99.92	99.99	2:08
c6288	12860056	56	99.97	100.00	0:43
c7552	19615716	226	99.89	99.99	33:42
s1196	721801	133	99.82	99.99	0:57
s1238	861328	171	99.75	99.99	2:00
s1423	936396	68	99.76	99.99	0:25
s13207	35781570	548	99.95	99.99	91:31
s1494	1721440	125	99.70	100.00	1:12
s15850	62445900	1004	99.95	99.99	65:49
s38417	380066235	1038	99.99	99.99	104:56
s386	99681	62	99.43	100.00	0:06
s5378	5987530	372	99.95	100.00	4:18
s9234	20656378	548	99.84	99.99	14:20
s953	378885	153	99.83	100.00	0:27

- In the first phase a large number of random patterns (approximately 10000) is simulated and a list of undetected faults is obtained. The faults not detected in this phase are the target faults for the deterministic phase.

- In the deterministic phase, tests for undetected faults are computed by targeting one fault at a time. For the target leakage fault an equivalent stuck-at fault is constructed. A general construction technique is discussed in Section 8.2.2.

  To compute a test for the stuck-at fault a combination of a branch-and-bound algorithm [106, 181] and a learning technique based on recursive learning [161, 284] is used. The branch-and-bound algorithm is useful in generating a test if a test exists for the target fault. The recursive learning technique is useful in determining undetectable faults.

  Two parameters are used to bound the computational resources used by the two algorithms. For the branch-and-bound algorithm this parameter is the **backtrack limit**: This limit is used to bound the amount of time spent by the algorithm to generate a test for the target fault. For recursive learning **maximum recursion depth** is the parameter used. The maximum value of this parameter, i.e., for exact learning, is the number of levels in the circuit.

  To obtain a balance between the computational effort spent in the branch-and-bound and recursive learning phases, multiple passes with different values of backtrack limit and maximum recursion depth are used. A maximum backtrack limit of 30 and a maximum recursion depth of 3 are used.

- Finally, the test set is compressed in two steps. First, the tests are simulated in the reverse order of their generation (**reverse-simulation**). A test T_i is deleted if and only if each fault detected by T_i is also detected by one of the vectors that has been simulated. In the next step the selection algorithm for extracted faults discussed in Section 7.3 is used.

Performance of CURRENT, for the ISCAS85 [34] and the combinational parts of ISCAS89 [32] circuits, is given in Table 8.7. Column "RP" gives the number of faults detected during the random phase, "DET" and "NOT DET" respectively gives the number of faults detected and proven irredundant during the deterministic phase, and "FC" gives the fault coverage. Fault efficiency for all circuits was 100%.

The CPU time, given in min:sec, is the total time for all three phases on an HP 9000/735 workstation. A disproportionate amount of time is spent in compressing the test set using the selection algorithm. To give an idea, for

Table 8.7 Performance results for CURRENT

Circuit	Total Faults	RP	DP DET	DP NOT DET	% FC	Size	CPU m:s
c432	8746	8633	113	0	98.71	52	0:04
c499	12958	12913	24	21	99.84	95	0:10
c880	16502	16459	17	26	99.84	65	0:07
c1355	24254	24178	45	31	99.87	96	0:08
c1908	33954	33772	108	74	99.78	171	0:28
c2670	48421	48202	23	196	99.60	73	0:37
c3540	66214	66026	2	186	99.72	85	0:50
c5315	101478	101369	0	109	99.89	87	1:12
c6288	110176	110046	0	130	99.88	50	0:55
c7552	138068	137397	498	173	99.87	131	2:05
s9234	163749	161108	2359	282	99.83	280	3:50
s13207	232269	229402	2671	196	99.92	383	7:29
s15850	279102	278141	677	284	99.90	317	6:49
s35932	639758	637970	0	1788	99.72	33	9:69
s38417	667346	666566	526	254	99.96	480	20:41
s38584	716616	712157	1011	3448	99.52	677	40:55

s38584 out of a total time of 40 minutes and 55 seconds, 26 minutes and 35 seconds were spent in this step. For more detailed data see [180].

8.4 ATPG FOR LEAKAGE AND METAL BFS USING CUTEGEN

CUTEGEN [139] is different from the other two classes of test generators in that it targets a combination of leakage faults and metal BFs. It uses only the gate-level description and targets leakage faults "indirectly".

CUTEGEN computes I_{DDQ} tests in two phases. It first targets leakage faults and computes a test set S_1. Next, I_{DDQ} tests for metal BFs not detected by S_1 are computed. The final I_{DDQ} test set is S_2. No compression of S_2 is done. For both phases it uses a stuck-at test generator. For the first phase, leakage faults are processed indirectly by computing pseudo stuck-at tests discussed in Chapter 5. All gate outputs are made observable. For the second phase, metal BFs are targeted using a stuck-at test generator. The experimental results are summarized in Table 8.8.

Column VEC gives the test set size, LEAK and MET give the fault coverage of leakage and metal BFs. There are three columns under CPU. All times are in minutes:seconds on a Sun Sparc 2 workstation. CUTEGEN has a preprocessing step that identifies sets of identical lines using an OBDD method [138]. The time taken for this is given in column EQUIV [138]. TG is the test generation time once the identical lines have been identified [139]. TOTAL is the total time for identifying redundant faults and computing tests. Stuck-at test generator HITEST [26] was used. Entries marked NA signify the unavailability of published data.

8.5 ANALYSIS OF ATPG FOR COMBINATIONAL CIRCUITS

Here and in the previous chapter we have seen a wide variety of techniques to compute I_{DDQ} tests for combinational circuits. For direct methods discussed in this chapter we ask ourselves the question: for a given fault model, based on available data, which is the best approach? We attempt an answer in Sec-

Table 8.8 Performance results for CUTEGENS

		Coverage		CPU		
Circuit	VEC	LEAK	MET	EQUIV	TG	TOTAL
		%	%	(m:s)	(m:s)	(m:s)
c880	23	100.00	99.93	0:45	0:38	1:23
c1355	92	100.00	99.93	31:25	2:55	34:20
c1908	116	99.96	99.78	49:07	5:09	54:16
c2670	32	99.95	99.91	12:29	2:37	15:06
c3540	72	99.58	99.85	22:02	9:29	31:31
c5315	38	99.99	99.95	45:35	6:27	52:02
c6288	42	99.74	99.99	NA	25:54	NA
c7552	54	99.91	99.96	125:03	9:07	134:10
s208	32	100.00	99.87	NA	0:29	NA
s298	23	100.00	99.64	NA	0:27	NA
s344	18	100.00	99.74	NA	0:25	NA
s349	17	99.57	99.74	NA	0:25	NA
s382	25	100.00	99.81	NA	0:37	NA
s1196	54	100.00	99.95	NA	1:36	NA
s1492	56	100.00	99.97	NA	1:46	NA
s5378	79	99.75	99.94	NA	9:18	NA
s9234	133	99.75	99.85	NA	29:07	NA

tion 8.5.1. For a given fault model, should one use the direct or the indirect method? We try to answer this in Section 8.5.2. While trying to answer these questions the test generation time, fault coverage and size of the computed tests have to be taken into consideration. We also address the issue of how much more difficult it is to compute tests for metal BFs as compared with extracted BFs. Like fault simulation and I_{DDQ} subset selection we will see that test generation for metal BFs is marginally more expensive than extractred BFs.

8.5.1 Direct Methods

For extracted BFs, CURRENT has the best reported result. For leakage faults, CURRENT will be very good and CUTEGEN, which uses pseudo stuck-at tests, may also be a useful approach. Since there is incomplete data on how much time CUTEGEN spent in identifying redundant leakage faults (which contributes very significantly to the total time), it is not clear whether CURRENT or CUTEGEN approachs will lead to a faster test generator. For all BFs, TC_ATPG is the only system reported in the literature.

For metal BFs, we need to compare CUTEGEN and TC_ATPG. Results are presented in Table 8.9. The fault coverages of the two systems are approximately the same. The size of the test sets computed by TC_ATPG is considerably smaller for all but two circuits - s5378 and s9234. Data on CUTEGEN for the larger circuits is not available. CPU times reported are run times on a SUN Sparc 2 workstation. Clearly, TC_ATPG is considerably faster. Thus, the approach used by TC_ATPG computes a smaller test set, most of the time, and is considerably faster. Since, CUTEGEN uses an OBDD method to identify redundant faults, we believe that CUTEGEN cannot be used for large circuits. Experimental results have already demonstrated that TC_ATPG can be used for the large benchmark circuits.

We have already seen that fault simulation, selection are only marginally more expensive for metal BFs when compared with extracted BFs. The same conclusion can be drawn for test generation from Table 8.10. This table is based on the performances of TC_ATPG and CURRENT.

Another important issue is the difference in the test set size when we target metal BFs, compared with the test set size when we target extracted BFs. For most circuits, the test set size for metal BFs is considerably smaller than the test set size for extracted BFs.

Table 8.9 Relative performance of TC_ATPG and CUTEGENS for metal BFs

	VECTORS		CPU	
Circuit	TC_ATPG	CUTEGEN	CUTEGEN	TC_ATPG
			(m:s)	(m)
c1355	92	68	34:20	1
c1908	116	44	54:16	6
c2670	32	29	15:06	8
c3540	72	61	31:31	12
c5315	38	36	52:02	14
c6288	42	28	NA	13
c7552	54	31	134:10	25
s298	23	13	NA	1
s344	18	14	NA	1
s382	25	15	NA	1
s1196	54	40	NA	4
s1494	56	39	NA	4
s5378	79	96	NA	28
s9234	133	162	NA	145

8.5.2 Direct and Indirect Methods

An important decision, given that we can apply test sets specifically for I_{DDQ} testing, is on using the direct or the indirect method to compute the tests. We believe that the direct method should be used primarily because of fault coverage reasons, and to some extent because of the smaller size of the test sets obtained. There is not much of an advantage in the time required. This conclusion is based on the performance comparison of TC_ATPG with the selection algorithm using the COVER heuristic of Chapter 7. The selection was done from a stuck-at test set computed using ATALANTA [169]. The results for metal BFs, leakage faults and extracted BFs is given respectively in Tables 8.11, 8.12 and 8.13.

In all cases tests were generated for the same fault model. Faults is the total number of faults targeted, FC is the fault coverage, FE is the fault efficiency, Size is the size of the I_{DDQ} test set, Left is the number of faults left undetected by the indirect method and Time is the CPU time.

Table 8.10 Comparison of test generation times for metal and extracted BFs (all time in CPU minutes)

	Metal				Extracted				Ratio	
Circuit	Faults	Tests	% FC	Time	Faults	Tests	% FC	Time	Faults	Adj Time
c5315	3086370	36	99.95	14	101478	49	99.89	1.0	30.4	3.1
c6288	2995128	28	99.99	13	110176	25	99.88	1.0	27.2	2.9
c7552	6913621	51	99.95	25	138068	70	99.87	2.0	50.0	2.8
s5378	4649725	96	99.93	28	91546	113	99.81	1.5	50.8	4.18
s9234	17073246	162	99.85	145	163749	127	99.83	3.5	104.2	9.2
s13207	37658181	243	99.94	501	232269	206	99.92	6.0	162.1	3.2
s15850	53929305	194	99.93	276	279102	172	99.90	6.5	193.2	9.4
s38417	286095160	257	99.98	1128	667346	250	99.96	19.5	428.7	13
s38584	217747146	315	99.54	1674	716616	344	99.52	39	303.8	9.6

Table 8.11 Tests for metal-BFs (comparison with indirect method)

	Indirect Method					Direct Method				
Circuit	Faults	Size	FC	Left	Time	Faults	Size	FC	FE	Time
			%		h:m			%	%	h:m
c1355	171991	29	99.01	1715	0:01	136503	68	100.0	100.00	0:01
c1908	416328	146	99.08	3813	0:02	191271	44	99.77	99.98	0:06
c2670	1016025	153	99.56	4380	0:27	627760	29	99.9	99.99	0:08
c3540	1476621	72	99.82	2603	0:12	668746	61	99.86	99.99	0:12
c432	19110	17	99.99	11	0:01	18145	19	99.96	100.00	0:01
c499	29403	16	98.92	317	0:01	22155	25	99.72	100.00	0:01
c5315	3086370	124	99.94	1737	0:55	1867278	36	99.95	99.99	0:14
c6288	2995128	57	98.68	39356	0:05	2956096	28	99.99	100.00	0:13
c7552	6913621	88	99.87	8975	0:46	3924201	51	99.95	99.99	0:25
s1423	279378	31	99.62	1071	0:01	179700	31	99.69	100.00	0:05
s13207	37658181	82	99.48	192679	2:49	8239770	243	99.94	100.00	8:21
s1494	218130	38	99.94	111	0:02	186355	39	99.97	100.00	0:04
s15850	53929305	94	99.82	94387	2:42	12819516	194	99.93	99.99	4:36
s5378	4649725	78	99.72	8323	0:32	1615503	96	99.93	100.00	0:28
s9234	17073246	68	99.65	59516	1:20	3823995	162	99.85	99.99	2:25

Table 8.12 Tests for leakage faults (comparison with indirect method)

		Indirect Method				Direct Method			
Circuit	Faults	Size	FC	Left	Time	Size	FC	FE	Time
			%		h:m		%	%	h:m
c1355	4796	35	93.01	335	0:03	84	100.00	100.00	0:03
c1908	5204	39	90.39	500	0:06	98	100.00	100.00	0:06
c2670	8973	46	94.72	473	0:23	36	99.44	99.93	0:12
c3540	11643	48	98.89	129	0:18	43	99.61	100.00	0:15
c432	1721	24	98.19	31	0:01	30	98.95	100.00	0:02
c499	3100	33	93.64	197	0:02	83	100.00	100.00	0:02
c5315	18933	86	99.41	111	1:01	46	99.98	100.00	0:26
c6288	22336	30	97.15	636	0:09	18	99.76	100.00	0:15
c7552	24972	70	97.08	728	1:04	65	99.87	100.00	0:51
s1423	5331	29	98.89	59	0:03	30	99.94	100.00	0:06
s13207	31505	82	92.04	2507	4:37	220	99.85	100.00	5:06
s1494	7765	48	99.48	40	0:06	44	100.00	100.00	0:04
s15850	42934	86	96.11	1668	3:25	164	99.83	99.99	5:13
s5378	12982	79	97.38	339	0:46	103	99.74	100.00	0:43
s9234	24462	64	94.63	1313	2:10	139	99.55	100.00	2:32

Table 8.13 Tests for extracted fault list (comparison with indirect method)

		Indirect method				Direct Method			
Circuit	Faults	Size	FC	Left	Time	Size	FC	Aborted	Time
			%		h:m	%		%	h:m
c1355	4356	15	98.76	54	0:01	10	98.76	0	0:03
c1908	4669	17	97.68	108	0:02	13	97.94	0	0:04
c2670	13589	134	93.53	878	0:55	23	97.71	0	0:10
c3540	16315	39	98.28	280	0:15	36	98.41	0	0:16
c432	1546	11	99.28	11	0:01	11	99.28	3	0:01
c499	2747	12	97.59	66	0:01	11	97.81	0	0:02
c5315	40143	118	96.92	1236	2:06	31	97.74	22	0:18
c6288	21475	37	94.50	1181	0:09	19	99.41	0	0:05
c7552	53439	76	95.99	2141	1:27	44	98.05	22	0:46
s1423	7500	25	98.53	110	0:02	25	99.10	0	0:06
s13207	65000	98	97.14	1858	4:12	162	99.95	0	7:15
s1494	10000	48	98.77	123	0:05	49	99.00	0	0:09
s15850	65000	85	98.98	663	2:48	132	99.82	0	3:43
s5378	20000	42	98.89	222	0:21	52	99.71	0	0:18
s9234	50000	64	98.29	854	1:43	106	99.86	0	2:16

Note that for metal BFs the size of the test set is usually much smaller. However, sometimes the test set computed by the direct method is larger. For example, for s13207 the test set computed by the direct method is more than double in size than that computed by the indirect method. This happens for circuits in which the indirect test set does not have good coverage. For s13207 the extra tests cover about 192,000 more metal BFs.

For leakage faults, the indirect method has very poor coverage and so it should not be used. For extracted BFs, the indirect method has better coverage. However, the coverage for the direct method is consistently better, with 100% fault efficiency for all circuits. In addition, for all circuits the test set is considerably smaller. The extracted set of BFs used is a random set of faults. Based on this, it appears that direct methods are better than indirect methods for computing I_{DDQ} tests for combinational circuits.

8.6 ATPG FOR SEQUENTIAL CIRCUITS

Unlike combinational circuits, direct methods for computing I_{DDQ} tests for faults in sequential circuits is still in its infancy. The only reported work is [170]. It uses genetic algorithm to compute test sequences. An important difference is that the population consists of sequences of vectors rather than individual vectors. Mutation, crossover and and reproduction operations are therefore defined over sets of sequences of vectors rather than over sets of vectors. The system does not have a deterministic part and therefore cannot identify irredundant faults.

Performance of the ATPG system on ISCAS89 [32] benchmark circuits is given in Table 8.14. For these experiments the following GA parameters were used: *Population_size*=16; *Num_generation*=8. Unlike the case of combinational circuits, each member of the population represents a sequence of ten tests. If the best sequence evolved after *Num_generation* generations does not detect additional faults, the process is stopped (**termination condition**). Note that the time taken is very small. However, we have to bear in mind that the system avoids the expensive step of identifying redundant faults.

Table 8.14 Sequential circuit test generation for ALL-BFs

Circuit	Total Faults	NumTests	FC (%)	CPU s
s1196	695610	276	99.81	1500
s1238	836571	358	99.722	2400
s1423	940506	253	98.654	1670
s1488	1615503	167	99.626	1960
s1494	1662576	169	99.612	1980
s5378	5380840	831	92.282	35200
s9234	19099290	229	5.566	3130
s13207	33215325	351	6.747	13900
s15850	55540530	739	23.618	102000
s35932	509044278	172	99.522	75800

8.7 SUMMARY

In summary, we note that much work has been done to directly compute I_{DDQ} tests for a variety of fault models that target bridges. On the other hand, no published work has discussed direct methods to compute I_{DDQ} tests for opens. We discussed fault collapsing techniques for extracted BFs, leakage faults, metal BFs and all BFs. Next we discussed TC_ATPG, that computes I_{DDQ} tests for metal BFs and all BFs, CURRENT, that computes I_{DDQ} tests for extracted BFs, and CUTEGEN, that computes I_{DDQ} tests for leakage faults and metal BFs. For extracted BFs CURRENT has the best reported results. We believe that for leakage faults CURRENT will outperform others. We argued that TC_ATPG has the best performance for metal BFs. TC_ATPG is the only system that targets all BFs. Note that all these systems assume the circuit to be combinational. There is very little published work on I_{DDQ} test generation for sequential circuits and, we believe that, more research is needed.

Finally, we addressed the issue of whether to use a direct or an indirect method to compute I_{DDQ} tests. Results show that, for a variety of fault models, the I_{DDQ} test set computed using a direct method is better because the test set size is smaller and the fault coverage is higher. In addition, the test generation time for direct methods is comparable to the "additional" test generation time required by indirect methods.

9

FAULT DIAGNOSIS

During testing faulty ICs are identified. These, along with ICs that fail in the field, are collected. To investigate causes of these failures, faulty ICs are examined off-line. This is referred to as **fault diagnosis** or **failure analysis.** Fault diagnosis provides information that is used to improve manufacturing yield.

The fault diagnosis process includes the computation of a **diagnostic test sequence** (DTS). We refer to this as **diagnostic test generation.** This is followed by **fault location.** In fault location, the diagnostic test sequence is applied to the faulty IC and the response is analyzed to determine the location of the defect. In the context of I_{DDQ} testing, the response used is a record of tests with high I_{DDQ}.

To locate the fault, the diagnostic test sequence should distinguish between faults. **Diagnostic measures** assess this quality of a diagnostic test sequence. **Diagnostic fault simulation** is the process of computing diagnostic measures for a given diagnostic test sequence.

In the sections that follow, we first discuss fault location techniques. In Section 9.2, we present a technique for diagnostic test generation. Next, we present a method for diagnostic fault simulation. We conclude this chapter with a novel approach to diagnosis, called dynamic diagnosis.

9.1 FAULT LOCATION

In fault location, a combinational circuit C, found to be faulty by an I_{DDQ} test set T is given. The objective is to identify the fault in C. It is assumed that the following are known: (i) the set of the input vectors T; (ii) for each input vector T_i in T whether or not C gave a faulty response, i.e., whether an abnormally high or a normal I_{DDQ} was measured. Furthermore, we assume that C has **only one** fault. Fault location for leakage and bridging faults is discussed in the next two sections.

9.1.1 Leakage Faults

One method for analyzing I_{DDQ} response uses I_{DDQ} **fault dictionaries**. A given circuit consists of a number of instances of **component types**. For example, a circuit made up of three 2-NAND gates and four 3-NOR gates has two component types, 2-NAND and 3-NAND, and seven gates in the circuit.

To construct I_{DDQ} fault dictionary, the I_{DDQ} **fault table** for each component type in the circuit is computed. The fault table has one row for each possible input pattern and one column for each leakage fault. A 1(0) in row i and column j of the fault table indicates that the input pattern corresponding to row i detects (does not detect) the leakage fault corresponding to column j. Table 9.1 is the I_{DDQ} fault table for the 2-NAND gate of Figure 9.1. In Table 9.1, columns $I1$ and $I2$ represent the inputs. Columns labeled bg represent the leakage fault between the base (b) and gate (g). The rest of the columns are similarly defined. In Table 9.1, columns bs and sd for $P1$ are such that every input pattern either detects both faults or does not detect any of them. Such faults that cannot be distinguished by I_{DDQ} tests are known as I_{DDQ} **indistinguishable faults**. Indistinguishable faults of a component are grouped together into I_{DDQ} **fault classes**. The I_{DDQ} fault classes of Table 9.1 are identified in the last row. Faults in the same fault class have the same fault class number.

The fault dictionary used consists of several parts, one for each gate in the circuit. We refer to the part of the dictionary for a given gate as the fault dictionary for that gate. The **fault dictionary for a gate**, for a given DTS, is computed as follows. Let $T = \{\langle 0,0,1,0\rangle, \langle 1,1,0,0\rangle, \langle 1,1,0,1\rangle, \langle 1,1,1,1\rangle\}$ be a DTS for the circuit C of Figure 9.2. Using logic simulation we can determine the input patterns that will appear at each gate on application of T. This is shown in Figure 9.2. For gate $G1$, they are $T0 = \langle 0,1\rangle, T1 = \langle 1,0\rangle$ and $T2 = \langle 1,1\rangle$.

Table 9.1 I_{DDQ} fault table of a two input NAND gate

		$P1$						$P2$					
$I1$	$I2$	bg	bs	bd	sd	gd	gs	bg	bs	bd	sd	gd	gs
		1	2	3	4	5	6	7	8	9	10	11	12
0	0	1	0	0	0	1	1	1	0	0	0	1	1
0	1	1	0	0	0	1	1	0	0	0	0	0	0
1	0	0	0	0	0	0	0	1	0	0	0	1	1
1	1	0	1	0	1	1	0	0	1	0	1	1	0
Fault class		2	6	-	6	8	2	4	6	-	6	10	4
		$N1$						$N2$					
$I1$	$I2$	bg	bs	bd	sd	gd	gs	bg	bs	bd	sd	gd	gs
		13	14	15	16	17	18	19	20	21	22	23	24
0	0	0	1	0	0	1	0	0	0	0	0	0	0
0	1	0	1	0	1	1	0	1	0	0	0	1	1
1	0	1	1	1	0	0	0	0	1	0	1	1	0
1	1	1	0	0	0	1	1	1	0	0	0	1	1
Fault class		9	5	3	1	8	6	7	3	-	3	11	7

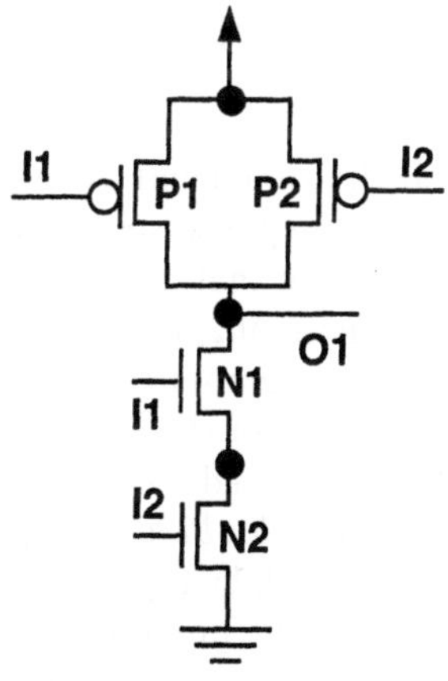

Figure 9.1 NAND gate example used to illustrate I_{DDQ} fault table

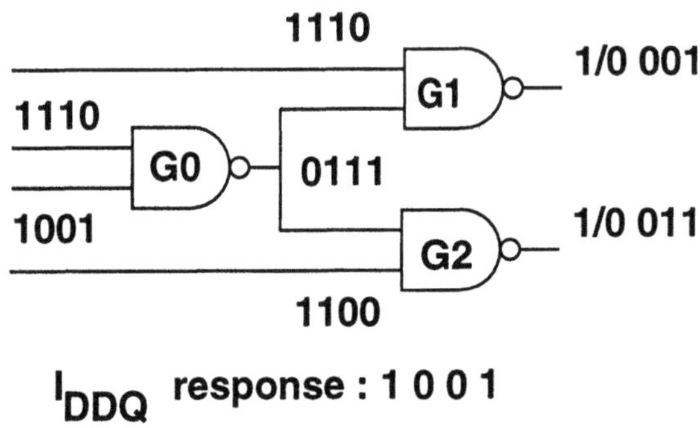

Figure 9.2 Example to illustrate the diagnosis algorithm

The fault dictionary of a gate consists of a table of ⟨response vector, list of fault classes⟩. The fault dictionary for $G1$ is given in Table 9.2. For each fault class of a gate we compute a response vector as follows. We use fault class 6 of gate $G1$ of Figure 9.2 as an example. Since, on application of the DTS, only three input patterns, $T0$, $T1$ and $T2$ can appear the **response vector** is of the form $\langle a0, a1, a2 \rangle$. If on appearance of Tj, $j = 0, 1, 2$, at the input of $G1$ faults in fault class 6 are activated then aj is 1 else it is 0. The reader can verify that the response vector for fault class 6 of $G1$ is $\langle 0, 0, 1 \rangle$. All fault classes with the same response vector are grouped together in the fault dictionary for that gate. In Table 9.2 the eight possible response vectors for gate $G1$ are listed in column 1. For each response we have also listed the fault classes of $G1$ that will lead to that response. Having computed I_{DDQ} fault dictionaries for all gates, faults can be located as follows.

In fault location we are given the **observed response vector** of the faulty circuit to the DTS. The observed response vector of the circuit is analyzed to locate the fault. For example assume that the observed response vector of the faulty instance of the circuit of Figure 9.2, to the example DTS T shown in the figure, is $R_c = \langle 1001 \rangle$. From that we compute the **observed response vector for each gate**. Since, on application of the DTS T, only patterns $T0, T1$ and $T2$ appear at the input of $G1$ the observed response vector R_{G1} for gate $G1$ is of the form $\langle b0, b1, b2 \rangle$. The values of $b0, b1, b2$ are computed as follows. When the input vector is $\langle 0, 0, 1, 0 \rangle$ then $T0 = \langle 0, 1 \rangle$ appears at the input of $G1$. Since the observed response vector of the circuit implies a high I_{DDQ} on application of $\langle 0, 0, 1, 0 \rangle$, $b0 = 1$. Similarly, on application of $\langle 1, 1, 0, 0 \rangle$ and $\langle 1, 1, 0, 1 \rangle$ pattern $T3 = \langle 1, 1 \rangle$ appears at the input of $G1$. Since the observed response vector of

Table 9.2 Example fault dictionary for gate G1 of Figure 9.2

Response vector	Fault classes
$\langle 0,0,0 \rangle$	Φ
$\langle 0,0,1 \rangle$	$\{6\}$
$\langle 0,1,0 \rangle$	$\{3,4\}$
$\langle 0,1,1 \rangle$	$\{9,10\}$
$\langle 1,0,0 \rangle$	$\{1,2\}$
$\langle 1,0,1 \rangle$	$\{7,8\}$
$\langle 1,1,0 \rangle$	$\{5\}$
$\langle 1,1,1 \rangle$	$\{11\}$

the circuit implies that a normal I_{DDQ} was observed on application of these two vectors, $b3 = 0$. The reader can verify that $R_{G1} = \langle 1,1,0 \rangle$.

The observed response vector R_G for gate G is used to identify the fault class of gate G that would result in the observed faulty response of the circuit. This is done by using R_G to index into the fault dictionary for G. From $R_{G1} = \langle 1,1,0 \rangle$ and the fault dictionary for gate $G1$ shown in Figure 9.2, we deduce the possible I_{DDQ} fault class contributed by $G1$ to be $\{5\}$. The union of the fault classes contributed by all gates is the final diagnosis.

We note that some gates may not contribute any fault. Consider gate $G2$ of Figure 9.2. On application of both the inputs $\langle 1100 \rangle$ and $\langle 0010 \rangle$, the pattern applied to $G2$ is $\langle 1,0 \rangle$. However, from the observed response vector of the circuit we note that on application of $\langle 1,1,0,1 \rangle$ I_{DDQ} is high while on application of $\langle 1,1,0,0, \rangle$ I_{DDQ} is normal. This inconsistency implies that no leakage fault in $G2$ can result in the observed response and the observed response vector for $G2$ in this case is defined to be inconsistent.

DIVA [11, 13], an experimental system for diagnosing leakage faults, uses an approach different from the one discussed above. It uses a combination of I_{DDQ} and logic measurement to locate leakage faults in combinational circuits. In DIVA, during testing, when high I_{DDQ} is detected the logic levels at the output and the sequence number of the corresponding input test pattern are stored. This information is then used for fault location as follows.

The method requires knowledge of the logic levels that can result at the output of a gate in the presence of leakage faults. This is precomputed for each component type and stored in a Logic/I_{DDQ} Fault Table (LIFT). Table 9.3 is the LIFT for the two input NAND gate of Figure 9.1. Each entry in the table can have one of the following values: H for high I_{DDQ}; X for unknown value at component output; and - for no change.

DIVA uses an iterative procedure. It starts with an initial fault list and continues until the list of potential faults is small. In each iteration, a block of 32 failing test inputs is processed as follows. A block of test patterns is simulated for each fault using a parallel pattern fault simulator and the logic fault table of the gates involved in the list of potential faults. This gives the logic level at the output in presence of faults. If the required memory is available, a dictionary containing predicted output for each fault could be precomputed. Next, the computed output pattern for each fault is compared to the observed output pattern and classified into fault groups as follows.

For output i and vector j, let O_i^j be the observed output value, $F_i^j(f)$ be the simulated faulty value in presence of fault f and S_i^j be the fault-free value. While comparing output patterns, the relation $=$ indicates *compatibility* as shown: $1 = 1$; $0 = 0$; $X = 1$; $X = 0$; and $X = X$. The relation $\neq$ is used to indicate incompatibility. Let G be the gate with fault f. Then, $Cone(f)$ is the set of primary outputs reachable from G.

- **Group 1:** Faults f such that for all test inputs j and all outputs i, $F_i^j(f) = O_i^j$.
- **Group 2:** Faults f such that for all test inputs j and all outputs $i \in Cone(f)$, $F_i^j(f) = O_i^j$.
- **Group 3:** Faults f such that for all test inputs j and all outputs $i \in Cone(f)$, $F_i^j(f) = O_i^j$ whenever $F_i^j(f) = S_i^j$.
- **Group 4:** Faults f such that for some test input j and some output $i \in Cone(f)$, $F_i^j(f) = O_i^j$ whenever $F_i^j(f) = S_i^j$.
- **Group 5:** Faults f such that for all test inputs j and all outputs i, $F_i^j(f) \neq O_i^j$.

If a fault can be placed in more than one group, it is placed in the group with the lowest index. Faults in group 1 are most likely to have caused circuit failure. Faults in remaining groups may cause failures with decreasing likelihood. After each iteration, if additional resolution is required, the most likely faults are placed in a potential list of faults. Using a new block of 32 test inputs and observed outputs, the process is then repeated.

Table 9.3 LIFT for a two input NAND gate

		P1						P2					
I1	*I2*	*bg*	*bs*	*bd*	*sd*	*gd*	*gs*	*bg*	*bs*	*bd*	*sd*	*gd*	*gs*
		1	2	3	4	5	6	7	8	9	10	11	12
0	0	HX	–	–	–	HX	HX	HX	–	–	–	HX	HX
0	1	HX	–	–	–	HX	HX	–	–	–	–	–	–
1	0	–	–	–	–	–	–	HX	–	–	–	HX	HX
1	1	–	HX	–	HX	HX	–	–	HX	–	HX	HX	–
Fault class		2	6	-	6	8	2	4	6	-	6	10	4
		N1						N2					
I1	*I2*	*bg*	*bs*	*bd*	*sd*	*gd*	*gs*	*bg*	*bs*	*bd*	*sd*	*gd*	*gs*
		13	14	15	16	17	18	19	20	21	22	23	24
0	0	–	HX	–	–	HX	–	–	–	–	–	–	–
0	1	–	HX	–	HX	HX	–	H-	–	–	–	H-	H-
1	0	HX	HX	HX	–	–	–	–	HX	–	HX	HX	–
1	1	HX	–	–	–	HX	HX	HX	–	–	–	HX	HX
Fault class		9	5	3	1	8	6	7	3	-	3	11	7

We illustrate this approach with the circuit in Figure 9.2. Four test vectors are applied as shown. Vector 1 corresponds to the input $\langle 0,0,1,0 \rangle$ and it resulted in the output pattern $\langle 1,1 \rangle$. High I_{DDQ} was observed for vectors 1 and 4, and faulty output values for vector 4. Using the observed output patterns and fault table given in Table 9.3 we can place leakage faults into the following groups:

- Group 1 $G0$: 8
- Group 2 $G1$: 9
- Group 3 $G0$: 1,6,11; $G1$: 1,5,10,11; $G2$: 1,11
- Group 4 $G0$: 2,3,4,7,9; $G1$: 2,3,4,7,8; $G2$: 2,3,4,5,7,8,9,10
- Group 5 $G0$: 5,10; $G1$: 6; $G2$: 6

To illustrate how the above groups were obtained, we consider the entry corresponding to leakage fault class 8 of gate $G0$ that is placed in group 1. First, from logic simulation of vectors 1-4, the expected I_{DDQ} and gate output values for faults in fault class 8 are obtained from Table 9.3. These are HX, B1, B1 and HX. For the vectors that resulted in high I_{DDQ}, the X values at the output of $G0$ is propagated to primary outputs. Hence, the output values on simulating the faults in fault class 8 of $G0$ for vectors 1 through 4 are: $G1$ output 1,0,0,X; and $G2$ output 1,1,0,X. The fault-free values are: $G1$ output 1,0,0,0; and $G2$ output 1,1,0,0. The observed values are: $G1$ output 1,0,0,0; and $G2$ output 1,1,0,0. Since for the class 8 fault in gate $G0$, the predicted and observed output patterns are compatible for all outputs and test inputs, it is placed in group 1.

DIVA [11, 13] was evaluated on a standard cell ASIC containing approximately 36k transistors. It was fabricated in a two-level metal 1μm CMOS process. The I_{DDQ} test set consisted of 140 tests. The chips selected (151 of them) for fault location were those that failed the I_{DDQ} test and passed all other production tests. Of the 151 chips selected, DIVA was able to diagnose 120 chips to one or more leakage faults in group 1. The DIVA algorithm was run on an HP 9000 series 700 workstation. Using a fault dictionary, the combined diagnosis time for all chips was less than 5 CPU minutes. In addition, the time required to compute the fault dictionary was approximately 5 CPU minutes.

9.1.2 Metal BFs and all BFs

A fault location technique for metal BFs and all BFs in combinational circuits is discussed [57, 51]. This technique does not require use of a fault dictionary. First we give some general rules for dropping faults. Next, we show how a judicious combination of these rules can be used for fault location.

In the algorithm, a set of BFs is represented by a set of SOPS of the form $\{\langle A_1, B_1\rangle, \langle A_2, B_2\rangle, \ldots, \langle A_k, B_k\rangle\}$ where A_i is a set of nodes in the circuit and B_i can be either a set of nodes in the circuit or the special symbol Ψ. SOPS are discussed in Section 6.2.1.

Fault Dropping

Let A be a set of nodes. On application of T_i, every node in A can have the value $0, 1$ or f. This can be derived by logic simulation. The following notation is used in subsequent discussions: $A^0 = \{n \mid T_i(n) = 0 \text{ and } n \in A\}$; $A^1 = \{n \mid T_i(n) = 1 \text{ and } n \in A\}$; and $A^f = \{n \mid T_i(n) = f \text{ and } n \in A\}$.

Our algorithm will serially process input vectors $T_1, T_2, \ldots, T_n$. After each vector is processed, faults will be dropped from the fault list using a set of rules described below. Let L_{i-1} be the SOPS representing the fault list after processing the input vectors $\{T_1 \cdots T_{i-1}\}$. Next, vector T_i is processed and faults from L_{i-1} are dropped using rules described below. The remaining set of faults is represented by the SOPS L_i. The fault dropping rules are described below.

Dropping faults detected by T_i. For each ordered pair $\langle A, B\rangle \in L_{i-1}$ faults detected by T_i are dropped by replacing $\langle A, B\rangle$ with a new SOPS as described below. We have two cases governed by the following rules:

- **Rule 1**. Let $\langle A, B\rangle, B = \Psi$. The following SOPS replaces $\langle A, B\rangle$: $\{\langle A^0, \Psi\rangle, \langle A^1, \Psi\rangle, \langle A^f, \Psi\rangle, \langle A^0, A^f\rangle, \langle A^1, A^f\rangle\}$. Rule 1 should drop BFs detected by T_i. Thus, all BFs $\langle x, y\rangle \in \langle A^0, A^1\rangle$ should be dropped. Note that this is exactly the set of faults dropped by Rule 1.

- **Rule 2**. Let $\langle A, B\rangle, B \neq \Psi$. The following SOPS replaces $\langle A, B\rangle$: $\{\langle A^0, B^0 \cup B^f\rangle, \langle A^1, B^1 \cup B^f\rangle, \langle A^f, B\rangle\}$. Rule 2 should drop BFs detected by T_i. Thus, all BFs $\langle x, y\rangle \in \langle A^0, B^1\rangle$ or $\langle A^1, B^0\rangle$ should be dropped. Note that this is exactly the set of faults dropped by Rule 2.

If set of bridging faults represented by the ordered pair $\langle A, B\rangle = \Phi$ then it is not added to the SOPS L_i.

Dropping faults not detected by T_i. On application of T_i, for an ordered pair $\langle A, B\rangle$, we again have two rules to drop faults not detected by T_i.

- **Rule 3.** $\langle A, B\rangle$ and $B = \Psi$. SOPS $\{\langle A^0, A^1\rangle\}$ replaces $\langle A, B\rangle$. Rule 3 should drop faults not detected by T_i. Thus, all BFs $\langle x, y\rangle \notin \langle A^0, A^1\rangle$ should be dropped. Note that this is exactly the set of faults dropped by Rule 3.
- **Rule 4.** $\langle A, B\rangle$ and $B \neq \Psi$. SOPS $\{\langle A^1, B^0\rangle, \langle A^0, B^1\rangle\}$ replaces $\langle A, B\rangle$. Rule 4 should drop faults not detected by T_i. Thus, all BFs $\langle x, y\rangle \notin \langle A^0, B^1\rangle$ or $\langle A^1, B^0\rangle$ should be dropped. Note that this is exactly the set of faults dropped by Rule 4.

Next, we state the algorithm [57, 51] and illustrate it with an example. Note that if only metal BFs are considered then in the above rules A^f and B^f are empty and hence the rules are simplified.

Algorithm DIAG

Notation. In addition to the notations used earlier, we use the following: $R = \{r_1, \cdots, r_N\}$ is the response vector where $\forall i, r_i \in \{0, 1\}$ and $r_i = 0(1)$ if there was not (there was) a large I_{DDQ} on application of T_i.

- Step 1. $L_0 = \{\langle X, \Psi\rangle\}$, is the set of all BFs.
- Step 2. For each T_i, $1 \leq i \leq N$ do Step 3.
- Step 3. Let $L_{i-1} = \{\langle A_1, B_1\rangle, \cdots, \langle A_q, B_q\rangle\}$. For $1 \leq j \leq q$ process $\langle A_j, B_j\rangle$ to form new ordered pairs as below. These new ordered pairs are added to L_i and $\langle A_j, B_j\rangle$ is deleted from L_{i-1}.
  - Compute using logic simulation, $A_j^0(T_i), A_j^1(T_i)$ and $A_j^f(T_i)$. If $B \neq \Psi$, compute $B_j^0(T_i)$, $B_j^1(T_i)$ and $B_j^f(T_i)$. We have four cases:
    1. if $B_j = \Psi$ and $r_i = 0$, use Rule 1.
    2. if $B_j = \Psi$ and $r_i = 1$, use Rule 3.
    3. if $B_j \neq \Psi$ and $r_i = 0$, use Rule 2.
    4. if $B_j \neq \Psi$ and $r_i = 1$, use Rule 4.

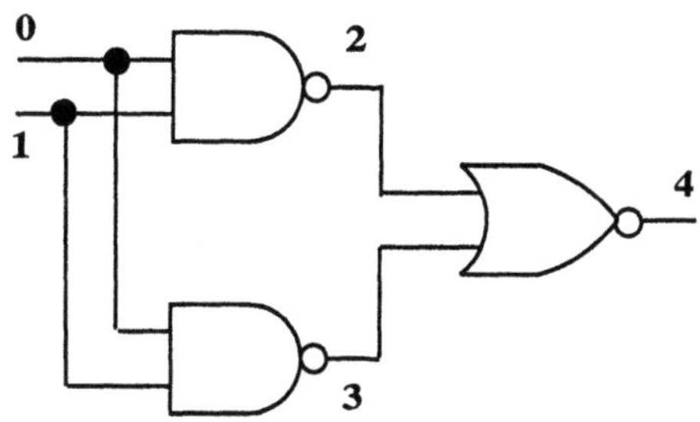

Figure 9.3 Example to illustrate fault location

Table 9.4 Example test set

Table 1		
	Node 0	Node 1
T_1	1	0
T_2	0	1

- Step 4: Let $L_N = \{\langle C_1, D_1\rangle, \cdots, \langle C_p, D_p\rangle\}$. The union of the set of BFs represented by these ordered pairs is the possible set of BFs. The number of undetected BFs is given by $\sum_{j=1}^{p} Count(\langle C_j, D_j\rangle)$, where $Count(\langle C_j, D_j\rangle$ is $\mid C_j \mid \times \mid D_j \mid$ if $D_j \neq \Psi$ else it is $\frac{|C_j| \times |C_j - 1|}{2}$.

Example: Consider the circuit of Figure 9.3, test set $T = \{T_1, T_2\}$ of Table 9.4 and the response vector $R = \{0, 1\}$.

Initially $L_0 = \{\langle\{0,1,2,3,4,5,6,7\}, \Psi\rangle\}$. Logic simulation of T_1 implies, $A^0 = \{1,2,3,5,6\}$, $A^1 = \{0,4\}$ and $A^f = \{7\}$. Since $B = \Psi$ and $r_1 = 0$, Rule 1 is used. Thus $L_1 = \{\langle\{1,2,3,5,6\}, \Psi\rangle, \langle\{0,4\}, \Psi\rangle, \langle\{1,2,3,5,6\}, \{7\}\rangle, \langle\{0,4\}, \{7\}\rangle\}$.

Next T_2 is processed and faults from L_1 are dropped for each ordered pair:

1. $\langle A_1, B_1\rangle = \langle\{1,2,3,5,6\}, \Psi\rangle$. Logic simulation implies $A_1^0 = \{2,3\}$, $A_1^1 = \{1,5,6\}$ and $A_1^f = \Phi$. Since $B_1 = \Psi$ and $r_2 = 1$, Rule 3 is used. Thus $\{\langle\{2,3\}, \{1,5,6\}\rangle\}$ replaces $\langle A_1, B_1\rangle$.

2. $\langle A_2, B_2\rangle = \langle\{0,4\},\Psi\rangle$. Logic simulation gives $A_2^0 = \{0\}, A_2^1 = \{1\}$ and $A_2^f = \Phi$. Since $B_2 = \Psi$ and $r_2 = 1$, Rule 3 is used. Thus $\{\langle\{0\},\{1\}\rangle\}$ replaces $\langle A_2, B_2\rangle$.

3. $\langle A_3, B_3\rangle = \langle\{1,2,3,5,6\},\{7\}\rangle$. Logic simulation implies $A_3^0 = \{2,3\}, A_3^1 = \{1,5,6\}, A_3^f = \Phi, B_3^0 = \Phi, B_3^1 = \Phi$ and $B_3^f = \{7\}$. Since $B_3 = \Psi$ and $r_2 = 1$, Rule 4 is used. Thus, $\langle A_3, B_3\rangle$ is replaced by the empty set.

4. $\langle A_4, B_4\rangle = \langle\{0,4\},\{7\}\rangle$. Logic simulation implies $A_4^0 = \{0\}, A_4^1 = \{4\}$, $A_4^f = \Phi, B_4^0 = \Phi, B_4^1 = \Phi$ and $B_4^f = \{7\}$. Here we use Rule 4 because $B_4 = \Psi$ and $r_2 = 1$ to obtain Φ.

From 1-4, $L_2 = \{\langle\{2,3\},\{1,5,6\}\rangle, \langle\{0\},\{1\}\rangle\}$. Since we have iterated through all the test vectors Step 4 is executed. The total number of possible BFs is 6+1 = 7. Note that the number of possible BFs reduces from 28 to 7.

Experimental Results

Experimental results obtained by running algorithm DIAG on combinational parts of ISCAS89 circuits are presented [57]. CPU times are for a Sun Sparc 2 workstation. For each circuit 25 random BFs were targeted. For each BF a random test set of size 100 was generated. For each test T_i, fault simulation was used to compute the response r_i. The results are tabulated in Table 9.5. Rows "Time" give the time taken in CPU seconds and rows "RF" give the average number of residual faults. The efficiency of this approach is clear from the observation that less than one hour of CPU time is required for the largest circuit. Experimental results for metal BFs can be found in [51].

9.2 DIAGNOSTIC TEST GENERATION

The goal of diagnostic test generation is to generate a set of test vectors of minimal cardinality for a given circuit such that it distinguishes all pairs of faults which are not equivalent. The diagnostic test generator we describe here can be used to target all BFs [108]. This test generation scheme consists of five phases. In the first phase, fault collapsing is performed to reduce the size of the initial test set. Next, using a simple genetic algorithm (GA), tests are generated. In the third phase, to distinguish fault pairs, tests are generated using a stuck-at test generator. Finally, we compact the computed diagnostic test set using reverse order simulation.

Table 9.5 Experimental results of fault diagnosis (random test sets)

Circuit	s13207	s13207.1	s15850	s15850.1	s35932
# of Faults	70383180	70383180	101609640	101609640	450945496
RF	1467	1743	1985	2183	2782
Time	567.12	884.5	832.64	1329.06	1845.78
Circuit	s38417	s38584	s38584.1	s5378	s9234.1
# of Faults	742376778	812952003	812952003	9792525	33763653
RF	3687	5643	6281	453	3852
Time	2173.04	3243.53	3539.66	131.70	445.60

The techniques used for fault collapsing are detailed in Section 8.1. These approaches can be used to reduce the size of the initial fault list. The GA phase is described in detail in Section 8.2.1. However, the GA parameters used are different. The fitness function used to evaluate a test v is given below:

$$FF(v) = \frac{1}{Q} \times \sum_{j=1}^{P}(Size_d(F_j) + Size_u(F_j)) \tag{9.1}$$

Where, assuming tests $T_1, \ldots, T_{i-1}$ have been selected, $F_1, \ldots, F_P$ are the P largest equivalence classes of faults in $EC(T_{i-1})$. Two faults belong to the same equivalence class in $EC(T_{i-1})$ if and only if they cannot be distinguished by $\{T_1, \ldots, T_{i-1}\}$. Methods to compute these classes are discussed in Section 9.3.2. $Size_d(F_j)$ is the size of subset of BFs detected in F_j and $Size_u(F_j)$ is the size of subset of BFs undetected in F_j. Q is the number of such non-empty subsets. The GA parameters, population size, number of generations and mutation probability are 20, 5 and 0.001, respectively. If fitness of the tests generated over five generations do not improve, we stop (**stopping criteria**).

Other fitness functions, like FF_1 and FF_2, can also be used. In equation 9.2 the larger the value of $FF_1(v)$, better is the vector. Whereas, in equation 9.3 the smaller the value of $FF_2(v)$, better is the vector.

$$FF_1(v) = \frac{1}{Q} \times \sum_{j=1}^{P}(Size_d(F_j) * Size_u(F_j)) \tag{9.2}$$

$$FF_2(v) = MAX_{j=1}^{P}\{(Size_d(F_j), Size_u(F_j))\} \tag{9.3}$$

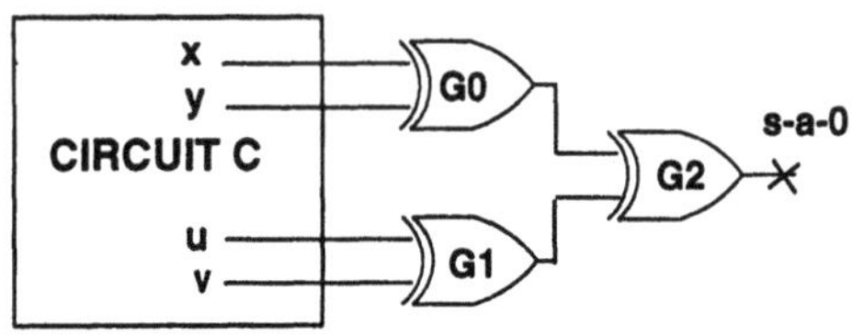

Figure 9.4 Example to illustrate use of stuck-at test generator

In the deterministic phase, we model the task of distinguishing fault pairs as a stuck-at fault. This is illustrated using Figure 9.4. Two BFs $\langle x, y\rangle$ and $\langle u, v\rangle$ can be distinguished by a test T, if T detects BF $\langle x, y\rangle$ and does not detect BF $\langle u, v\rangle$ or vice versa. This condition is satisfied if the XOR gate $G2$ is set to logic value "1". Hence, the stuck-at fault *G2 s-a-0* is the fault targeted by the stuck-at test generator to generate a test that distinguishes BFs $\langle x, y\rangle$ and $\langle u, v\rangle$ from each other.

9.2.1 Experimental Results

In this section we present experimental results using the GA-based diagnostic test generation scheme. The circuits used in the experiments are the ISCAS85 and combinational parts of the ISCAS89 benchmark circuits. All times are in CPU seconds on a SUN 4. For our experiments a subset of metal BFs were targeted and target fault list was obtained by randomly generating 15000 BFs. In Table 9.6, columns "Size" and "Comp" are sizes of the generated test set before and after compression, respectively.

9.3 DIAGNOSTIC FAULT SIMULATION

Analogous to detection test sets, diagnosis requires **diagnostic test sets** (DTS). **Diagnostic fault simulation** is the process of evaluating the capability of a DTS to diagnose a circuit. Like fault coverage and fault efficiency computed by fault simulation algorithms diagnostic fault simulation algorithms compute quantitative measures [40, 49, 50, 151, 160, 270, 320], called **diagnostic measures**, as defined below.

Table 9.6 Experimental results using GA-based diagnostic test generator

Circuit	Size	Comp	Time
C432	42	38	122.5
C499	49	43	1425
C880	50	44	1061
C1355	97	80	9014
C1908	92	63	1293952
S208	48	41	159.5
S208.1	41	35	664.8
S298	31	29	139.3
S344	33	32	1709
S349	34	32	1553
S382	38	35	685.8
S386	49	45	99.86
S400	38	34	699.4
S420	76	67	1226
S420.1	78	62	3269
S444	39	34	903.7
S510	58	55	141.6
S526	40	38	2972
S526n	40	36	2335
S641	48	42	1296
S713	49	46	10028
S838	125	115	3063
S838.1	127	117	19951
S953	117	102	3326
S1196	93	84	2249
S1238	102	88	17158
S1423	70	61	103390

The diagnostic measures assume a set F of faults like stuck-at, etc. First consider only combinational circuits. Two faults f_1 and $f_2 \in F$ are **equivalent**, with respect to a set of input vectors T, iff for each input vector $T_i \in T$ the responses of the two faulty circuits, respectively for f_1 and f_2, are the same. For combinational circuits this binary relation on F is an equivalence relation and partitions F into **equivalence classes** (EC). Let $E_1, \ldots, E_m$ be the equivalence classes of F with respect to the test set T, R be the number of faults in F and P be the number of equivalence classes of size 1. After each diagnosis using DTS T we are left with a set of possible faults from F that are equivalent with respect to T. This set is the **residual set.**

Diagnostic Resolution (DR) of DTS T is the percentage of faults in F that can be completely distinguished from all other faults using T. More formally, $DR = \frac{P}{R} \times 100$.

Expected Residual Size (ERS) of DTS T is the expected value of the size of the residual set when the circuit is diagnosed using T. More formally, $ERS = \sum_{i=1}^{m} \frac{|E_i| \times |E_i|}{R}$, where $| A |$ denotes the cardinality of set A.

For sequential circuits, when we assume an unknown initial state, the logic level of some of the primary outputs may not be defined. For this reason the binary relation defined above on S need not be an equivalence relation [270]. So, $E_1, \ldots, E_m$, may not be mutually disjoint and are more appropriately called **Indistinguishable Classes** (IC) [270]. Beside this difference, the definition of the above diagnostic measures remain unchanged.

Diagnostic simulation algorithms for stuck-at faults have been discussed in [49, 50, 160, 270, 271]. Not withstanding recent advances to speed up such algorithms [151, 320] diagnostic fault simulation remains a computationally difficult problem. The primary reason is the need to perform fault simulation without fault dropping. The trend to diagnose larger circuits and consider fault models like bridging faults, leakage faults, etc., tend to increase the number of faults to be analyzed. This adds to the difficulty of diagnostic fault simulation very significantly. To cope with this increase in complexity "sampling" is used.

A similar approach, called "fault sampling" [8, 37, 322], has been used very successfully for fault simulation. We briefly review "fault sampling" in the next section. "Fault sampling" cannot be used for diagnostic fault simulation. We propose using "EIC Sampling" that samples the set of ECs of combinational circuits and ICs of sequential circuits for diagnostic fault simulation.

One difficulty of sampling ECs or ICs is that no explicit enumeration of the set of ECs or ICs is available as is possible for the set of faults in "fault sampling". In fact, if an explicit enumeration of the ECs or ICs exists the diagnostic measures can be computed very easily. The method described samples the set of ECs or ICs without explicitly computing the entire set of ECs or ICs. This technique for evaluating I_{DDQ} diagnostic tests sets is presented. The exact diagnostic simulation algorithm is discussed in Section 9.3.2 and the approximate diagnostic simulation algorithm in Section 9.3.3. Experimental results are presented in Section 9.3.4. These results are encouraging as the approximation algorithm is quite fast and the error is small.

9.3.1 Sampling

Fault Sampling is a statistical sampling technique that has been used to estimate the value of the fault coverage of detection test sets. Let T be the detection test set, F be the set of faults and D the subset of F detected by T. Then, **fault coverage** $C = \frac{|D|}{|F|}$.

In *fault sampling*, to reduce the complexity of fault simulation, a small subset S of F is selected at random without replacement. Next, the subset Q of S detected by T is computed. Then, estimated fault coverage is: $C_{app} = \frac{|Q|}{|S|}$. It has been shown that $| C_{app} - C |$ is small, with a very high probability. This approach has been used very effectively.

The question we ask is: how can "sampling" be used for diagnostic simulation? This question is important because diagnostic simulation is computationally more difficult than fault simulation. The primary reason being that, except in some special cases [320], no fault dropping is possible. Therefore, there is an even greater need for approximate diagnostic fault simulation algorithms [47].

Note that for diagnostic simulation the sample space to be used is the set of ECs or ICs of the set of faults F with respect to the test set T. So, the sampling technique used for fault simulation cannot be used for this purpose. A difficulty in sampling the sample space of equivalence classes is that no explicit enumeration of this set exists. In addition, this set is as difficult to compute as diagnostic simulation itself. This leaves us with no choice but to **sample the set of ECs or ICs implicitly** [47]. This immediately raises the question: how can the set of ECs or ICs be sampled implicitly? We suggest the following technique.

EIC Sampling.

Notations: F is the set of modeled faults, T is the diagnostic test set being evaluated, $E = \{E_1, E_2 \ldots, E_m\}$ is the set of ECs or ICs of F with respect to the test set T, $P(f_i)$ is the equivalence class, with respect to T, containing f_i, n is the number of sampled faults, $S = \{f_1, f_2, \ldots, f_n\}$ is the set of sampled faults and $M = | \{P(f_i) :| P(f_i) |= 1\} |$.

Step 1. Randomly select, without replacement, a set of faults S.

Step 2. For each fault $f_i \in S$, compute $P(f_i)$. Let $P(f_1), P(f_2), \ldots P(f_n)$ be the list of equivalence classes so computed. Note that if two faults f_i and f_j are equivalent (or indistinguishable) with respect to T and both f_i and $f_j \in S$ then $P(f_i)$ and $P(f_j)$ are identical. However, we retain both of them in the list. This is equivalent to sampling with replacement.

Step 3. Compute the approximate diagnostic measures as follows.

$$DR_{app} = \frac{M}{n} \quad ; \quad ERS_{app} = \frac{\sum_{i=1}^{n} | P(f_i) |}{n}$$

There are two issues that need to be addressed. (i) In Step 2 why do we perform sampling with replacement? (ii) How do we compute $P(f_i)$?

The reason for (i) is that if we carefully examine the definition of ERS we note that it is equivalent to the expected value of the size of an equivalence class when one is selected at random. In this case the larger equivalence classes have larger weight. We use sampling with replacement in our approximation to simulate this effect.

The second issue is considerably more difficult to answer in general. Methods to compute $P(f_i)$ will depend on the fault model and the type of circuit - sequential or combinational - under investigation. We present one method to compute $P(f_i)$ selectively.

9.3.2 Exact Diagnostic Simulation

The exact diagnostic simulation algorithm consists of two basic steps, **UpdatePartitions** and **UpdateEquivClasses**. The algorithm iterates through these two basic steps for each input vector. The working of the algorithm is illustrated using Figure 9.5 and Table 9.7. For a discussion on node partitions

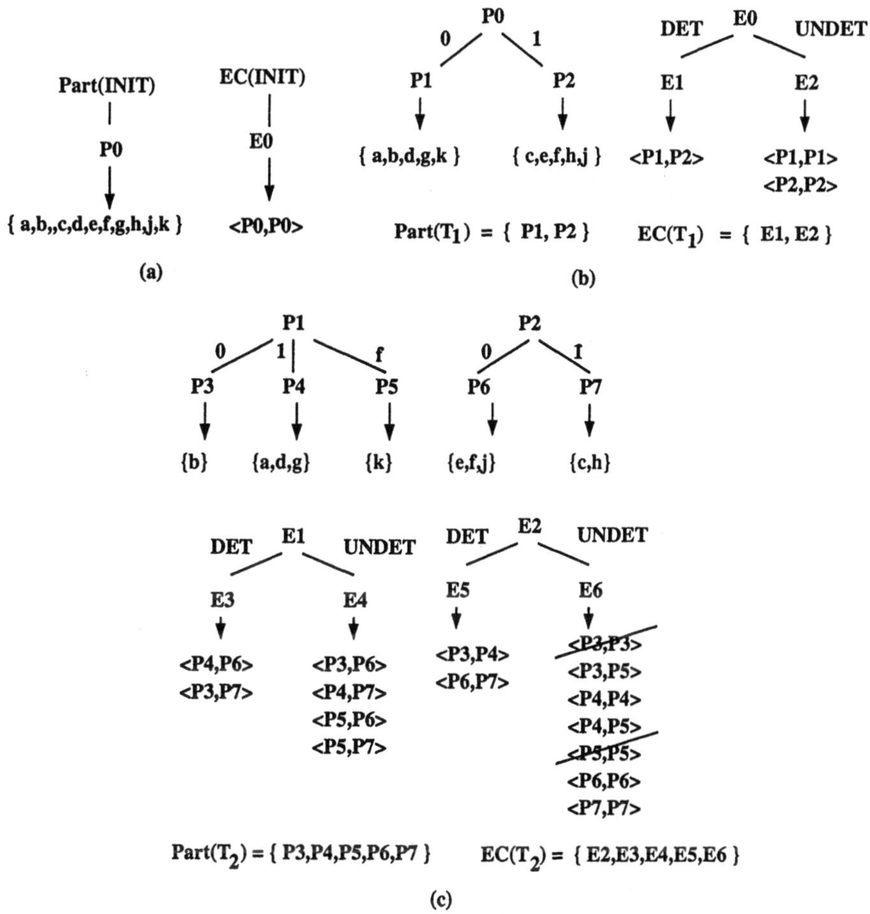

Figure 9.5 Example illustrating UpdateEquivClass

and procedure **UpdatePartitions**, see Section 6.2.2. We get $Part(T_2) = \{P3, P4, P5, P6, P7\} = \{\{b\}, \{a, d, g\}, \{k\}, \{e, f, j\}, \{c, h\}\}$.

Equivalence classes are represented using the SOPS representation. For details on the SOPS representation and its implementation, see Section 6.2.1. Equivalence classes are computed as follows. The working of our example is shown in Figure 9.5.

Let $EC(INIT)$ be the initial set of ECs and $EC(T_i)$ be the set of ECs with respect to $\{T_1, \ldots, T_i\}$. To start with, we have $Part(INIT) = \{P0\}$ and $EC(INIT) = \{\langle P0, P0\rangle\} = \{E0\}$.

Next, we process T_1. *Node partition* $Part(T_1) = \{P1, P2\} = \{\{a, b, d, g, k\}, \{c, e, f, h, j\}\}$ is computed. $P1(P2)$ is the subset of $P0$ set to 0(1) by $T1$. Thus, $\{\langle P1, P2\rangle\}$ is the subset of $E0$ detected by T_1. Similarly, $\{\langle P1, P1\rangle, \langle P2, P2\rangle\}$ is the subset of $E0$ not detected by T_1. Hence, $E0$ is split into two ECs: (i) the subset of $E0$ detected by T_1; and (ii) the subset of $E0$ not detected by T_1. The new set of equivalence classes with respect to $\{T_1\}$ is $EC(T_1) = \{\{\langle P1, P2\rangle\}, \{\langle P1, P1\rangle, \langle P2, P2\rangle\}\} = \{E1, E2\}$. After computing $EC(T_1)$ the space used by $EC(INIT)$ and $Part(INIT)$ is released.

Next T_2 is processed as shown in Figure 9.5(c). We first compute $Part(T_2) = \{\{b\}, \{a, d, g\}, \{k\}, \{e, f, j\}, \{c, h\}\} = \{P3, P4, P5, P6, P7\}$. Next each of $E1$ and $E2$ in $EC(T_1)$ is split into two Equivalence Classes as was done for $E0$. Note that $\langle P3, P3\rangle$ and $\langle P5, P5\rangle$ do not represent any BF and are deleted. At the end of this we have:

$$\begin{aligned} EC(T_2) &= \{\{\langle P4, P6\rangle, \langle P3, P7\rangle\}, \{\langle P3, P6\rangle, \langle P4, P7\rangle, \\ &\quad \langle P5, P6\rangle, \langle P5, P7\rangle\}, \{\langle P3, P4\rangle, \langle P6, P7\rangle\}, \\ &\quad \{\langle P3, P5\rangle, \langle P4, P4\rangle, \langle P4, P5\rangle, \langle P6, P6\rangle, \langle P7, P7\rangle\}\} \\ &= \{E3, E4, E5, E6\} \end{aligned} \tag{9.4}$$

More formally, the rules for computing $EC(T_i)$ from $EC(T_{i-1})$ follow. Let $EC(T_{i-1})$ be $\{E_0, \ldots, E_M\}$. First, using UpdatePartition, $Part(T_i)$ is computed from $Part(T_{i-1})$. Next, each $E_j \in EC(T_{i-1})$ is split into two sets of BFs $E_j(Det)$ and $E_j(UnDet)$ where $E_j(Det)$ ($E_j(UnDet)$) is the subset of E_j detected (not detected) by T_i.

Let $E_j \in EC(T_{i-1})$ be represented by the SOPS $\{\langle A_1, B_1\rangle, \ldots, \langle A_Q, B_Q\rangle\}$ where the A_k's and B_k's are pointers to partitions in the collection of partitions $Part(T_{i-1})$. $E_j(Det)$ is computed as follows. For each $\langle A_k, B_k\rangle \in E_j$ we compute $Det(k)$, the subset of BFs represented by this pair of sets that is detected by T_i, as follows. Let $A_k(0), A_k(1), A_k(f)$ be pointers to partitions in $Part(T_i)$ such that for all $t \in \{0, 1, f\}$ $A_k(t)$ points to the subset of nodes in A_k set to t by T_i. Notations $B_k(0), B_k(1)$ and $B_k(f)$ are similarly defined. Then,

$$Det(k) = \{\langle A_k(0), B_k(1)\rangle, \langle A_k(1), B_k(0)\rangle\} \tag{9.5}$$

$$E_j(Det) = \bigcup_{k=1}^{Q} Det(k) \tag{9.6}$$

Let $UnDet(k)$ be the subset of BFs represented by $\langle A_k, B_k\rangle$ that are not detected by T_i. Then,

$$\begin{aligned} UnDet(k) &= \{\langle A_k(0), B_k(0)\rangle, \langle A_k(0), B_k(f)\rangle, \\ &\quad \langle A_k(1), B_k(1)\rangle, \langle A_k(1), B_k(f)\rangle, \langle A_k(f), B_k(0)\rangle, \\ &\quad \langle A_k(f), B_k(1)\rangle, \langle A_k(f), B_k(f)\rangle\} \qquad (9.7) \\ E_j(UnDet) &= \bigcup_{k=1}^{Q} UnDet(k) \qquad (9.8) \end{aligned}$$

In both equation (9.5) and equation (9.7) if any of the pairs of pointers denote an empty set of BFs it is deleted from the list. Note that, in equation (9.5), if A_k is the same as B_k then the two pairs denote the same set of BFs. Only one of them is retained. Similarly, in equation (9.7), if A_k is the same as B_k then $\langle A_k(1), B_k(f)\rangle$ is same as $\langle A_k(f), B_k(1)\rangle$ and only one is retained. We achieve this by adding two other rules for pairs of the form $\langle A_k, A_k\rangle$ which are simplification of equation (9.5) and equation (9.7). Using these additional rules we guarantee that every pair of partition pointers represents disjoint sets of BFs and, by definition, the ECs represent disjoint set of BFs. The use of equation (9.5) to equation (9.8) is referred to as **UpdateEquivClasses** and uses "list splitting", a paradigm used earlier for diagnostic simulation of stuck-at faults [49, 50, 271, 320].

Outline of the Algorithm

Note that as the number of vectors simulated increases the number of pairs of partition pointers in each SOPS representing an EC increases. However, it will never be worse than explicit enumeration of all fault in the EC. In the initial stages of the simulation the use of SOPS results in a considerable speedup. To improve the speed of simulation during later stages we use the following strategy. Note that once an EC has reduced to size 1 we can delete it and keep track of the number of ECs of size 1. The better the diagnostic test set the more effective this strategy will be. The diagnostic simulation algorithm is outlined below.

ExactIddqDiagSim.
/* Let $\{T_1, \ldots, T_N\}$ be the given diagnostic test set. */
BEGIN
  Compute $Part(INIT) = \{P0\}$. /* It will contain only one partition pointed to by $P0$ */
  Set $EC(INIT)$ to $\{\langle P0, P0\rangle\}$.

SINGLETONS = 0;
For each test T_i
BEGIN
  $Part(T_i)$ = UpdatePartition($Part(T_{i-1})$);
  $EC(T_i)$ = UpdateEquivClass($EC(T_{i-1})$);
  Traverse $EC(T_i)$ and delete ECs of size 1. When such an EC is deleted increment SINGLETONS.
  Release Space used by $Part(T_{i-1}), EC(T_{i-1})$.
END
Traverse $EC(T_N)$ and compute DR, ERS, MAXECS defined in Section 9.3.1. The value of SINGLETONS have to be taken into account in this calculation.
END(* **IddqDiagSim** *)

To complete the example we started in the previous subsection refer to Figure 9.6. For now, ignore the numbers within [ ] and { }. Figure 9.6(a) shows how each partition is split into its subpartitions using UpdatePartitions. In Figure 9.6(b) we show how each E_j in $EC(T_2)$, from Figure 9.5(c), is split into its two subsets - $E_j(Det)$ and $E_j(UnDet)$.

To compute the diagnostic measures first compute the size of each EC. This is done by computing the number of BFs represented by each pair of partition pointers. For example, since both $P10$ and $P12$ point to partitions of size 2 $\langle P10, P12\rangle$ represents 4 BFs. In Figure 9.6(b) the number of BFs represented by each pair of partition pointers is shown next to it, within [ ]. Note that $\langle P12, P12\rangle$ represents 1 not 4 BFs. $\langle P9, P9\rangle$ and $\langle P12, P12\rangle$ represent empty sets and are deleted.

Recall that in SOPS a pair of partition pointers represents a set of BFs. Consider two pairs of partition pointers such that they differ in at least one of the two pointers. Since the node partitions are disjoint the two sets of BFs represented by these pairs of partition pointers are disjoint. In addition, each EC is represented by a pair of partition pointers and a BF occurs in an EC at most once. From this we deduce that size of an EC is the sum of the sizes of the sets of BFs represented by the pairs of partition pointers constituting an EC. For each EC of Figure 9.6(b) its size is shown within { }. Thus SINGLETONS = 0, the total number of faults is 45, $DR = \frac{0}{45} \times 100 = 0\%$, MaxECS is 7 and

$$ERS = \frac{(49 + 16 + 49 + +49 + 9 + 36 + 49 + 16)}{45} = 6.07.$$

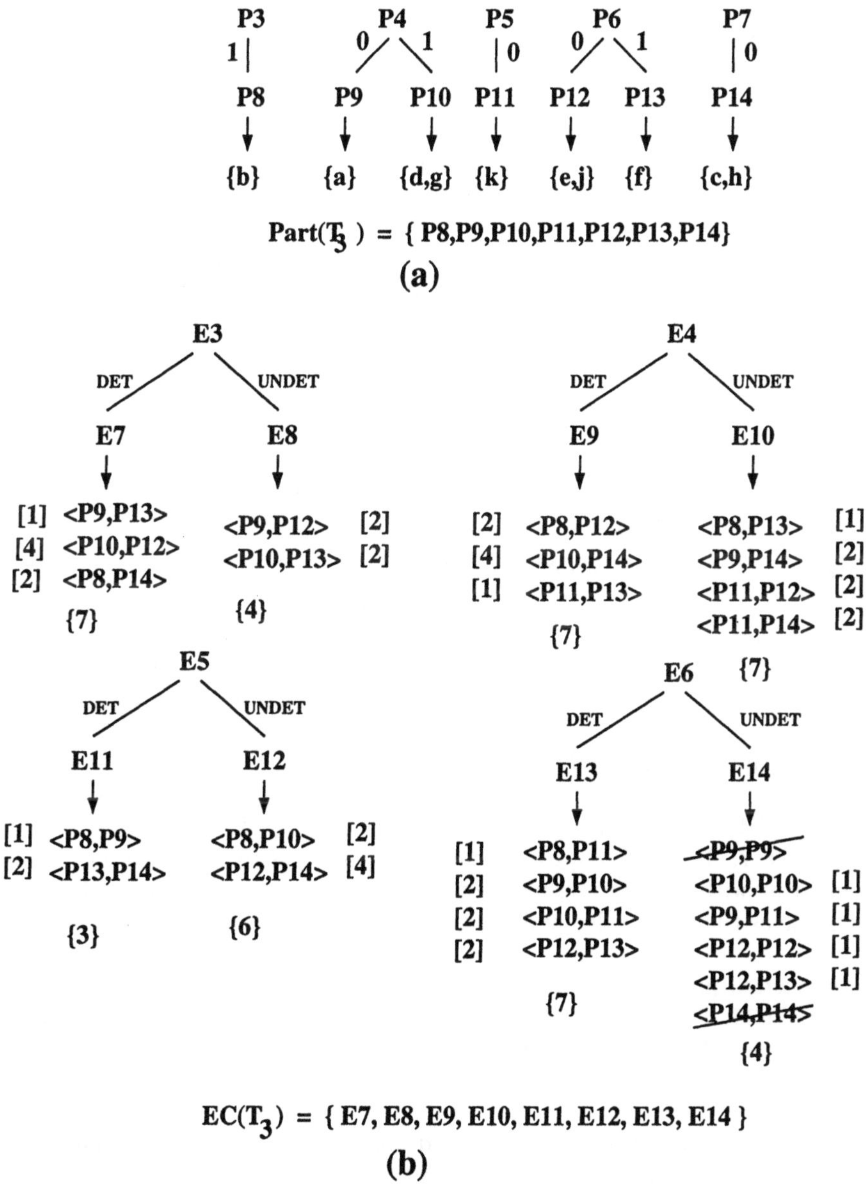

Figure 9.6 Example illustrating UpdateEquivClass (contd)

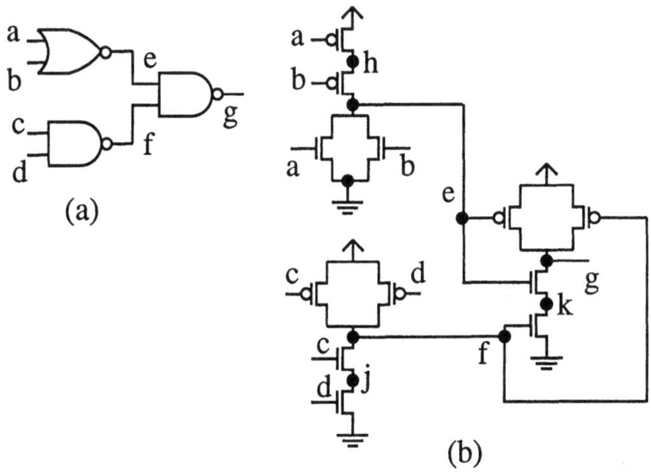

Figure 9.7 Example used for diagnostic fault simulation

9.3.3 Approximate Diagnostic Simulation

For approximate diagnostic simulation we use the EIC Sampling technique outlined in Section 9.3.1. We first illustrate the process using the example of Figure 9.7 and Table 9.7. For this example, we assume the all BF fault model.

Let the sample of faults be $S = \{f_1 = \langle e, h\rangle, f_2 = \langle b, g\rangle, f_3 = \langle a, b\rangle, f_4 = \langle d, k\rangle\}$. From Figure 9.6 we note that $P(f_1) = E12, P(f_2) = E12$, $P(f_3) = E11$ and $P(f_4) = E13$. Therefore, $\mid P(f_1) \mid= 6$, $\mid P(f_2) \mid= 6$, $\mid P(f_3) \mid= 3$, and $\mid P(f_4) \mid= 7$. Hence, $DR_{app} = 0$ and $ERS_{app} = \frac{1}{4} \times (6 + 6 + 3 + 7) = 5.5$. Comparing with the exact values we note that the approximate value of DR is the same as the exact value and the error for ERS is about 10%.

An important point is that our approximation is based on a subset of ECs consisting of $E11, E11, E12$ and $E13$ instead of the entire set of ECs $\{E7, \ldots, E14\}$. This complete list of ECs is never computed. We next illustrate how this is done.

Let the sample of faults S be as defined above. The idea is to compute the equivalence classes of S with respect to the test set of Table 9.7. For each equivalence class A of S we also compute the equivalence class B of F such that $A \subset B$. Unnecessary equivalence classes of F, as soon as they are recognized, are deleted.

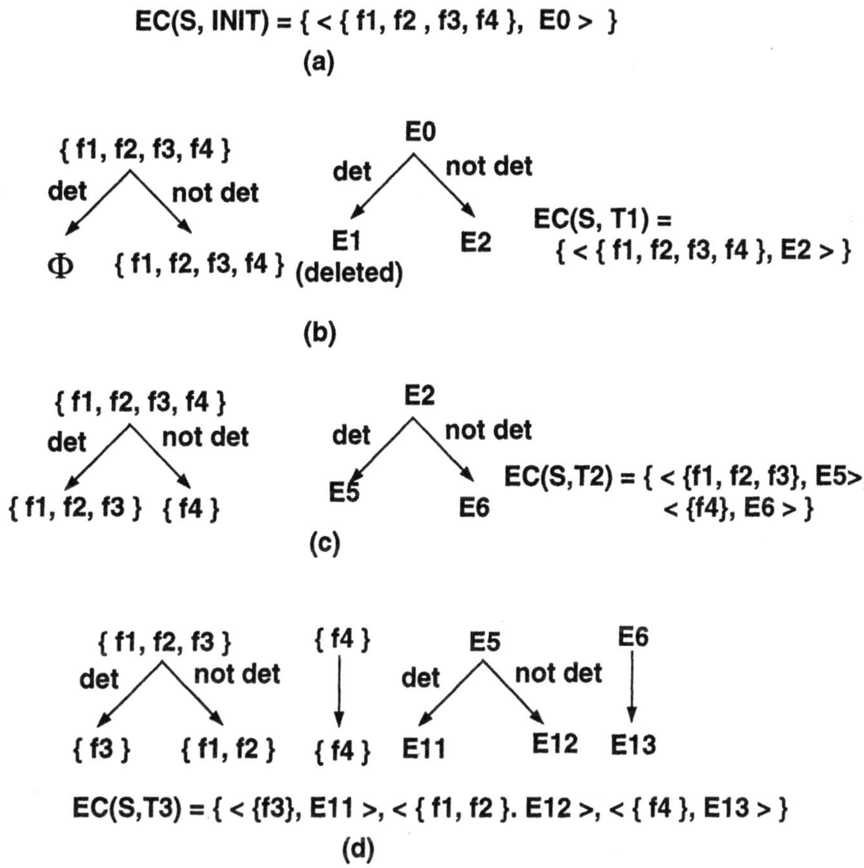

Figure 9.8 Example illustrating approximate diagnostic simulation

Let $EC(S, INIT)$ be the initial sample of equivalence classes of F. It consists of an ordered pair $\langle A, B \rangle$ where A is the set of sampled faults S and B is $E0$, the set of all faults being modeled. $E0$ is defined in Figure 9.5 and $EC(S, INIT)$ is shown in Figure 9.8(a).

Let $EC(S, T_i)$ be the sample of the equivalence classes of F, i.e., a sample of $EC(T_i)$ (see Section 9.3.2 on computing equivalence classes) with respect to the fault sample S. Each member of $EC(S, T_i)$ consists of an ordered pair $\langle A, B \rangle$. A is an equivalence class of S with respect to $\{T_1, \ldots, T_i\}$. B points to the member of $EC(T_i)$ that A is a subset of. All members of $EC(T_i)$ such that they are not a subset of any equivalence class of S are not included in the sample.

First, $T1$ is processed and $EC(S,T1)$ is computed from $EC(S,INIT)$. None of f_1, f_2, f_3, f_4 is detected by T_1. So, $\{f_1, f_2, f_3, f_4\}$ is split as shown in Figure 9.8(b). From Figure 9.5 we see that $E0$ is split into $E1$ and $E2$ where $E1$ ($E2$) is the subset of $E0$ detected (not detected) by $T1$. We therefore delete $E1$ and $EC(S,T1)$ as shown in Figure 9.8(b).

Next, $T2$ is processed and $EC(S,T2)$ is computed from $EC(S,T1)$. For each ordered pair $\langle A, B\rangle$ of $EC(S,T1)$ we split A into the subset A_d (A_u) of faults in A detected (not detected) by $T1$. Similarly, we split B into the subset B_d (B_u) of faults in B detected (not detected) by $T1$. Replace $\langle A, B\rangle$ by $\langle A_u, B_u\rangle$ and $\langle A_d, B_d\rangle$. If $A_u(A_d)$ is empty then $\langle A_u, B_u\rangle$ ($\langle A_d, B_d\rangle$) is deleted. Continuing with our example, we have only one ordered pair $\langle A, B\rangle$. From Figure 9.8(c), $A_u = \{f_4\}, A_d = \{f_1, f_2, f_3\}$, $B_d = E5$ and $B_u = E6$. ECs $E5$ and $E6$ are defined in Figure 9.6(b). Hence, $EC(S,T2)$ is as shown in Figure 9.8(c). The processing of $T3$ is shown in Figure 9.8(d). This completes the example. The list of sampled ECs is read off from $EC(S,T3)$.

Note that the approximation algorithm is a modification of *UpdateEquivClasses*. The equivalence classes of the set of sampled faults S is computed. This is done by splitting each equivalence class A of S into two subsets: the subset A_u (A_d) of A not detected (detected) by the current vector T_i being processed (**SplitFaultClass**). Let B be the set of equivalence classes of the modeled faults such that $A \subset B$. Then, B is split into its detected (B_d) and undetected (B_u) subsets (**SplitEquivClass**). If A_d (A_u) is empty then B_d (B_u) is eliminated and not processed any further. This elimination of the equivalence classes of the modeled faults, as soon as they are found to be irrelevant, is what contributes to the speed up of the process. It also contributes to inaccuracies in the values computed. A high level description of the above process follows. A similar method can be used for other fault models.

ApproxIddqDiagSim.
/* Let $\{T_1, \ldots, T_N\}$ be the given diagnostic test set. */
BEGIN
  Let $S = \{f_1, \ldots, f_n\}$ be the set of sampled faults.
  Compute $Part(INIT) = \{P0\}$. /* It will contain only one partition pointed to by $P0$ */
  Set $E0$ to $\langle P0, P0\rangle$ and $EC(S,INIT)$ to $\{\langle S, E0\rangle\}$.
  For each test T_i
  BEGIN

```
    Part(T_i) = UpdatePartition(Part(T_{i-1}));
    /* Compute EC(S,T_i) from EC(S,T_{i-1}). EC(S,T_{i-1}) = {⟨A_1,B_1⟩, ...,
                                                            ⟨A_P,B_P⟩}. */
    For each ⟨A,B⟩ ∈ EC(S,T_{i-1})
    BEGIN
      SplitFaultClass(A) to get A_u, A_d.
      SplitEquivClass(B) to get B_u, B_d.
      If A_u (A_d) is empty then delete B_u (B_d).
      Replace ⟨A,B⟩ by non-empty pairs from {⟨A_u,B_u⟩,⟨A_d,B_d⟩}.
    END
  END
  Traverse EC(S,T_N) and compute DR_app, ERS_app, MAXECS_app defined in
   Section 9.3.1.
END (*ApproxIddqDiagSim*)
```

9.3.4 Experimental results

Table 9.8 lists ISCAS85 [34] and ISCAS89 [32] circuits used in our experiment [47]. Only combinational parts of ISCAS89 circuits were used. *Vectors* is the size of the stuck-at detection test set from [169] used in our experiment. *Sample* is the number of fault sample used. *Gates* (*Nodes*) is the number of gates (nodes) in the circuit. The number of faults in the sample equals the number of nodes in the circuit. *Metal BF* (*all BF*) is the number of faults in metal BFs (all BF). No fault collapsing was used. Note that the sample of faults is very small compared to the total number of faults.

We first compare the values of DR and ERS computed by ExactIddqDiagSim and ApproxIddqDiagSim. Table 9.9 is for metal BFs. The column labels are self-explanatory. The ERROR column was computed as follows: $ERROR = \frac{|EXACT-APPROX|}{EXACT} \times 100$. For some circuits, like s9234 onward, exact diagnostic simulation had to be aborted for lack of resources. The average error for DR is 3.5% and the standard deviation is low. For ERS the average error is 7.2% and the standard deviation is 4.91. Similar results are tabulated in Table 9.10 for all BFs. The average error for DR is about 5.33% with a standard deviation of 5.19. The approximate values of ERS are not as good as for metal BFs.

Table 9.11 tabulates the CPU time for metal BFs (in column labeled metal BF) and all BFs (in column labeled all BF). EXACT and APPROX are the CPU seconds for exact and approximate diagnostic simulation on a SUN 4/30 workstation. RATIO ($= \frac{EXACT}{APPROX}$) is the speedup factor. Note that for both

Table 9.7 An example test set

	a	b	c	d	e	f	g	h	j	k
$T1$	0	0	1	0	1	1	0	1	1	0
$T2$	1	0	1	1	0	0	1	1	0	f
$T3$	0	1	0	1	0	1	1	0	0	0

fault models the speedup is between 14 and 15, on an average. The approximation algorithm is also less memory intensive making it possible to run for some of the larger circuits like S13207.1, S15850. The speedup is better for larger circuits, for which data are available, like S5378 and C7552.

Note that for some circuits the error is quite large. This is bound to happen because sampling is a random process. To reduce the error bound the algorithm can be executed a couple of times and the average of the values computed is taken as an approximation. Alternatively, we can use a larger sample size. If either of the above alternatives is used, the speed up factor will be reduced.

9.4 DYNAMIC DIAGNOSIS

In the diagnosis paradigm discussed so far, a diagnostic test sequence (DTS) targeting a set of faults is computed using a diagnostic test generator (DTG). The quality of the DTS is evaluated using diagnostic simulation. Finally, a fault dictionary is computed either during diagnostic simulation or independently. All above steps are completed before the actual diagnosis and hence we refer this paradigm as **static diagnosis**.

Static diagnosis has the following disadvantages: diagnostic fault simulation requires considerable space and time; fault dictionaries require large amounts of space and time to compute; and DTS can be large particularly for large circuits resulting in storage problems. To overcome the above disadvantages, several alternatives to static diagnosis have been studied [1, 2, 48, 52, 109, 242, 272]. One such alternative is **dynamic diagnosis** [109]. Here, the computationally expensive steps of fault dictionary computation and diagnostic fault simulation are eliminated.

In dynamic diagnosis, the DTS is generated during fault location. Portions of the fault dictionary, if required, are computed during fault location. For I_{DDQ}

Table 9.8 Circuit characteristics

Circuit	Vectors	Sample	Gates	Nodes	METAL	ALL
C432	47	466	196	466	19110	108345
C499	53	1027	243	1027	29403	526851
C880	53	935	443	935	97903	436645
C1355	85	1163	587	1163	171991	675703
C1908	120	1594	913	1594	416328	1269621
C2607	105	2795	1502	2795	1127251	3904615
C3540	153	3579	1719	3579	1476621	6402831
C5315	113	5496	2485	5496	3086370	15100260
C6288	28	5088	2448	5088	2995128	12941328
C7552	210	7371	3719	7371	6913621	27162135
S208.1	34	234	122	234	7381	27261
S382	36	365	182	365	16471	66430
S386	71	478	172	478	14706	114003
S420.1	75	494	252	494	31626	121771
S444	32	403	205	403	20910	81003
S510	62	512	236	512	27730	130816
S526	59	553	217	553	23436	152628
S526n	59	552	218	552	23653	152076
S641	59	696	433	696	93528	241860
S713	55	756	447	756	99681	285390
S820	114	916	312	916	48516	419070
S298	33	308	136	308	9180	47278
S838.1	103	1014	512	1014	130816	513591
S953	86	873	440	873	96580	380628
S1196	114	1260	561	1260	157080	793170
S1238	149	1319	540	1319	145530	869221
S1423	71	1589	748	1589	279378	1261666
S1488	127	1951	667	1951	222111	1902225
S1494	119	1965	661	1965	218130	1929630
S5378	251	4665	2993	4665	4477528	10878780
S344	22	346	184	346	16836	59685
S349	23	351	185	351	17020	61425
S382	36	365	182	365	16471	66430
S9234	367	9604	5844	9604	17073246	46113606
S13207.1	469	13491	8651	13491	37415575	90996795
S15850	440	16585	10383	16585	53898153	137522820

Table 9.9 Comparison of exact and approximation results for metal BF

Circuit	DR			ERS		
	EXACT	APPROX	ERROR	EXACT	APPROX	ERROR
432	78.76	75.97	3.54	1.41	1.46	3.55
499	66.45	64.76	2.54	3.07	2.99	2.61
880	58.82	60.53	2.91	2.07	1.91	7.73
1355	65.36	66.55	1.82	2.03	1.97	2.96
1908	7.47	7.15	4.28	23.16	21.03	9.20
2670	13.45	12.77	5.06	9.36	9.21	1.60
3540	15.53	15.48	0.32	20.81	19.59	5.86
5315	27.32	27.13	0.70	8.56	7.15	16.47
6288	77.11	76.30	1.05	1.71	1.73	1.17
7552	19.92	19.54	1.91	12.64	11.90	5.85
s208.1	43.04	39.74	7.67	2.96	2.84	4.05
s298	29.53	32.79	11.04	3.41	2.86	16.13
s344	43.93	41.62	5.26	3.50	3.80	8.57
s349	44.44	43.59	1.91	3.69	3.88	5.15
s382	38.08	38.36	0.74	2.99	2.92	2.34
s386	56.80	62.97	10.86	1.86	1.68	9.68
s420.1	46.78	44.33	5.24	2.81	2.51	10.68
s444	41.01	39.95	2.58	3.15	2.95	6.35
s510	91.82	87.89	4.28	1.12	1.12	0.00
s526	40.78	40.51	0.66	2.91	2.97	2.06
s526n	38.94	37.50	3.70	3.02	2.73	9.60
s641	7.70	7.90	2.60	16.57	13.24	20.10
s713	7.21	8.33	15.53	19.77	17.20	13.00
s820	53.49	53.82	0.62	2.17	2.23	2.76
s832	59.42	62.10	4.51	1.89	1.80	4.76
s838.1	13.77	14.20	3.12	251.96	239.46	4.96
s953	81.79	81.44	0.43	1.27	1.21	4.72
s1196	60.84	61.82	1.61	2.19	2.01	8.22
s1238	82.45	80.90	1.88	1.25	1.35	8.00
s1423	47.52	46.63	1.87	10.99	9.17	16.56
s1488	64.80	63.60	1.85	1.74	1.66	4.60
s1494	70.06	71.25	1.70	1.60	1.52	5.00
s5378	11.79	11.60	1.61	18.06	16.72	7.42
s9234		41.62			3.80	
s13207.1		6.58			68.03	
s15850		5.53			396.66	
AVE			3.50			7.02
STD DEV			3.40			4.91

Table 9.10 Comparison of exact and approximation results for all BF

Circuit	DR			ERS		
	EXACT	APPROX	ERROR	EXACT	APPROX	ERROR
432	49.63	51.07	2.90	20.48	16.36	20.12
499	28.86	28.72	0.49	33.13	45.94	38.67
880	40.44	41.28	2.08	4.30	3.48	19.07
1355	30.47	31.04	1.87	6.71	7.84	16.84
1908	6.70	5.90	11.94	37.30	29.34	21.34
2670	12.38	12.70	2.58	24.61	21.37	13.17
3540	16.04	15.93	0.69	30.39	20.03	34.09
5315		22.40			19.05	
6288	24.95	24.33	2.48	5.71	6.60	15.59
s208.1	16.49	15.39	6.67	10.35	9.10	12.08
s344	16.43	15.32	6.76	7.76	5.45	29.77
s349	17.68	19.94	12.78	8.61	8.90	3.37
s382	29.01	27.94	3.69	5.33	6.02	12.95
s386	21.12	20.92	0.95	10.04	10.30	2.59
s420.1	20.83	20.85	0.10	13.14	12.64	3.81
s444	26.78	29.78	11.20	9.84	12.06	22.56
s510	45.62	48.24	5.74	7.03	6.45	8.25
s526	34.72	36.17	4.18	7.96	5.79	27.26
s526n	35.24	30.98	12.09	7.83	9.66	23.37
s641	10.37	12.64	21.89	18.56	15.95	14.06
s713	9.28	8.33	10.24	30.47	30.60	0.43
s820	34.04	30.90	9.22	26.61	20.59	22.62
s832	34.72	33.73	2.85	27.94	20.01	28.38
s838.1	10.95	11.24	2.65	2014.14	2122.20	5.37
s953	41.78	37.46	10.34	10.05	12.60	25.37
s1196	38.27	38.49	0.57	8.05	6.05	24.84
s1238	42.79	40.86	4.51	13.25	16.81	26.87
s1423	25.40	25.61	0.83	17.68	16.03	9.33
s1488	23.00	22.91	0.39	29.99	31.69	5.67
s1494	23.60	24.02	1.78	30.06	33.21	10.48
s5378		21.35			15.64	
s9234		3.63			5032.54	
s13207.1		7.46			67.63	
s15850		6.62			1753.60	
AVE			5.33			17.18
STD DEV			5.19			10.13

Table 9.11 Comparison of time required for exact and approximate computation

Circuit	METAL			ALL		
	EXACT	APPROX	RATIO	EXACT	APPROX	RATIO
C432	7.98	1.98	4.03	62.97	7.97	7.90
C499	9.99	2.97	3.36	350.99	45.98	7.63
C880	28.99	4.98	5.82	256.98	21.99	11.69
C1355	53.94	9.98	5.40	668.98	38.97	17.17
C1980	264.97	66.98	3.96	1241.97	17.98	69.08
C2670	732.95	32.96	22.24	3251.99	120.97	26.88
C3540	1074.96	49.99	21.50	8344.97	254.97	32.73
C5315	2385.95	93.99	25.39		607.96	
C6288	979.96	87.98	11.14	4229.99	360.97	11.72
C7552	11089.98	204.99	54.10			
S208.1	1.98	0.03	66.00	11.98	1.98	6.05
S298	1.96	0.02	98.00	20.98	2.97	7.06
S344	3.97	0.98	4.05	14.97	1.96	7.64
S349	3.98	0.95	4.19	18.97	2.97	6.39
S382	5.96	0.98	6.08	33.98	3.99	8.52
S386	7.96	1.97	4.04	122.98	9.98	12.32
S420.1	17.98	2.98	6.03	109.97	12.99	8.47
S444	7.98	0.99	8.06	33.96	3.98	8.53
S510	11.98	1.99	6.02	104.98	12.96	8.10
S526	9.97	1.97	5.06	108.99	11.97	9.11
S526n	10.97	1.97	5.57	124.98	10.99	11.37
S641	25.95	3.96	6.55	114.97	9.98	11.52
S713	24.96	3.98	6.27	113.94	10.98	10.38
S820	48.97	6.96	7.04	702.98	58.98	11.92
S832	44.99	7.99	5.63	661.97	65.99	10.03
S838.1	55.99	6.99	8.01	264.99	50.97	5.20
S953	70.98	6.98	10.17	396.96	36.98	10.73
S1196	116.99	15.98	7.32	1455.98	86.98	16.74
S1238	120.97	16.97	7.13	1777.96	113.99	15.60
S1423	112.98	14.99	7.54	942.97	51.99	18.14
S1488	142.97	19.97	7.16	3701.99	179.98	20.57
S1494	141.98	19.98	7.11	3929.98	196.00	20.05
S5378	5326.99	153.98	34.60		702.00	
S9234		472.97			2427.97	
S13207.1		1433.98			4528.98	
S15850		2176.95			7472.97	
AVE			14.68			14.31
STD DEV			20.67			12.10

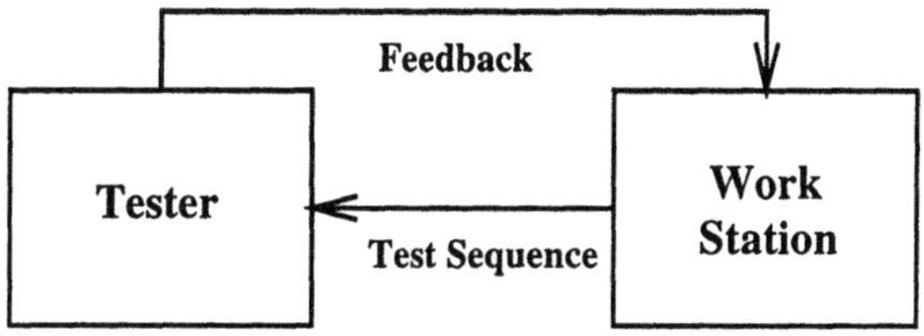

Figure 9.9 Dynamic diagnosis system

measurement based diagnosis of BFs, as we have already seen, no dictionary is required.

Figure 9.9 shows the system setup. It is a multi-phase iterative process. Each phase consists of two parts. In the first part, a small DTS is generated using a DTG to distinguish targeted faults. This DTS is next applied to the circuit under diagnosis (CUD). The responses (i.e., logic/I_{DDQ}) of the CUD are stored. In the second part of this phase the stored response is used for fault dropping. This reduced fault list is used as target fault list at the start of the following phase. This process continues until no fault dropping is possible, i.e., all faults in the fault list are equivalent. Note that the system setup required for dynamic diagnosis is the same as that for static diagnosis.

In each phase, if the DTS generated is a single vector the DTG phase can be considered "well-informed" about the status of diagnosis. But this implies a huge communication overhead. To avoid this, we generate the DTS in sets and each set of vectors is processed in one phase. A potential overhead of this approach is the increased usage of tester time due to increased time required by the DTG in one or more phases.

An experimental system using dynamic diagnosis has been implemented [109]. It completely diagnoses metal BFs in combinational circuits. In addition, it assumes that the information from I_{DDQ} measurements is to be used for diagnosis. In this system, the DTS was generated in three phases. In the first phase, $log(n)$ random vectors were generated. Using this DTS, and the observed response of the CUD to these tests, faults were dropped. In the second phase, faults involving nodes that are hard to set to 0(1) are targeted. These are the lines with either low or high controllability. BFs between pairs of lines such that both have either low or high controllability are enumerated. A stuck-at test generator is used to compute a detection test set for these BFs. This DTS is now applied to the CUD and the response is obtained. This response is used to drop faults from the set of faults left after the first stage. In the third phase,

Table 9.12 Comparison with static system (metal BF)

Circuit	Vectors			CPU seconds		
	AD		SD	AD		SD
	Ave	Max		Ave	Max	
C1355	40.83	83	78	0.780	2.80	668.43
C1908	36.62	69	74	3.691	83.18	1116.73
C2670	27.33	54	36	8.095	124.70	1331.03
C3540	37.84	106	65	4.248	171.20	2246.38
C5315	29.98	61	45	68.68	29950.00	3415.43
C6288	28.05	47	53	3.392	125.30	3104.72
C7552	33.24	68	58	7.365	87.55	8230.52
S1196	44.79	69	55	0.999	7.30	626.58
S1238	33.51	68	70	0.972	2.76	703.80
S1423	32.56	52	40	0.803	14.51	687.45
S5378	26.91	140	113	1.029	5583.00	10152.90
S13207.1	72.16	250	NA	141.6	55650.00	NA
S15850.1	59.07	259	NA	52.39	9846.00	NA
S35932	35.91	49	NA	20.63	64.05	NA
S38417	59.00	306	NA	58.26	854.60	NA
S38584.1	95.74	436	NA	855.1	211900.00	NA

pairs of faults left after phase two are enumerated. Tests to distinguish such pairs of faults are computed using a deterministic stuck-at test generator. For details the reader may refer to Section 9.2.

Experiments were performed using ISCAS85 and combinational parts of ISCAS89 benchmark circuits. In experiments, 500 faults were randomly generated to simulate 500 faulty circuits. Table 9.12 gives the comparison between the dynamic DTG and static DTG approaches. The data under "AD" are the average and maximum over 500 fault location experiments. The residual fault set size is not reported because the dynamic approach will always outperform the static approach. On an average the dynamic system requires fewer measurements. However, the big difference is in worst case time. Also, while static DTG could not be used for circuits larger than c7552, the dynamic DTG could be used effectively for all benchmark circuits.

It is important to ask why dynamic diagnosis performs so much better. Let r_1 and r_2 be the responses of the two faulty circuits, respectively, with faults f_1 and f_2, to the input sequence S. The sequence S **distinguishes** faults f_1 and f_2 iff $r_1 \neq r_2$. Two faults f_1 and f_2 are **distinguishable** iff there exist an input sequence S that distinguishes them. A **complete static DTS** distinguishes between every distinguishable fault pair. A static DTG (SDTG) computes a complete static DTS.

SDTG requires enourmous resources because for some distinguishable fault pairs there are very few vectors that distinguish them. In addition, most of these vectors distinguish none or only a small number of other fault pairs. Such fault pairs are known as **hard-to-distinguish-fault-pairs** (HDFP). HDFPs are analogous to hard-to-detect fault vis-a-vis detection. Analogous to hard-to-detect faults HDFPs are the main cause of the difficulty of computing static DTS.

In practice, a weaker notion of a complete DTS is more useful. Faults f_1 and f_2 are **equivalent w.r.t. a DTS** S iff $\forall S_i \in S$, S_i cannot distinguish f_1 and f_2. A DTS is **complete w.r.t. a fault** f_i iff the following condition holds. For each fault f_j, f_i and f_j are equivalent iff they are equivalent w.r.t. S.

For a given fault location experiment we only need to completely distinguish the fault in the faulty circuit from all other possible faults in the circuit. That is, a complete DTS w.r.t. the fault f_i in the CUD is required. This is considerably easier to compute than a complete DTS. This is what is done in dynamic diagnosis.

10

INSTRUMENTATION FOR I_{DDQ} MEASUREMENT

KENNETH M. WALLQUIST
Philips Semiconductors
Albuquerque, NM USA
Email: wallqst@abqn02.scs.philips.com

10.1 INTRODUCTION

I_{DDQ} is a strong indicator of inherent weaknesses and defects in CMOS circuitry. However, once this is accepted a new journey must begin. The goal is to actually implement I_{DDQ} testing in production, so that the manufacturer and user can reap the benefits of reduced failure rates. Unfortunately this is much harder than it would appear. The question of production I_{DDQ} testing is not simply a lack of hardware. In fact, there are many resources available to the test engineer for this task. The challenge lies in how to determine which solution effectively balances such things as accuracy, speed, and resolution. Each of the many methods available for measuring I_{DDQ} contains inherent advantages and disadvantages. It is up to the informed engineer to determine which best suits the needs for device testing. This chapter will attempt to describe the many types of instrumentation available to the test engineer. The pros and cons of each method are described. It is hoped that this will prove useful in the selection of a proper instrumentation setup.

10.2 THE INSTRUMENTATION PUZZLE

The problem is not that I_{DDQ} is difficult to measure. In fact, setting up a measurement can be as simple as connecting a precision ammeter to a device's power supply on the bench. Ammeters measuring down in the pA range are readily available.

The value of such setups is seen in low-level device characterization. For SSI devices, this is perfectly valid. However, in order to properly design a production test screen for VLSI devices, it is important that the device characteristics be well-known. This often requires that the device be placed into several known states, each of which tests a number of potential defects. This device testing does not come easily to the bench setup, but rather requires that an ATE system be put into place to drive the device. Only then can true I_{DDQ} coverage be attempted, and only then can the production tests be implemented.

The basic problem is that often the ATE systems of choice do not provide the high-throughput I_{DDQ} testing desired. It is sometimes necessary to augment the tester's capabilities in this area, either with external circuitry or by adding optional boards to the tester itself. Complex circuitry can be designed into the loadboard for characterization purposes, but the reality is that simpler designs are required in production. They are much simpler to maintain, and are less likely to be damaged.

I_{DDQ} monitors can be classified according to their location. Each has its own inherent strengths and weaknesses. There are three different locales where monitors can reside: inside the tester, on the loadboard, or inside the device under test (DUT).

Tester based monitors can be quite elaborate in instrumentation, since consumed area is not at as high a premium as the other locations. As a general rule, tester-based monitors tend to be slower, but can provide more functionality and greater accuracy. As discussed later, these monitors can (and even must) provide this functionality because of their location.

Loadboard monitors are presently designed with discrete components. They are typically located as close to the DUT socket or probe card as possible. Because they are close to the DUT, they have some speed advantages over the tester monitors. However, they suffer from tighter area constraints. These monitors attempt to take the best characteristics from the other two monitor types.

Monitors located inside the DUT are often called **Built-In Current Sensors** (BICS). They can target specific internal circuitry within the device, avoiding areas with high dc currents or high current switching. Typically, they are only designed to return a pass/fail reading, since consumed area must be kept to a minimum. Because of their location, they can offer speeds in the MHz range. However, such flexibility comes at a price, namely silicon area consumption and a potential reduction in the operating speed of the device. Since designs of BICS are often proprietary in nature, and can often vary significantly between implementations, these monitors will not be covered in this chapter. Papers have also been written on techniques and strategies for placing the BICS sensors appropriately on the DUT landscape to provide the best defect coverage [38]. This chapter will instead concentrate on how to implement I_{DDQ} testing, assuming the DUT must be tested externally.

10.3 THINGS TO CONSIDER

When selecting a monitor, several factors must be considered. Each is important; it is the job of the informed engineer to decide how to weigh them to decide on the type of monitor to use.

Perhaps the most desired attribute of a monitor is **accuracy**. The results obtained from the monitor must be guaranteed to be within a reasonable margin of error from the true value. Notice that the term "reasonable" is used; a monitor which is required to set a limit of 100 μA does not necessarily need to have a 1 nA accuracy. Relaxation in accuracy, when appropriate, can allow other important factors to take a larger role.

Another important factor is **sensitivity** or **resolution**. The monitor should be able to discern small changes in current. Sometimes the difference between good and bad devices is a microampere; in this case it is essential that the monitor of choice be able to have enough resolution to make that distinction. For less constraining cases, where current limits may be set at 100 μA or so, specifications may be relaxed but not forgotten.

For production testing, it is also required that the monitor be **repeatable**. It should give the same results the first time and every time. Readings should not vary from system to system. If temperature extremes will be encountered, the monitor should work at all temperatures with repeatable results.

In the production environment, **speed** is a significant concern. If the DUT is to have several measurements taken, then the monitor must be as quick as possible in order to avoid impacts on throughput. Currently available monitors allow I_{DDQ} testing in the upper kilohertz range, making I_{DDQ} testing economical, particularly when its benefits are considered. Since production testing is concerned mainly with good or bad devices, the pass/fail types of monitors are more useful than value monitors which take absolute measurements of current.

Of course, a monitor is of practically no use unless it is **easy to use**. The monitor should be relatively easy to install or design into a test setup. It is of even greater benefit if software exists to aid in the setup and use of the monitor. A turnkey solution is highly valued in the production environment, as it takes less resources to maintain.

The monitor should be **flexible** with application across many different products and hardware setups. The monitor should be cheap enough to build and replicate. Having a universal I_{DDQ} monitor (not one that is specifically designed for one of many products) offers many benefits. However, what is best for ICs at one end of the spectrum may not be particularly useful for those at the other end. Sometimes the requirements of one device family may be so at odds with another that one cannot help but use two different monitors. All of the above factors are important. Predictably enough, however, they tend to be at odds with one another. A good example of this is speed and sensitivity, which often form an inverse relationship; the faster a monitor is run, the less sensitive it becomes. And, as if to add insult to injury, several additional factors act in concert to eat away at the ability to operate an I_{DDQ} monitor at optimum performance.

10.4 DEGRADING FACTORS

I_{DDQ} measurements are particularly sensitive to environmental concerns due mainly to the small currents and high resolution that is demanded. Many individual factors can degrade the operation of an I_{DDQ} monitor. Usually, with adequate compensation, the monitor can take these factors into account, but only with a sacrifice in operating speed.

A serious detriment to I_{DDQ} measurement is **noise**. Noise can appear from any number of sources, but its effect is the same: measurements which jump around significantly. Some origins of noise may include thermal noise, cross-coupled signals, improper grounding, or inherent noise in the DUT itself. Since I_{DDQ} monitors can sometimes be required to differentiate currents on the order of 1-5μA, noise becomes a great concern. Low-pass filters can be helpful in suppressing excess noise, but again these filters tend to slow down the measurement circuitry.

Another factor that affects I_{DDQ} monitor performance is **capacitance**. Since high frequency devices tend to consume large currents with significant loading on the power supply, it is often necessary to bypass the supply pins on the device with extra capacitance. The magnitude of this capacitor can be in the $10\mu F$ range. It is impossible to simply disconnect the capacitor while the device is in operation; if it remains connected, it will most likely interfere with the I_{DDQ} monitor, causing oscillations or highly degraded response. The capacitor may in fact be so large as to render the monitor completely useless. With clever compensation schemes in the monitor, as well as innovative current measurement methods, the monitor can compensate for large capacitors effectively.

This often comes at the cost of reduced operating speed. Therefore, bypass capacitance must be kept to a minimum or a monitor which can sufficiently deal with it must be found.

Another important concern is the **distance** between the monitor and the DUT. BICSs mentioned earlier in this chapter can often run at speeds in the MHz range; this is a strong indication that the closer the monitor is to the circuit it is measuring, the faster it will be able to run. Shorter distances reduce the risk of noise entering the system. Monitors mounted inside the test head often need to compensate in some way for the increased noise pick up.

It is important that the I_{DD} **characteristics** of the device be known. This can determine exactly how fast - or even how slow - the I_{DDQ} measurement can be made. Devices with dynamic nodes may remain stable for a certain period of time, after which the node starts losing too much charge and internal gates start drawing excessive current. The measurement is also limited by the transient portion of the curve. All transient currents must be dissipated (at least beyond the resolution of the monitor) before measurements can be performed. In addition, states in which the DUT is drawing excessive current must be avoided; sometimes test modes are added to shut off parts of the DUT which prevent I_{DDQ} testing from being performed. The DUT itself, then, plays a very important role in determining the type of monitor to implement.

One of the more esoteric influences on monitor speed or operation is the **tester** itself. The current loading on all output pins must be disabled during measurement, all active pullup or pulldown resistors must be disabled if drawing current, and all input pins on the device must be driven to an appropriate state. On some test systems, the tester comparators themselves can be a source of concern. It was observed during testing I_{SSQ}[1] on devices on one test system where current was actually flowing into the ground pin [327]! In this case, ECL comparators draw current from each output pin to a negative potential, thus seriously impacting the measurements. In addition to leakage, tester channels can have associated capacitance which can affect how fast output drivers can switch logic values.

One final influence on the circuitry is the I_{DDQ} **limit**. The desired limit will have a significant impact on the type of monitor that is needed. If a relatively high test limit is desired, say in the 100μA range, then the monitor constraints can be relaxed to allow it to run slightly faster. However, if the monitor is to

[1]When measurement is made at the ground rather than the V_{DD} node the steady state supply current is denoted by I_{SSQ}.

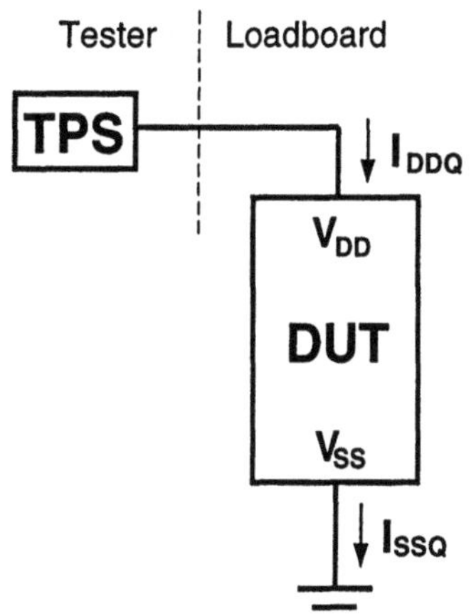

Figure 10.1 I_{DDQ} vs. I_{SSQ} measurement

be applied across a product line with wide variations in I_{DDQ} test limits, then the monitor constraints can become quite stringent. In fact, I_{DDQ} limits may vary with different tests on the same device, or even vector to vector. Such demands on the monitor require a high degree of flexibility.

10.5 I_{DDQ} VS. I_{SSQ} MEASUREMENT

In theory, it should not matter whether current going into the V_{DD} pin of the DUT (I_{DDQ}) or current leaving the V_{SS} pin (I_{SSQ}) is measured (see Figure 10.1). Defects in the core logic will elevate equally the values for I_{DDQ} or I_{SSQ}. Certain defects in the I/O or protection circuits will specifically elevate I_{DDQ} or I_{SSQ}. These defects are typically detected during I/O pin testing and are not an issue for I_{DDQ}/I_{SSQ} discrimination. In most cases, the magnitude of the two currents should be the same. In reality, however, there are other circumstances which can make one method desirable over the other. Some situations may actually prevent the use of one of the methods. One of the most notable examples of this are pullup or pulldown resistors. Internal pullup resistors on input pins will draw current from V_{DD} when the pin is driven low, but will not affect I_{SSQ}. Internal pulldown resistors on input pins will force current through V_{SS} when the pin is driven high, but will not affect I_{DDQ}. Unless the input level of the device remains in a state which does not activate the resistor, the type of test to be done has already been decided.

When testing multiple sites on a probing system, it is important to recognize the effects of common substrates. Two different die, while electrically isolated on the V_{DD} node, may be connected with the V_{SS} node through the substrate. This will prevent independent measurement of the I_{SSQ} current from each device.

A strong argument for the use of I_{SSQ} testing is bypass capacitance. While capacitance on the V_{DD} pin can be quite high, there is typically no capacitance on the V_{SS} pin of the device. This in effect removes the capacitor from affecting measurements, significantly reducing the oscillations present in the monitor. In fact, large bypass capacitance on the V_{DD} pin can sometimes provide more stable results in I_{SSQ} testing.

Perhaps the strongest case made against I_{SSQ} testing is the interruption of the ground on the DUT. Since the monitor must be introduced between the V_{SS} of the device and ground, the device no longer has as solid a ground as may be desired. This can increase ground bounce on the device, resulting in loss of device state or decreased device performance [327]. For mixed-signal ICs, this may be completely unacceptable. Often, given a choice, test engineers prefer to use the V_{DD} pin instead because of these factors.

It should be noted, however, that I_{SSQ} testing has been used successfully for many years, even though I_{DDQ} is the more "popular" of the two. Some built-in sensors are in fact I_{SSQ} monitors [189]. It is really left up to the test engineer which type of monitor to choose, given the requirements of the DUT. In the description of the monitors to follow, only I_{DDQ} types will be discussed, although the techniques can usually be applied to I_{SSQ} with a little modification.

10.6 HISTORICAL METHODS OF MEASURING I_{DDQ}

I_{DDQ} testing has been around for a long time. This section deals with some of the methods that have been used in the past. The drawbacks and advantages of each technique are discussed. This is intended as background information for the next section, which will discuss some of the recent advancements and innovations in I_{DDQ} instrumentation. In fact, advanced monitors sometimes borrow from some of the older techniques and apply some new improvements. In any case, it is useful to review the simpler methods, if for no other reason than

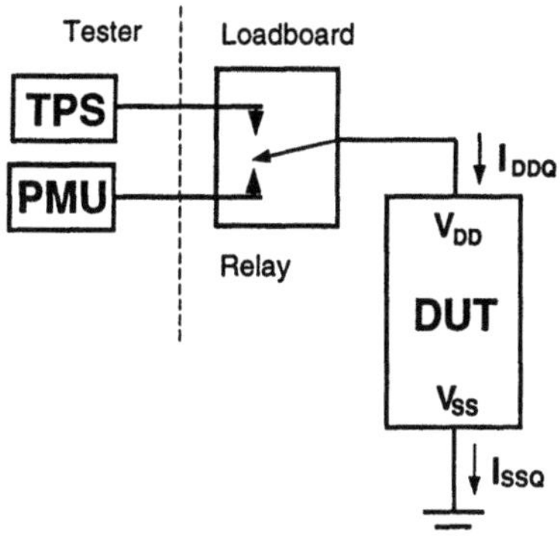

Figure 10.2 Simple PMU setup for I_{DDQ} measurement

to appreciate the frustration with using traditional methodology for testing I_{DDQ} in production.

10.6.1 The Parametric Measurement Unit (PMU) [154]

Nearly every test system in place today contains some type of PMU. This is a flexible precision dc measurement tool which can be used to measure voltage or current in one of several ranges. It can be used normally to measure pin input leakage, I/O voltage levels, or output resistance of drivers. In this case, the PMU can be used as a simple and accurate measurement tool for I_{DDQ} testing (see Figure 10.2). It can measure low nA currents.

Typically, the power supply drives the V_{DD} pin of the device which is placed into a known logic state; PMUs usually cannot offer the quick current response required during transient phases. When the desired state is achieved where I_{DDQ} is to be measured, all clocks are stopped; the PMU is switched into the V_{DD} node while the tester power supply (TPS) is disconnected, and the current is measured by the PMU. The process is then reversed to replace the device under power supply control so that it can move to the next I_{DDQ} state.

While this is a perfectly valid technique and easy to implement, it does have several drawbacks:

- Power supply capacitance may be of concern. Large capacitance may cause oscillation in the PMU that can be circumvented by switching out any large capacitance with the power supply. Theoretically, the part is already in a static state so that switching out the capacitor should have little effect.
- PMUs are designed to make precision dc measurements. This takes longer than a simple comparison. This is desired in engineering modes, but in production testing a measurement may take more time than desired. PMU I_{DDQ} test rates are on the order of 5 - 10 Hz.
- Each reading requires stopping the clock on the device and performing setup and measurements. If the device is not fully static (but still I_{DDQ} testable), this may cause loss of state in the device.
- Because of the amount of time required for a single measurement, only a few select measurements may be taken on the device in order to avoid impact on test time. If the DUT has many transistors, a few measurements may not be sufficient to guarantee high I_{DDQ} fault coverage. The point should be made, however, that 1 to 10 measurements are better than none.

The unfortunate conclusion is that if a high defect coverage is desired, PMUs will almost certainly not get you there. However, if 1 to 10 measurements are adequate, then this solution may be the simplest and most readily available one.

10.6.2 Resistor Method [154]

The current drawn by the device can be easily converted into a voltage by inserting a resistor into the path of the power supply (Figure 10.3). This is more easily handled by the tester equipment. The resistor can be made relatively small and the voltage drop across it can be amplified to the point where standard tester instrumentation can make a measurement. While this technique is very simple, it has drawbacks.

If the measurement is to be made quickly, then the resistor itself must be made to be part of the power supply path (otherwise, you might just as well use a PMU). Therefore, any high-current transient states that occur when device inputs change state will be forced through the resistor. This has serious potential for causing loss of device state, as V_{DD} drops instantaneously well below the input levels on the DUT. Also, input protection circuit can be damaged if V_{DD} voltage drops below the device input voltage level. Clearly, this is undesirable.

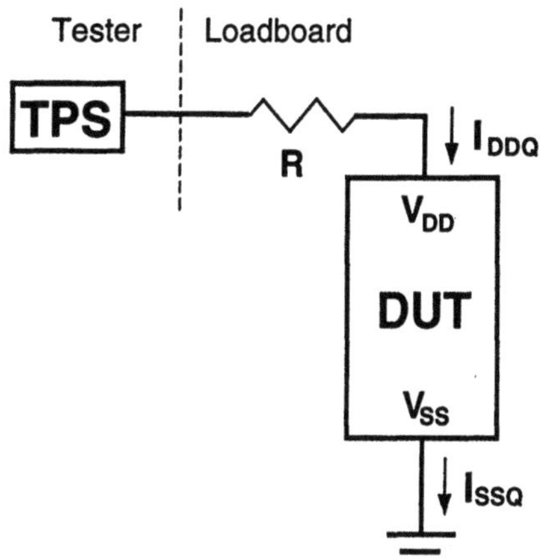

Figure 10.3 The resistor method

This can be alleviated by lowering the resistance value. However, in doing so, the amplification necessary to make a decent voltage measurement becomes intractable. Also, since noise is directly amplified along with the power supply current, the measurement becomes more unstable.

10.6.3 Resistor-Diode Method [154]

A simple way to avoid the above situation is to place a diode in parallel with the resistor (Figure 10.4). When large currents are present, the diode turns on and provides a lower resistance path between the power supply and DUT. This requires that the drop across the resistor during testing must be lower than the diode's turn-on voltage, or diode leakage may affect measurements. While the voltage drop from the true V_{DD} is limited to a diode drop, this may still be unacceptable. During normal operating mode, there will be a constant drop across the diode, which may affect operating speed of the device. Diode leakage must also be taken into consideration.

10.6.4 FET-Resistor Method [154]

The FET-Resistor method replaces the diode with a low-resistance FET switch (Figure 10.5). This can reduce the resistance in the V_{DD} line significantly. Mod-

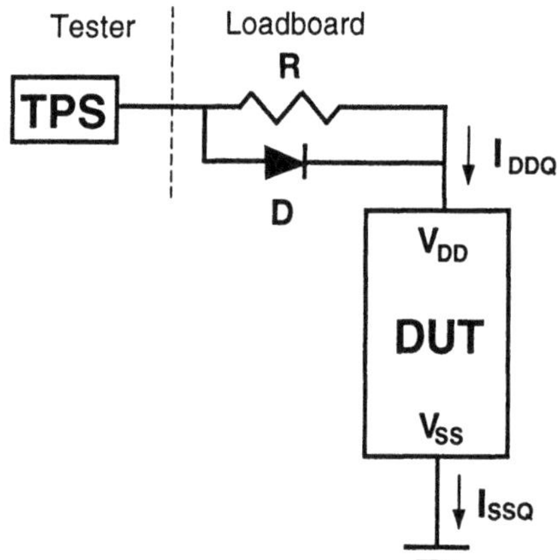

Figure 10.4 The resistor-diode method

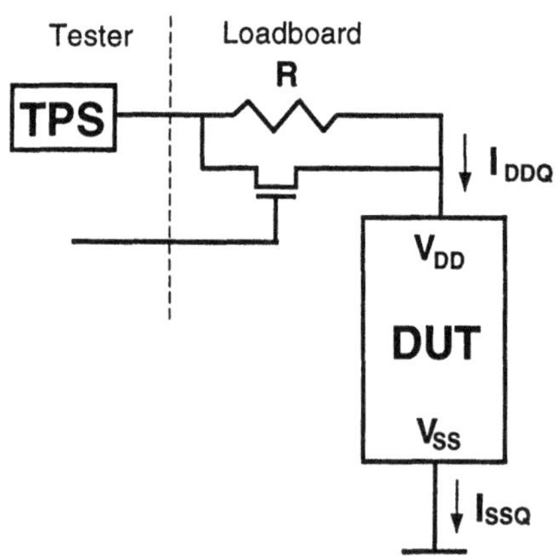

Figure 10.5 The FET-resistor method

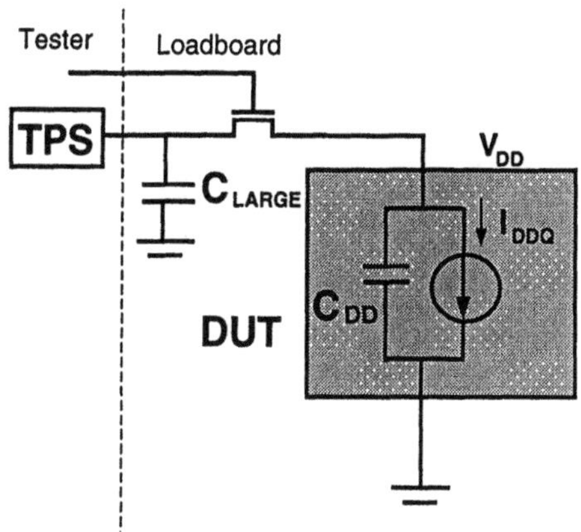

Figure 10.6 Model of Keating-Meyer technique

ern power FET transistors can achieve resistance below 1 ohm when activated, yet still provide good off-state leakage characteristics. When measurements are to be taken, the FET switch is turned off and the voltage drop across the resistor begins to build up. While this method offers some improvement, it has limiting factors. First, by switching from a lower to a higher resistance, there will be a new time constant imposed. The circuit will need to settle down based on the RC circuit formed by the higher resistance and any bypass and stray capacitances. Secondly, the FET switch can inject charge onto the V_{DD} node, which can affect some measurements. Lastly, the drop across the resistor is no longer limited, so that the V_{DD} voltage at the device can potentially drop dangerously low during measurement periods.

10.6.5 Keating-Meyer Method [154]

At the 1987 International Test Conference, Keating and Meyer proposed a new technique for I_{DDQ} measurement which has shown great promise [154]. It uses the capacitance of the DUT's power supply node (C_{DD}) to measure I_{DDQ} quickly. It removes the resistor from the FET-resistor method, allowing the V_{DD} node of the device to float (Figure 10.6).

Figure 10.7 shows typical waveforms from a Keating-Meyer circuit. During periods of DUT activity, such as when inputs are switching, the excessive transient currents are drawn from the power supply through the FET switch. Once these transients have dissipated, the FET switch is deactivated, so that the

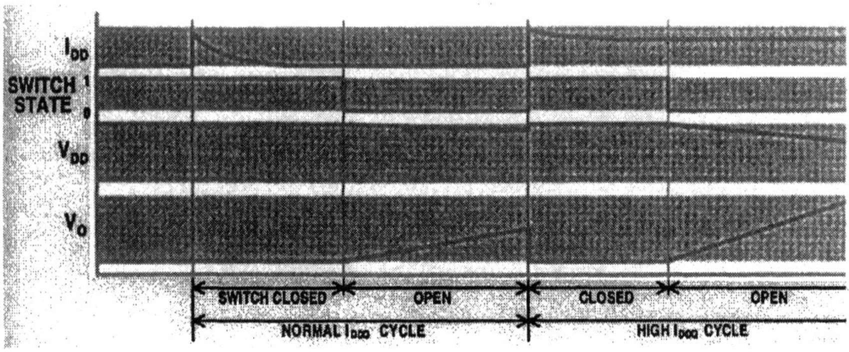

Figure 10.7 Typical Keating-Meyer waveforms

V_{DD} node of the DUT is floating. The current required by the device is then drawn from its own capacitance (and any extra capacitance which is still connected to the DUT). This total capacitance is lumped into a single term C_{DD}. As current is drawn from C_{DD}, the voltage at the V_{DD} node decreases. The relationship between I_{DDQ} and voltage drop is

$$I_{DDQ} = C_{DD}\frac{dV_{DD}}{dt} \tag{10.1}$$

This equation shows that, at least for small drops in V_{DD}, the voltage decrease will be constant for a constant I_{DDQ}. Thus, after a predetermined period of time, the voltage drop from nominal value can be measured and converted directly into current. Larger voltage drops would be indicative of larger currents, and therefore defective devices (Figure 10.8). A similar situation applies at the V_{SS} node of the DUT, where capacitance C_{SS} is seen from the V_{SS} pin to the V_{DD} pin (Figure 10.9). In this case, the DUT current I_{SSQ} causes the voltage at V_{SS} to rise when the switch is opened. While this method has proved to be very fast, it has a number of fundamental downfalls:

- The V_{DD} node of the device is floating. If not properly monitored or limited, this can lead to loss of state in the device.
- In order to discern small currents quickly, C_{DD} must necessarily be as small as possible. This means that bypass capacitance may severely limit the speed of the method.

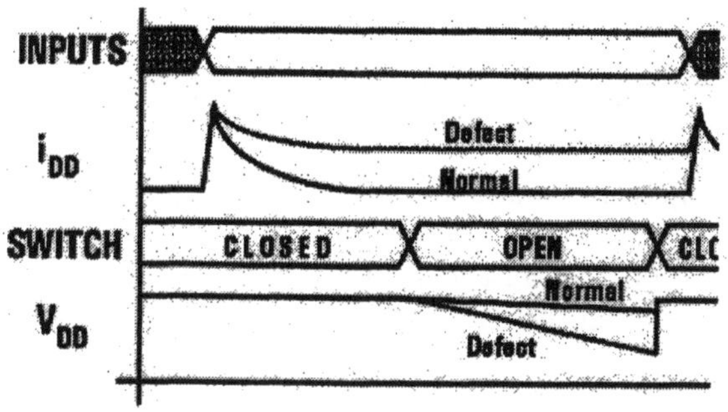

Figure 10.8 Keating-Meyer example showing normal and defective devices

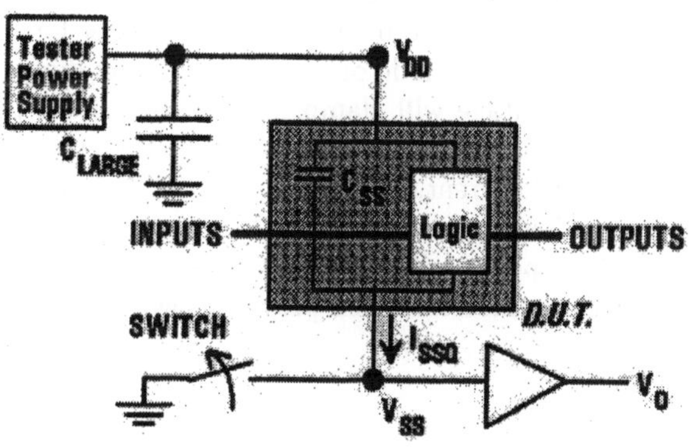

Figure 10.9 I_{SSQ} model for the Keating-Meyer method

- The method requires precise discharge times in order to be accurate. This includes any delay in switching of the FET transistor, any delays in the amplifiers, and strict control over timing resources.
- The method requires knowledge of C_{DD}, which can be strongly affected by the device's own capacitance. C_{DD} can vary between devices or even on a vector-to-vector basis, depending on the device in question [327].
- The charge injection caused by switching off the FET switch must be known or compensated for. This typically causes a constant voltage offset, depending on the voltage swing on the FET gate and the ratio of capacitances between the FET and C_{DD}.
- Any preexisting voltage drop across the FET switch must be accounted for. Such voltage drops are the result of the dc current flowing through the switch at the time it is opened.

It may seem that this method has too much going against it, but all of the above effects can actually be compensated for. A circuit operating on the next generation of this principle will be described later in this chapter.

10.6.6 Diode-Capacitor Method

This method was pioneered by Sandia Labs in the late 1970s for low gate count devices [269]. When the current is to be measured, the tester forces the desired I_{DDQ} current limit into the V_{DD} node of the device. Depending on the actual device current, the V_{DD} voltage will clamp high or low to the power supply. Then, a simple tester comparator can decide whether the current is higher or lower than the PMU-forced current. While this method is simple in practice, in order to work for larger devices, low resistance bypass is required. In addition, the method does result in potentially large voltage swings on the V_{DD} node of the DUT. However, this method has been proven to work with 50μA limits at 100 kHz.

10.7 ADVANCED TESTER-BASED INSTRUMENTATION

Monitors in this class are designed to offer the benefits of I_{DDQ} measurement from inside the test head offering several advantages to the user. The most obvious of these is the turnkey nature of the solution; typically, the user can program the instrument using built-in tester commands and get reasonably quick measurements off the system. In addition, because it is located inside the test head, area consumed by the monitor is not as important an issue, and many more enhancements can be added. Such monitors can offer highly accurate readings, and are typically geared to return real measurements of current back to the test program for comparison. A secondary advantage is that since hardware is located inside the test head, the monitor can be used across many different hardware setups and product types, assuming the monitor's usage constraints are met.

On the other hand, these types of instruments have disadvantages. One drawback is their location; they are relatively far away from the DUT and, as such, must contend with cross-coupled signals, noise, and capacitive effects due to that environment. Because of this, these types of monitors tend to be slow, in an effort to cancel out the noise and oscillations.

Of the three types of monitors, this is arguably the most desirable. If a measurement solution is offered by an ATE manufacturer, then there is less effort that needs to be put into the design and verification process. However, the most potent argument to date against the currently available options is that most are just not fast and sensitive enough for achieving high defect coverage for a VLSI device.

In addition to the generic dc measurement tools available on a test system, the company manufacturing it may offer as a separate feature a circuit specially designed to make power supply current measurements, or even I_{DDQ} measurement in particular. For example, Advantest has an optional Bit Power Supply Current Measurement Option for their T3340 and T3320 test systems [7]. This monitor is capable of being programmed to take power supply current measurements on a per-vector basis while supplying the device current during transient phases. It can operate at speeds of 15 kHz at an accuracy of 1μA. As another example, Teradyne has proposed a new technique for performing I_{DDQ} measurements in the test head with good sensitivity results [148].

The final barriers between widespread creation of tester-based I_{DDQ} instrumentation seems twofold. While people seeking to implement I_{DDQ} test desire a turnkey solution, a satisfactory one does not exist, so they go about designing one. Hence, the tester manufacturer never enters the loop. Add on circuits could be relatively cheap relative to the rest of the test system, but may take up some tester resources. The second hurdle seems to be the proliferation of loadboard monitors themselves; they often have better characteristics than tester-based schemes at a lower cost.

10.8 LOADBOARD-MOUNTED INSTRUMENTATION

Often, the only choice left to the engineer is to use a loadboard mounted monitor. Mounting additional hardware onto the loadboard offers advantages over other I_{DDQ} measurement schemes. Loadboard monitors can offer the best of both worlds by providing a single monitor to test many different parts, while offering speed and sensitivity.

10.8.1 Advantages

Loadboard monitors are often the most cost effective monitors, costing very little in parts to create and maintain relative to the other schemes. Monitors can range from a single chip to a collection of discrete components, all of which are easy to replace on a loadboard.

The monitor is also very flexible in location. It can be placed as close to or as far away from the DUT as possible. If it is made from surface mount components, it is even possible to place it directly underneath the test socket. Monitors benefit greatly from being designed directly into the loadboard; this can offer much greater control over the environment the monitor has to deal with.

Because the monitor is designed into the loadboard, it can be used to test many devices. This argument also applies to wafer testing. The monitor can be designed into the probe card or tester-prober interface loadboards. The reusability can be extended further if the test setup consists of a generic "motherboard" into which many different loadboards (or probe cards) are installed. In this case, installing the monitor onto the motherboard allows the monitor to be used more generically. This can degrade the monitor's performance, but if cost and reusability are a concern, then the reduced performance may be acceptable.

The speed of loadboard monitors has been improved significantly over the past few years. Loadboard monitors can now approach the speeds of BICS monitors, although they cannot often reach the sensitivity of a tester-based measurement, such as a PMU. This is due mainly to the fact that the extra circuitry required to fine tune the measurement to that degree cannot be reliably maintained on a loadboard.

10.8.2 Disadvantages

Of course, the loadboard monitor does not come without its own set of challenges. Perhaps the most threatening of these is the very location of the monitor itself. Because it is designed into the loadboard, it runs significant risk of being either mechanically or electrically damaged when the loadboard is handled. In setups which require frequent docking and undocking of test hardware and frequent board swapping, this risk becomes even greater. It is therefore essential that some kind of regular circuit check or preventive maintenance be performed on the board on a regular basis. This raises the effective cost of the board, particularly if the monitor is not sufficiently protected from damage and must have components replaced.

Such concerns can be minimized with careful design. Low profile components will be less likely to be mechanically damaged. If the monitor consists of transistors or resistors, it is preferable to use surface mount components over leaded components for this reason. Electrically, the circuit must be well protected from electrostatic discharge (ESD) events. It is also a good idea to have the tester perform a sanity check on the circuit prior to device testing, which will determine how well the monitor is performing. Failure to pass this loadboard check should prevent the test program from proceeding and alert the operator to the condition.

Another disadvantage to the loadboard monitor is the need to duplicate it for additional setups. While this may not be a significant concern for volume production where loadboards may be used to test many thousands of units, production lines where lots of parts are small may incur a larger cost. One such case would be an ASIC supplier, where different loadboard setups may be required for a large number of low volume products. In such a case, a loadboard monitor needs to be redesigned into each loadboard. This may turn out to be a costly situation, but may pay for itself if the test time saved by a faster loadboard monitor can counteract the design cost.

10.8.3 Quality Test Action Group (QTAG)

The proliferation of loadboard based monitors led to the creation of an industry-wide committee of test professionals called the Quality Test Action Group (QTAG) [15]. The goal of this organization, which was founded by Keith Baker of Philips Labs, at the 1993 International Test Conference in Baltimore, Maryland is the eventual standardization of loadboard I_{DDQ} or I_{SSQ} monitors as well as other topics related to I_{DDQ}. The committee consists of members from the industry test community, automatic test equipment (ATE) vendors, and other experts in the testing field.

The standard for monitors seeks to integrate all aspects of using the monitor in production. Compatible monitors would not only have a standardized pinout, but would have included software information [17] which would be used by the tester compiler for pattern generation and monitor control. The monitors would be designed to a variety of device needs by IC manufacturers.

Currently, the proposed standard provides for three monitor classes [15].

Class 1 monitors are single chip monitors designed into a single 8-pin SSOP package. This is a highly desirable configuration because of the flexibility to place the monitor virtually anywhere on the board, including as close to the DUT as possible. The package has a standardized pinout, so that the engineer can pick and choose from a variety of available monitors and not worry about design considerations. With the software format, the introduction and testing of the monitors become nearly "plug & play" [17]. Two of the Class 1 monitors currently under development are described later in this section.

Class 2 monitors are basically one step up from the Class 1. In theory, these monitors can be used in either production or engineering mode. They can take up more space on the loadboard with more available interface pins. They can be made of discrete components, as long as the total area falls within certain constraints. One of the more popular Class 2 monitors called QuiC-Mon will be discussed later in this chapter.

Class 3 monitors are larger and capable of more advanced engineering tasks. The footprint allowed is much larger. While these are unwieldy for production testing, they serve an important role in device characterization. They can offer highly accurate measurements of I_{DDQ} for use in histograms and limit calculations.

10.8.4 Examples of Loadboard Monitors

IDUNA. IDUNA[2] was initially created by Philips Semiconductors as an on-chip sensor for SRAMs [15]. Since then, a more recent version (IDUNA-2) was created as a stand-alone IC manufactured to be compatible with the provisional QTAG standard. IDUNA operates on a current-mirror principle. Current drawn by the DUT is replicated and sent to a thresholding circuit designed to send a pass/fail value to the tester. In addition to the I_{DDQ} measurement circuitry, IDUNA-2 contains a special bypass transistor designed to supply the DUT during transient current stages.

OCIMU. At the 1994 International Test Conference, a group working in conjunction with Alcatel-Bell in Belgium presented results on an off-chip measurement circuit designed with the QTAG standards in mind [178]. This circuit, dubbed Off-Chip I_{DDQ} Current Measurement Unit (OCIMU), is capable of driving significant capacitive loads with its Stabilized Voltage Drop technique. This circuit is also a single-chip implementation.

QuiC-Mon. QuiC-Mon is an advanced adaptation based on the Keating-Meyer method [16, 328]. Instead of using a standard amplifier to measure voltage drops [329], QuiC-Mon uses a "derivative amplifier" as a first stage. QuiC-Mon is composed of several discrete components, but has been successfully designed into a small area (see Figure 10.10). Figure 10.11 shows a simplified model of QuiC-Mon during I_{DDQ} measurement. I_{DDQ} is drawn from the two capacitors C_{DD} and C1. This creates a current divider with part of the current being drawn from the amplifier and part from the device capacitance. With C1 $\gg C_{DD}$, i.e., at least an order of magnitude, the transfer function is:

$$V_S = \left(1 + \frac{R_3}{R_2}\right) \times R_1 \times I_{DDQ} \qquad (10.2)$$

Figure 10.12 shows the full schematic for the I_{DDQ} version of QuiC-Mon. Power MOSFET transistor M2 is an Si9947DY from Siliconix, which has a very low on-resistance (0.3 Ω) and moderate gate capacitance (1 nF). This transistor is used as a circuit bypass providing a high current path for DUT switching during normal functional testing. This switch cannot be used as the I_{DDQ} switch because testing revealed a large amount of oscillation when the transistor was switched off. Thus, transistor M1 is used in I_{DDQ} testing. It has characteristics (R_{ON} = 4Ω, C_G = 80 pF) to which the QuiC-Mon amplifiers respond more favorably. Diode D_1 prevents large

[2]Name of a Norse Goddess who guards the apples of eternal life.

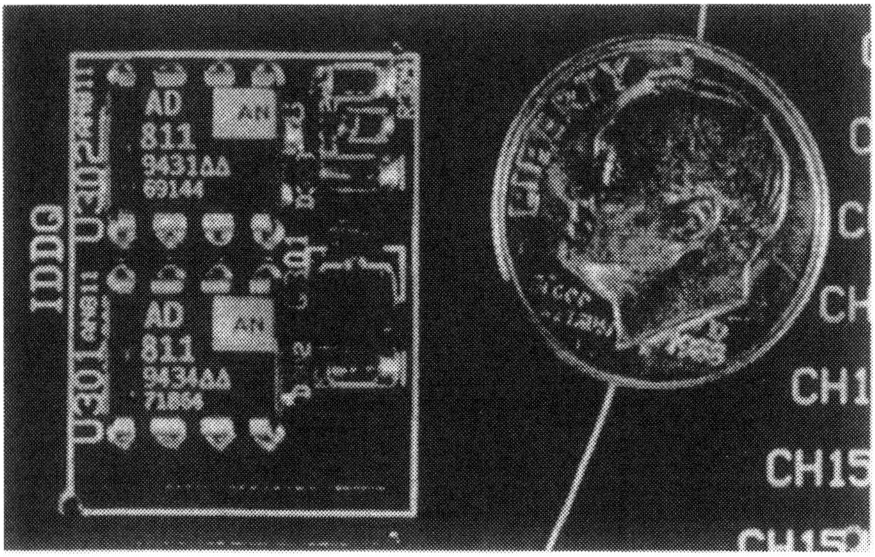

Figure 10.10 Photograph of QuiC-Mon v5.0 implementation

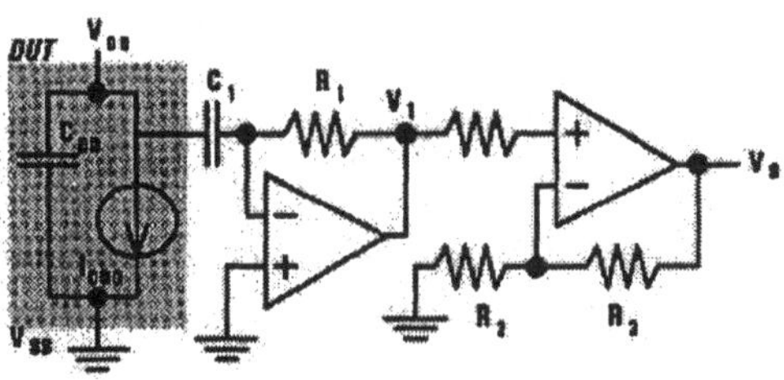

Figure 10.11 Model for QuiC-Mon v5.0

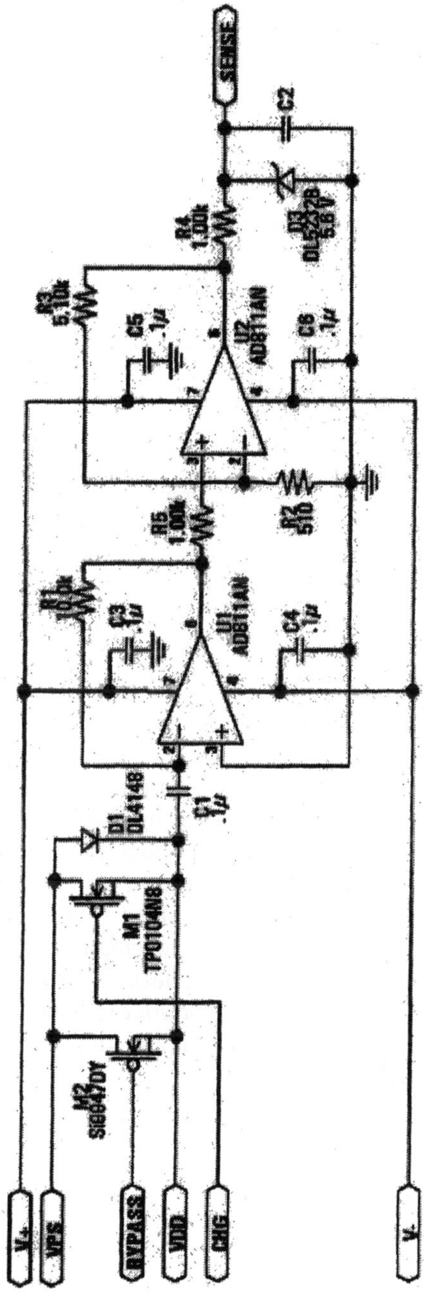

Figure 10.12 Complete schematic for QuiC-Mon v5.0

current transients from sending the DUT into latchup by holding V_{DD} to a diode drop below the supply level.

Measurements are begun by switching off M_1, with M_2 in its inactive state. This forces I_{DDQ} through the amplifier. The gain as shown in the schematic is about 110 $\frac{mV}{\mu A}$.

Resistor R_5 is essential, acting both as a low pass filter between stages and as an impedance matching element for the second-stage amplifier. Without it, serious oscillations prevent any meaningful measurements. The value of 1.0 kΩ was chosen primarily to match the inverting node's resistance and has little effect on operating speed.

The network formed by R_4, D_3, and C_2 clamps and filters the output. Clamping was added to protect tester channels from voltages outside of their operating range; a surface mount DL5232B provides a nice 5.6 V clamp, keeping the output in normal logic levels. The low pass filter is essential in reducing any noise the circuit picks up; this can be substantial, particularly if V_{DD} trace length is long or if the DUT exhibits current noise. The value of C_2 should be experimentally determined to provide optimum balance between noise reduction and desired circuit speed.

10.9 CONCLUSION

There are many ways that I_{DDQ} measurements can be taken from engineering to production. The implementation chosen by the user will depend upon several factors, namely desired performance specifications and design freedom. Each of the three classes of monitors offer their own set of advantages and disadvantages. The characteristics of each make them tailored for different sets of situations. The unfortunate reality is that the engineer is often forced into taking matters into his or her own hands, so to speak, in creating a specific load-board monitor. Choosing the right monitor for your specific situation depends heavily on the needs of the device under test and the speed and sensitivity desired in production testing.

REFERENCES

[1] M. Abramovici and M. A. Breuer, "Multiple Fault Diagnosis of MOS Combinational Networks," *IEEE Transactions on Computers*, Vol. C-29, No. 6, pp. 451-460, June 1980.

[2] M. Abramovici and M. A. Breuer, "Fault Diagnosis in Synchronous Sequential Circuits Based on an Effect-Cause Analysis," *IEEE Transactions on Computers*, Vol. C-31, No. 12, pp. 1165-1172, Dec. 1982.

[3] M. Abramovici, M. A. Breuer and A. D. Friedman, *Digital Systems Testing and Testable Design.* Computer Science Press, 1990.

[4] M. Abramovici, J. J. Kulikowski and R. K. Roy, "The Best Flip-Flops to Scan," *Proceedings of IEEE International Test Conference*, 1991, pp. 166-173.

[5] J. M. Acken, "Testing for Bridging [Shorts] in CMOS Circuits," *Proceedings of ACM/IEEE Design Automation Conference*, 1983, pp. 717-718.

[6] J. M. Acken, *Deriving Accurate Fault Models.* PhD Thesis, Department of Electrical Engineering, Stanford University, 1988.

[7] Advantest Corporation, Advantest T3381/T3342 Bit Current Measurement Option Reference Manual, 1987.

[8] V. D. Agrawal, "Sampling Techniques for Determining Fault Coverage in LSI Circuits," *Journal of Digital Systems*, Vol. 5, No. 3, pp. 189-202, Fall 1981.

[9] V. D. Agrawal, S. C. Seth and P. Agrawal, "Fault Coverage Requirements in Production Testing of LSI Circuits," *IEEE Journal of Solid State Circuits*, Vol. SC-17, pp. 57-61, 1982.

[10] A. V. Aho, J. E. Hopcroft and J. Ullman, *The Design and Analysis of Computer Algorithms.* Addison-Wesley, Reading, 1974.

[11] R. C. Aitken, "Fault Location with Current Monitoring," *Proceedings of IEEE International Test Conference*, 1991, pp. 623-632.

[12] R. C. Aitken, "A Comparison of Defect Models for Fault Location with IDDQ Measurements," *Proceedings of IEEE International Test Conference*, 1992, pp. 778-787

[13] R. C. Aitken, "Diagnosis of Leakage Faults with IDDQ," *Journal of Electronic Testing: Theory and Applications*, Vol. 3, No. 4, pp. 367-375, Dec. 1992.

[14] Y. Aizenbud, et al., "AC Test Quality: Beyond Transition Fault Coverage," *Proceedings of IEEE International Test Conference*, 1992, pp. 568-577.

[15] K. Baker, "QTAG: A Standard for Text Fixture Based I_{DDQ}/I_{SSQ} Monitors," *Proceedings of IEEE International Test Conference*, pp. 194-202, 1994. Further QTAG info on the World Wide Web at ftp://ftp.polito.it/pub/etttc/qtag/.

[16] K. Baker et al., "Development of a Class 1 QTAG Monitor," *Proceedings of IEEE International Test Conference*, 1994, pp. 213-222.

[17] K. Baker et al., "Plug & Play I_{DDQ} Monitoring with QTAG," *Proceedings of IEEE International Test Conference*, 1995, pp. 739-749.

[18] K. Baker and B. Verhelst, "Iddq Testing Because Zero Defects is Not Enough: A Philips Perspective," *Proceedings of IEEE International Test Conference*, 1990, pp. 253-254

[19] K. Baker, "Philips Draft Proposal for a Quiescent Current Monitor Standard under QTAG," Dec. 1993.

[20] T. Barrette, V. Bhinde, K. De, M. Stover and E. Sugasawara, "Evaluation of Early Failure Screening Method," *Proceedings of IEEE International Workshop on IDDQ Testing*, 1996, pp. 14-19.

[21] Z. Barzilai and B. K. Rosen, "Comparison of AC Self-Testing Procedures," *Proceedings of IEEE International Test Conference*, 1983, pp. 89-94.

[22] J. Beasley, H. Ramamurthy, J. Ramirez-Augulo and M. De Yong, "Idd Pulse Response Testing of Analog and Digital CMOS Circuits," *Proceedings of IEEE International Test Conference*, 1993, pp. 626-634

[23] B. Beck, "Practical Application of Statistical Process Control in Semiconductor Manufacturing," *Proceedings of IEEE International Test Conference*, 1993, pp. 99-107.

[24] I. M. Bell, D. A. Camplin, G. E. Taylor and B. Banister,"Supply Current Testing of Mixed Analog and Digital ICs," *IEE Electronics Letters*, Vol. 27, No. 17, pp. 1581-1583, Aug. 1991.

[25] R. Bencivenga, T. J. Chakraborty and S. Davison, "The Architecture of the Gentest Sequential Test Generator," *Proceedings of IEEE Custom Integrated Circuits Conference*, 1991, pp. 17.1.1-17.1.4.

[26] M. Bending, "HITEST: A Knowledge-Based Test Generation System," *IEEE Design & Test of Computers*, Vol. 1, No. 3, pp. 83-93, May 1984.

[27] M. Bershteyn, "Calculation of Multiple Sets of Weights for Weighted Random Testing," *Proceedings of IEEE International Test Conference*, 1993, pp. 1031-1040.

[28] D. Bhattacharjee, P. Agrawal and V. D. Agrawal, "Delay Fault Test Generation for Scan/Hold Circuits using Boolean Expressions," *Proceedings of ACM/IEEE Design Automation Conference*, 1992, pp. 159-164.

[29] S. Bhawmik, C. J. Lin, K. T. Cheng and V. D. Agrawal, "PASCANT: A Partial Scan and Test Generation System," *Proceedings of IEEE Custom Integrated Circuits Conference*, 1991, pp. 17.3.1-17.3.4.

[30] S. W. Bollinger and S. F. Midkiff, "On Test Generation for Iddq Testing of Bridging Faults in CMOS Circuits," *Proceedings of IEEE International Test Conference*, 1991, pp. 598-607.

[31] S. Bose, P. Agrawal and V. D. Agrawal, "Generation of Compact Delay Tests by Multiple-Path Activation," *Proceedings of IEEE International Test Conference*, 1993, pp. 714-723.

[32] F. Brglez, D. Bryant and K. Kozminski, "Combinational Profiles of Sequential Benchmark Circuits," *Proceedings of IEEE International Symposium on Circuits and Systems*, 1989, pp. 1929-1934.

[33] R. E. Bryant, "An Algorithm for MOS LSI Logic Simulation," *LAMBDA*, Fourth Quarter, 1980, pp. 46-53.

[34] F. Brglez and H. Fujiwara, "A Neutral Netlist of 10 Combinational Benchmark Circuits and a Target Translator in Fortran," Special session on ATPG and fault simulation, *Proceedings of IEEE International Symposium on Circuits and Systems*, 1985.

[35] R. Burch, F. Najm, P. Yang and D. Hocevar, "Pattern Independent Current Estimation for Reliability Analysis of CMOS Circuits," *Proceedings of ACM/IEEE Design Automation Conference*, 1988, pp. 294-299.

[36] D. Burns, "Locating High Resistance Shorts in CMOS Circuits by Analyzing Supply Current Measurements Vectors," *Proceedings of International Symposium on Testing and Failure Analysis*, 1989, pp. 231-237.

[37] T. T. Butler, T. G. Hallin, J. J. Kulzer and K. W. Johnson, "LAMP: Application to Switching System Development," *Bell Systems Technical Journal*, Vol. 53, pp. 1535-1555, Oct. 1974.

[38] T. Calin, F. L. Vargas and M. Nicolaidis, "Upset-Tolerant CMOS SRAM Using Current Monitoring: Prototype and Test Experiments," *Proceedings of IEEE International Test Conference*, 1995, pp. 45-53.

[39] A. M. Campbell, E. I. Cole, C. L. Henderson and M. R. Taylor, "Case History: Failure Analysis of a CMOS SRAM with an Intermittent Open Contact," *Proceedings of International Symposium for Testing and Failure Analysis*, 1991, pp. 261-269.

[40] P. Camurati, D. Medine, P. Prinetto and M. S. Reorda, "A Diagnostic Test Pattern Generation Algorithm," *Proceedings of IEEE International Test Conference*, 1990, pp. 52-58.

[41] J. L. Carter, V. S. Iyengar and B. K. Rosen, "Efficient Test Coverage Determination for Delay Faults," *Proceedings of IEEE International Test Conference*, 1987, pp. 418-427.

[42] G. R. Case, "Analysis of Actual Fault Mechanisms in CMOS Logic Gates," *Proceedings of ACM/IEEE Design Automation Conference*, 1976, pp. 265-270.

[43] T. J. Chakraborty, S. Bhawmik, R. Bencivenga and C. J. Lin. "Enhanced Controllability for Iddq Test Sets Using Partial Scan," *Proceedings of ACM/IEEE Design Automation Conference*, 1991, pp. 278-281.

[44] S. Chakravarty, "A Characterization of Robust Test-Pairs for Stuck-Open Faults," *Journal of Electronic Testing: Theory and Applications*, Vol. 1, No. 4, pp. 275-286, 1991.

[45] S. Chakravarty, "Heuristics for Computing Robust Tests for Stuck-open Faults from Stuck-at Test Sets," *Proceedings of IEEE European Design Automation Conference*, 1992, pp. 416-420.

[46] S. Chakravarty, "A Study of Theoretical Issues in the Synthesis of Delay Fault Testable Circuits," *IEEE Transaction on Computers*, Vol. 45, No. 8, pp. 985-991, August 1996.

[47] S. Chakravarty, "A Sampling Technique for Diagnostic Fault Simulation" *Proceedings of IEEE VLSI Test Symposium*, 1996, pp. 192-197.

[48] S. Chakravarty and Y. Gong, "An Algorithm for Diagnosing Two-Line Bridging Faults in Combinational Circuits," *Proceedings of ACM/IEEE Design Automation Conference*, 1993, pp. 520-524.

[49] S. Chakravarty and Y. Gong, "Diagnostic Simulation of Stuck-at Faults in Combinational Circuits," *Proceedings of IEEE VLSI Test Symposium*, 1994, pp. 128-133.

[50] S. Chakravarty, Y. Gong and S. Venkataraman, "Diagnostic Simulation of Stuck-at Faults in Combinational Circuits," *Journal of Electronic Testing: Theory and Applications*, Vol. 8, No. 1, pp. 87-98, February 1996.

[51] S. Chakravarty and M. Liu, "Algorithms for Current Monitor Based Diagnosis of Bridging and Leakage Faults," *Proceedings of ACM/IEEE Design Automation Conference*, 1992, pp. 353-356.

[52] S. Chakravarty and M. Liu, "Algorithms for I_{DDQ} Measurement Based Diagnosis of Bridging Faults," *Journal of Electronic Testing: Theory and Applications*, Vol. 3, No. 4, pp. 377-395, Dec. 1992.

[53] S. Chakravarty and A. Shekhawat, "Parallel and Serial Heuristics for the Minimum Set Cover Problem," *Journal of Supercomputing*, No. 5, pp. 331-345, 1992.

[54] S. Chakravarty and P. J. Thadikaran, "Simulation and Generation of I_{DDQ} Tests for Bridging Faults in Combinational Circuits," *Proceedings of IEEE VLSI Test Symposium*, 1993, pp. 25-32.

[55] S. Chakravarty and P. J. Thadikaran, "A Study of I_{DDQ} Subset Selection Algorithms for Bridging Faults," *Proceedings of IEEE International Test Conference*, 1994, pp. 403-412.

[56] S. Chakravarty and P. J. Thadikaran, "Simulation and Generation of I_{DDQ} Tests for Bridging Faults in Combinational Circuits," *IEEE Transactions on Computers*, Vol. 39, No. 10, pp. 1131-1140, Oct. 1996.

[57] S. Chakravarty and S. Suresh, "Iddq Measurement Based Diagnosis of Bridging Fault in Full Scan Circuits," *Proceedings of Seventh International Conference on VLSI Design*, 1994, pp. 179-182.

[58] V. H. Champac, R. Rodrigues-Montanes, J. A. Seguera, J. Figueras and A. Rubio, "Fault Modeling of Gate-Oxide Shorts, Floating Gate Shorts and Bridging Failures in CMOS Circuits," *Proceedings of IEEE European Test Conference*, 1991, pp. 143-148.

[59] V. H. Champca, A. Rubio and J. Figueras, "Electrical Model of Floating Gate Defect in CMOS IC's: Implications on I_{DDQ} Testing," *IEEE Transactions on Computer-Aided Design of Integrated Circuits and Systems*, Vol. 13, No. 3, pp. 359-369, 1994.

[60] H. A. Chan, P. J. Englert, M. A. Oien and S. Rajaraman, "Environmental Stress Testing," *AT&T Technical Journal*, Vol. 73, pp. 77-85, April 1994.

[61] R. Chandramouli, "On Testing Stuck-Open Faults," *Proceedings of IEEE 13th International Fault-Tolerant Computing Symposium*, 1983, pp. 258-265.

[62] J. Chang and E. McCluskey, "Detecting Delay Flaws by Very-Low-Voltage Testing," *Proceedings of IEEE International Test Conference*, 1996, pp. 367-376.

[63] J. Chang and E. J. McCluskey, "SHOrt Voltage Elevation (SHOVE) Testing," *Proceedings of IEEE International Workshop on IDDQ Testing*, 1996, pp. 45-49.

[64] V. Chvatal, "A Greedy Heuristic for the Set-Covering Problem," *Mathematics of Operations Research*, Vol. 4, pp. 233-235, 1979.

[65] C. Chen and J. A. Abraham, "High Quality Tests for Switch Level Circuits Using Current and Logic Test Generation Algorithms," *Proceedings of IEEE International Test Conference*, 1991, pp. 615-622.

[66] C. Chen and J. A. Abraham, "Generation and Evaluation of Current and Logic Tests for Switch Level Sequential Circuits," *Journal of Electronic Testing: Theory and Applications*, Vol. 3, No. 4, pp. 359-366, Dec. 1992.

[67] K. Cheng, "Transition Fault Simulation for Sequential Circuits," *Proceedings of IEEE International Test Conference*, 1992, pp. 723-731.

[68] K. T. Cheng and V. D. Agrawal, "A Partial Scan Method for Sequential Circuits with Feedback," *IEEE Transaction on Computers*, Vol. 39, No. 4, pp. 544-548, April 1990.

[69] K. Cheng and H. Chen, "Delay Testing for Non-Robust Untestable Circuits," *Proceedings of IEEE International Test Conference*, 1993, pp. 954-961.

[70] K. T. Cheng, S. Devadas and K. Keutzer, "Robust Delay Fault Test Generation and Synthesis for Testability under a Standard Scan Design Methodology," *Proceedings of ACM/IEEE Design Automation Conference*, 1991, pp. 80-86.

[71] W. T. Cheng and T. J. Chakraborty, "GENTEST: An Automatic Test Generation System for Sequential Circuits," *Computer*, Vol. 22, No. 4, pp. 43-49, April 1989.

[72] K. W. Chiang and Z. G. Vranesic, "On Fault Detection in CMOS Logic Networks," *Proceedings of ACM/IEEE Design Automation Conference*, 1983, pp. 50-56.

[73] C. Crapuchettes, "Testing CMOS Iddq on Large Devices," *Proceedings of IEEE International Test Conference*, 1987, pp. 310-315.

[74] V. P. Dabholkar, *Optimization Problems in Low Power and Stress Testing.* PhD Thesis, Aug. 1986, Department of Computer Science, State University of New York, Buffalo, NY.

[75] R. G. Daniels, "The Changing Demands of Microprocessor Testing," Keynote address, *Proceedings of IEEE International Test Conference*, 1990, p. 21.

[76] D. V. Das, S. C. Seth, P. T. Wagner, J. C. Anderson and V. D. Agrawal, "An Experimental Study on Reject Ratio Prediction for VLSI Circuits: Kokomo Revisited," *Proceedings of IEEE International Test Conference*, 1990, pp. 712-720.

[77] B. Davari, R. H. Dennard and G. G. Shahidi, "CMOS Scaling for High Performance and Low Power-The Next Ten Years," *Proceedings of the IEEE*, Vol. 83, No. 4, pp. 595-606, April 1995.

[78] W. Debany, "Coverage of Node Shorts Using Internal Access Methods," *Proceedings of IEEE VLSI Test Symposium*, 1991, pp. 215-220.

[79] W. H. Debany, "Coverage of Node Shorts Using Internal Access and Equivalence Classes," *International Journal of Computer Aided VLSI Design*, Vol. 1, No. 1, pp. 71-86, September 1993.

[80] S. Dasgupta, R. G. Walters, T. W. Williams and E. B. Eichelberger, "An Enhancement to LSSD and Some Applications of LSSD in Reliability, Availability and Servicebility," *Proceedings of IEEE 11th Fault-Tolerant Computing Symposium*, 1981, pp. 32-34.

[81] R. Dekker et al., "Fault Modeling and Test Algorithm Development for Static Random Access Memories," *Proceedings of IEEE International Test Conference*, 1988, pp. 343-352.

[82] S. J. Duey, A. R. Harvath and P. G. Kowalczyk, "Improved Quality and Reliability Using Operating Extremes Test Methods," *Proceedings of IEEE Customs Integrated Circuits Conference*, 1993, pp. 30.5.1-30.5.4.

[83] Y. M. El-Ziq, "Automatic Test Generation for Stuck-Open Faults in CMOS VLSI," *Proceedings of ACM/IEEE Design Automation Conference*, 1981, pp. 347-354.

[84] Y. M. El-Ziq and R. J. Cloutier, "Functional Level Test Generation for Stuck-Open Faults in CMOS VLSI," *Proceedings of IEEE International Test Conference*, 1981, pp. 536-546.

[85] F. Fantini and C. Morandi, "Failure Modes and Mechanisms for VLSI ICs - A Review," *IEE Proceedings: Part G*, Vol. 132, No. 3, pp. 74-81, June 1985.

[86] M. Favalli, M. Dalpasso, P. Olivo and B. Ricco, "Analysis of Dynamic Effects of Resistive Bridging Faults in CMOS and BiCMOS Digital ICs," *Proceedings of IEEE International Test Conference*, 1993, pp. 865-874.

[87] D. B. I. Feltham, P. J. Nigh, L. R. Carley and W. Maly, "Current Sensing for Built-in Testing of CMOS Circuits," *Proceedings of ACM/IEEE International Conference on Computer-Aided Design*, 1988, pp. 454-457.

[88] A. Ferre and J. Figueras, "On Estimating Bounds on the Quiescent Current for I_{DDQ} Testing," *Proceedings of IEEE VLSI Test Symposium*, 1996, pp. 106-111.

[89] F. J. Ferguson and J. P. Shen, "A CMOS Fault Extractor for Inductive Fault Analysis," *IEEE Transaction on Computer-Aided Design of Integrated Circuits and Systems*, Vol. 7, No. 11, pp. 1181-1194, Nov. 1988.

[90] F. J. Ferguson, M. Taylor and T. Larrabee, "Testing for Parametric Faults in Static CMOS Circuits," *Proceedings of IEEE International Test Conference*, 1990, pp. 436-443.

[91] F. J. Ferguson and T. Larrabee, "Test Pattern Generation for Current Testable Faults in Static CMOS Circuits," *Proceedings of IEEE VLSI Test Symposium*, 1991, pp. 297-302.

[92] F. J. Ferguson and T. Larrabee, "Test Pattern Generation for Realistic Bridge Faults in CMOS ICs," *Proceedings of IEEE International Test Conference*, 1991, pp. 492-499.

[93] J. Flood, "Fundamentals of Testing COS/MOS Integrated Circuits," RCA Solid-State Integrated Circuits Databook, ICAN-6532, 1978, pp. 622-630.

[94] J. F. Frenzel and P. N. Marinos, "Power Supply Current Signature (PSCS) Analysis: A New Approach to System Testing," *Proceedings of IEEE International Test Conference*, 1987, pp. 125-129.

[95] R. R. Fritzemeier, J. M. Soden, R. K. Treece and C. F. Hawkins, "Increased CMOS IC Stuck-at Fault Coverage with Reduced Iddq Test Sets," *Proceedings of IEEE International Test Conference*, 1990, pp. 427-435

[96] R. R. Fritzemeier, C. F. Hawkins and J. M. Soden, "CMOS IC Fault Models, Physical Defect Coverage and Iddq Testing," *Proceedings of IEEE Custom Integrated Circuits Conference*, 1991, pp. 13.1.1-13.1.8.

[97] K. Fuchs, F. Fink and M. H. Schultz, "DYNAMITE: An Efficient ATPG System for Path Delay Faults," *IEEE Transaction on Computer-Aided Design of Integrated Circuits and Systems*, Vol. 10, No. 10, pp. 1323-1335, Oct. 1991.

[98] H. Fujiwara and T. Shimono, "On the Acceleration of Test Generation Algorithms," *IEEE Transactions on Computers*, Vol. C-32, No. 12, pp. 1137-1144, December 1983.

[99] J. Galiay, Y. Crouzet and M. Vergniault, "Physical versus Logical Fault Models in MOS LSI Circuits, Impact on Their Testability," *IEEE Transactions on Computers*, Vol. C-27, No. 6, pp. 527-531, June 1980.

[100] R. Gayle, "The Cost of Quality: Reducing ASIC Defects with Iddq, At-Speed Testing and Increased Fault Coverage," *Proceedings of IEEE International Test Conference*, 1993, pp. 285-292.

[101] A. Gattiker and W. Maly, "Current Signatures," *Proceedings of IEEE VLSI Test Symposium*, 1996, pp. 112-117.

[102] A. Gattiker, P. Nigh, D. Grosch and W. Maly, "Current Signatures for Production Testing," *Proceedings of IEEE International Workshop on IDDQ Testing*, 1996, pp. 25-28.

[103] M. A. Gharaybeh, M. L. Bushnell and V. D. Agrawal, "Classification and Test Generation for Path-Delay Faults Using Single Stuck-Fault Tests," *Proceedings of IEEE International Test Conference*, 1995, pp. 139-148.

[104] P. Girard, C. Landrault and S. Pravossoudovich, "An Implicit Delay Fault Simulation Method with Approximate Detection Threshold Calculation," *Proceedings of IEEE International Test Conference*, 1993, pp. 705-713.

[105] C. T. Glover and M. R. Mercer, "A Method for Delay Fault Test Generation," *Proceedings of ACM/IEEE Design Automation Conference*, 1988, pp. 90-95.

[106] P. Goel, "An Implicit Enumeration Algorithm to Generate Tests for Large Combinational Circuits," *IEEE Transactions on Computers*, Vol. 30, No. 3, pp. 215-292, March 1981.

[107] D. E. Goldberg, *Genetic Algorithms in Search, Optimization & Machine Learning*. Addison Wesley, Reading, MA, 1989.

[108] Y. Gong, *Exploiting Dynamic Computation in Diagnosis of Bridging Faults*. PhD Thesis, Aug. 1995, Department of Computer Science, State University of New York at Buffalo.

[109] Y. Gong and S. Chakravarty, "On Adaptive Diagnostic Test Generation," *Proceedings of ACM/IEEE International Conference on Computer-Aided Design*, 1995, pp. 181-184.

[110] R. K. Gulati, W. Mao and D. Goel, "Detection of 'Undetectable' Faults Using Iddq Testing," *Proceedings of IEEE International Test Conference*, 1992, pp. 770-777.

[111] H. Gundlach, B. Koch and K. D. Muller-Glasser, "On Selection of a Partial Scan Path with Respect to Target Faults," *Proceedings of European Conference on Design Automation*, 1991, pp. 219-223.

[112] R. Gupta and M. Breuer, "Ordering Storage Elements in a Single Scan Chain," *Proceedings of IEEE International Conference on Computer-Aided Design*, 1991, pp. 408-411.

[113] R. Gupta and M. A. Breuer, "A Ballast Methodology for Structured Partial Scan," *IEEE Transaction on Computers*, Vol 39, No. 4, pp. 538-544, April 1990.

[114] A. Hales, "Texas Instruments Draft Proposal for the QTAG Off-Chip IDDQ/ISSQ Monitor Standard," Jan. 1994.

[115] H. Hao and E. J. McCluskey, "Resistive Shorts within CMOS Gates," *Proceedings of IEEE International Test Conference*, 1991, pp. 292-301.

[116] M. Hashizume, et al., "Fault Detection of Combinational Circuits Based on Supply Current," *Proceedings of IEEE International Test Conference*, 1988, pp. 374-380.

[117] C. F. Hawkins and J. M. Soden, "Electrical Characteristics and Testing Considerations for Gate-Oxide Shorts in CMOS ICs," *Proceedings of IEEE International Test Conference*, 1985, pp. 544-555.

[118] C. F. Hawkins and J. M. Soden, "Reliability and Electrical Properties of Gate-Oxide Shorts in CMOS ICs," *Proceedings of IEEE International Test Conference*, 1986, pp. 443-451.

[119] C. F. Hawkins, J. M. Soden, R. R. Fritezemeier and L. K. Horning, "Quiescent Power Supply Current Measurement for CMOS IC Defect Detection," *IEEE Transactions Industrial Electronics*, Vol. 36, No. 2, pp. 211-218, May 1989.

[120] C. F. Hawkins, J. M. Soden, A. W. Righter and F. J. Ferguson, "Defect Classes: An Overdue Paradigm for CMOS IC Testing," *Proceedings of IEEE International Test Conference*, 1984, pp. 413-425.

[121] S. Hellebrand, S. Tarnick, J. Rajski and B. Courtois, "Generation of Vector Patterns Through Reseeding of Multiple-Polynomial Linear Feedback Shift Registers," *Proceedings of IEEE International Test Conference*, 1992, pp. 120-129.

[122] S. Hellebrand, B. Reeb, S. Tarnick and H. Wunderlich, "Pattern Generation for a Deterministic BIST Scheme," *Proceedings of ACM/IEEE International Conference on Computer-Aided Design*, 1995, pp. 88-94.

[123] C. L. Henderson, J. M. Soden and C. F. Hawkins, "The Behavior and Testing Implications of CMOS IC Logic Gate Open Circuits," *Proceedings of IEEE International Test Conference*, 1991, pp. 302-310.

[124] T. H. Henry and T. Soo, "Burn-In Elimination of a High Volume Microprocessor Using I_{DDQ}," *Proceedings of IEEE International Test Conference*, 1996, pp. 242-249.

[125] Y. Higami, S. Kajihara and K. Kinoshita, "Reduced ScanShift: A New Testing Methodology for Sequential Circuits," *Proceedings of IEEE International Test Conference*, 1994, pp. 624-630.

[126] D. Hochbaum, "Approximation Algorithms for the Set Covering and Vertex Covering Problems," *SIAM Journal of Computing*, Vol. 11, pp. 555-556, 1982.

[127] E. Hnatek, "Thoughts on VLSI Burn-in," *Proceedings of IEEE International Test Conference*, 1984, pp. 531-535.

[128] S. Holland et al., "On Physical Models for Gate-Oxide Breakdown," *Electron Device Letters*, Vol. 5, Aug. 1984.

[129] L. K. Horning, et al., "Measurements of Quiescent Power Supply Current for CMOS ICs in Production Testing," *Proceedings of IEEE International Test Conference*, 1987, pp. 300-309.

[130] E. P. Hsieh, R. A. Rasmussen, L. J. Vidunas and W. T. Davis, "Delay Test Generation," *Proceedings of ACM/IEEE Design Automation Conference*, 1977, pp. 486-491.

[131] C. Hsue and C. J. Lin, "Built-in Current Sensor for Iddq Test in CMOS," *Proceedings of IEEE International Test Conference*, 1993, pp. 635-641.

[132] U. Hubner and H. T. Vierhaus, "Built-in Current Testing versus Delay Fault Testing, a Case Study," *Proceedings of CompEuro*, 1992, pp. 268-273.

[133] S. Hwang and R. Rajsuman, "Effectiveness of Stuck-at Test Sets to Detect Bridging Faults in Iddq Environment," *Proceedings of Second IEEE Asian Test Symposium*, 1993, pp. 249-254.

[134] S. Hwang and R. Rajsuman, "VLSI Testing for High-Reliability: Mixing Iddq and Logic Testing," *International Journal on VLSI*, Vol. 2, 1994.

[135] S. Hwang, R. Rajsuman and S. Davidson, "Detect Efficiency of Stuck-at Tests on Bridging Faults in Iddq Environment," *Proceedings of Seventh International Conference on VLSI Design*, 1994, pp. 183-186.

[136] Y. C. Hsu and S. K. Gupta, "A Simulator for At-Speed Robust Testing of Path Delay Faults in Combinational Circuits," *IEEE Transactions on Computers*, Vol. 45, No. 11, pp. 1312-1318, Nov. 1996.

[137] M. Inoue, T. Yamada and A. Fujiwara, "A New Testing Acceleration Chip for Low-Cost Memory Test," *IEEE Design & Test of Computers*, Vol. 10, pp. 15-19, March 1993.

[138] E. Isern and J. Figueras, "Analysis of Redundant Structures in Combinational Circuits," *Proceedings of IEEE VLSI Test Symposium*, 1993, pp. 21-23.

[139] E. Isern and J. Figueras, "Test Generation with High Coverages for Quiescent Current Test of Bridging Fault in Combinational Circuits," *Proceedings of IEEE International Test Conference*, 1993, pp. 73-82.

[140] V. S. Iyengar, B. K. Rosen and I. Spillinger, "Delay Test Generation 1 - Concepts and Coverage Metrics," *Proceedings of IEEE International Test Conference*, 1988, pp. 857-866.

[141] V. S. Iyengar, B. K. Rosen and I. Spillinger, "Delay Test Generation 2 - Algebra and Algorithms," *Proceedings of IEEE International Test Conference*, 1988, pp. 867-876.

[142] V. S. Iyengar, B. K. Rosen and J. A. Waicukauski, "On Computing the Sizes of Detected Delay Faults," *IEEE Transaction on Computer-Aided Design of Integrated Circuits and Systems*, Vol. 9, pp. 299-312, 1990.

[143] S. K. Jain and V. D. Agrawal, "Modeling and Test Generation Algorithms for MOS Circuits," *IEEE Transactions on Computers*, Vol C-34, No. 5, pp. 426-433, May 1985.

[144] A. Jee and F. J. Ferguson, "CARAFE: An Inductive Fault Analysis Tool for CMOS VLSI Circuits," *Proceedings of IEEE VLSI Test Symposium*, 1993, pp. 92-98.

[145] N. K. Jha and Q. Tong, "Detection of Multiple Input Bridging and Stuck-on Faults in CMOS Logic Circuits Using Current Monitoring," *Proceedings of IEEE European Design Automation Conference*, 1990, pp. 350-354.

[146] D. S. Johnson, "Approximation Algorithms for Combinatorial Problems," *Journal of Computing and System Sciences,* 1974, pp. 256-278.

[147] S. Johnson, "Residual Charge on a Faulty Floating Gate MOS Transistor," *Proceedings of IEEE International Test Conference*, 1994, pp. 555-561.

[148] G. H. Johnson and J. B. Wilstrup, "A General Purpose ATE Based I_{DDQ} Measurement Circuit," *Proceedings of IEEE International Test Conference*, 1995, pp. 97-105.

[149] D. D. Josephson, D. J. Dixon and B. J. Arnold, "Test Features of the HP PA7100LC Processor," *Proceedings of IEEE International Test Conference*, 1993, pp. 763-772.

[150] D. Josephson, M. Storey and D. Dixon, "Microprocessor I_{DDQ} Testing: A Case Study," *Design & Test of Computers*, Vol. 12, No. 2, pp. 42-52, 1995.

[151] J. M. Jou and S. C. Chen, "A Fast and Memory Efficient Diagnostic Fault Simulation Algorithm for Sequential Circuits," *Proceedings of ACM/IEEE International Conference on Computer-Aided Design*, 1994, pp. 723-726.

[152] R. Kapur, S. Patil, T. J. Snethen and T. W. Williams, "Design of an Efficient Weighted Random Pattern Generation System," *Proceedings of IEEE International Test Conference*, 1994, pp. 491-500.

[153] R. Kawahara, O. Nakayama and T. Kurasawa, "The Effectiveness of IDDQ and High Voltage Stress for Burn-in Elimination," *Proceedings of IEEE International Workshop on IDDQ Testing*, 1996, pp. 9-13.

[154] M. Keating and D. Meyer, "A New Approach to Dynamic IDD Testing," *Proceedings of IEEE International Test Conference*, 1987, pp. 316-321.

[155] B. Koenemann, et al., "Delay Test: The Next Frontier for LSSD Test Systems," *Proceedings of IEEE International Test Conference*, 1992, pp. 578-587.

[156] S. Koeppe, "Modeling and Simulation of Delay Faults in CMOS Logic Circuits," *Proceedings of IEEE International Test Conference*, 1986, pp. 530-536.

[157] J. B. Khare and W. Maly, *From Contamination to Defects, Faults and Yield Loss.* Kluwer Academic Publishers, Boston, 1996.

[158] J. Khare, S. Griep, H. D. Oberle, W. Maly, D. S. Landsiedel, U. Kollmer and D. M. H. Walker, "Key Attributes of an SRAM Testing Strategy Required for Effective Process Monitoring," *Proceedings of IEEE International Workshop on Memory Testing*, 1993, pp. 84-89.

[159] E. E. King and G. P. Nelson, "Radiation Testing of an 8-bit CMOS Microprocessor," *IEEE Transaction on Nuclear Science*, Vol. NS-22, No. 5, pp. 2120-2120, Oct. 1975.

[160] K. Kubiak, S. Parkes, W. K. Fuchs and R. Saleh, "Exact Evaluation of Diagnostic Test Resolution," *Proceedings of ACM/IEEE Design Automation Conference*, 1992, pp. 347-352.

[161] W. Kunz and D. Pradhan, "Recursive Learning: An Attractive Alternative to the Decision Tree for Test Generation in Digital Circuit," *Proceedings of IEEE International Test Conference*, 1992, pp. 816-825.

[162] A. Kunzmann and H. J. Wunderlich, "An Analytical Approach to the Partial Scan Design Problem," *Journal of Electronic Testing: Theory and Applications*, Vol. 1, pp. 163-174, May 1990.

[163] T. Larrabee, "Test Pattern Generation Using Boolean Satisfiability," *IEEE Transactions on Computer-Aided Design of Integrated Circuits and Systems*, Vol. 11, No. 1, pp. 4-15, January 1992.

[164] K. J. Lee and M. A. Breuer, "On Detecting Single and Multiple Bridging Faults in CMOS Circuits Using the Current Supply Monitoring Method," *Proceedings of IEEE International Symposium on Circuits and Systems*, 1990, pp. 5-8.

[165] K. J. Lee and M. A. Breuer, "A Universal Test Sequence for CMOS SCAN Register," *Proceedings of IEEE Custom Integrated Circuits Conference*, 1990, pp. 28.5.1-28.5.4.

[166] K. J. Lee and M. A. Breuer, "Constraints for Using Iddq Testing to Detect CMOS Bridging Faults," *Proceedings of IEEE VLSI Test Symposium*, 1991, pp. 303-308.

[167] K. J. Lee and M. A. Breuer, "Design and Test Rules for CMOS Circuits to Facilitate Iddq Testing for Bridging Faults," *IEEE Transactions on Computer-Aided Design of Integrated Circuits and Systems*, Vol. 11, No. 5, pp. 659-670, May 1992.

[168] H. K. Lee and D. S. Ha, "HOPE: A Fast Parallel Fault Simulator for Sequential Circuits," *Proceedings of ACM/IEEE Design Automation Conference*, 1992, pp. 336-340.

[169] H. K. Lee and D. S. Ha, "ATALANTA: Deterministic Stuck-at Test Generation," Technical Report, Electrical Engineering Department, Virginia Ploytechnique Institute and State University, VA.

[170] T. Lee, I. N. Hajj, E. M. Rudnick and J. H. Patel, "Genetic-Algorithm-Based Test Generation for Current Testing of Bridging Faults in CMOS VLSI Circuits," *Proceedings of IEEE VLSI Test Symposium*, 1996, pp. 456-462.

[171] D. H. Lee and S. M. Reddy, "On Determining Scan Flip-Flops in Partial-Scan Designs," *Proceedings of ACM/IEEE International Conference on Computer-Aided Design*, 1990, pp. 322-325.

[172] S. Y. Lee and K. K. Saluja, "An Algorithm to Reduce Test Application Time in Full Scan Design," *Proceedings of IEEE International Conference on Computer-Aided Design*, 1992, pp. 17-20.

[173] Y. Levendel and P. R. Menon, "Transition Faults in Combinational Circuits: Input Transition Test Generation and Fault Simulation," *Proceedings of IEEE Fault Tolerant Computing Symposium*, 1986, pp. 278-283.

[174] M. W. Levi, "CMOS is Most Testable," *Proceedings of IEEE International Test Conference*, 1981, pp. 217-220.

[175] C. C. Liaw, S. Y. Su and Y. K. Malaiya, "Test Generation for Delay Faults Using Stuck-at Fault Test Set," *Proceedings of IEEE International Test Conference*, 1980, pp. 167-175.

[176] C. J. Lin and S. M. Reddy, "On Delay Fault Testing in Logic Circuits," *IEEE Transactions on Computer-Aided Design of Integrated Circuits and Systems*, Vol. 6, pp. 694-703, 1987.

[177] J. C. Lo, J. C. Daly and M. Nicolaidis, "Design of Static CMOS Self-Checking Circuits Using Built-in Current Sensing," *Proceedings of International Symposium on Fault-Tolerant Computing*, 1992, pp. 104-111.

[178] H. A. R. Manhaeve et al., "An Off-Chip I_{DDQ} Current Measurement Unit for Telecommunications ASICs," *Proceedings of IEEE International Test Conference*, 1994, pp. 203-212.

[179] U. Mahlstedt, "DELTEST: Deterministic Test Generation for Gate-Delay Faults," *Proceedings of IEEE International Test Conference*, 1993, pp. 972-980.

[180] U. Mahlstedt, J. Alt and M. Heinitz, "CURRENT: A Test Generation System for I_{DDQ} Testing," *Proceedings of IEEE VLSI Test Symposium*, 1995, pp. 317-323.

[181] U. Mahlstedt, T. Grunning, C. Ozcan and W. Daehn, "CONTEST: A Fast ATPG Tool for Very Large Combinational Circuits," *Proceedings of ACM/IEEE International Conference on Computer-Aided Design*, 1990, pp. 222-225.

[182] Y. K. Malaiya and S. Y. H. Su, "A New Fault Model and Testing Technique for CMOS Devices," *Proceedings of IEEE International Test Conference*, 1982, pp. 25-34.

[183] Y. K. Malaiya, "Testing Stuck-on Faults in CMOS Integrated Circuits," *Proceedings of International Conference on Computer-Aided Design*, 1984, pp. 248-250.

[184] Y. K. Malaiya, A. P. Jayasumana and R. Rajsuman, "A Detailed Examination of Bridging Faults," *Proceedings of IEEE International Conference on Computer Design*, 1986, pp. 78-81.

[185] Y. K. Malaiya, A.P. Jayasumana, Q. Tong and S. M. Menon, "Enhancement of Resolution in Supply Current Based Testing for Large ICs," *Proceedings of IEEE VLSI Test Symposium*, 1991, pp. 291-296.

[186] Y. K. Malaiya and R. Rajsuman, *Bridging Faults and I_{DDQ} Testing.* IEEE Computer Society Press, Los-Alamitos, CA, 1992.

[187] W. Maly, "Modeling of Lithography Related Yield Losses for CAD for VLSI Circuits," *IEEE Transaction on Computer-Aided Design of Integrated Circuits and Systems*, Vol. CAD-4, No. 4, pp. 166-177, 1985.

[188] W. Maly, "Realistic Fault Modeling for VLSI Testing," *Proceedings of ACM/IEEE Design Automation Conference*, 1987, pp. 173-180.

[189] W. Maly and P. Nigh, "Built-in Current Testing-Feasibility Study," *Proceedings of ACM/IEEE International Conference on Computer-Aided Design*, 1988, pp. 340-343.

[190] W. Maly, P. K. Nag and P. Nigh, "Testing Oriented Analysis of CMOS ICs with Opens," *Proceedings of ACM/IEEE International Conference on Computer-Aided Design*, 1988, pp. 344-347.

[191] W. Maly and S. B. Naik, "Process Monitoring Oriented IC Testing," *Proceedings of IEEE International Test Conference*, 1989, pp. 527-532.

[192] W. Maly and M. Patyra, "Built-in Current Testing," *IEEE Journal Solid State Circuits*, Vol. 27, No. 3, pp. 425-428, March 1992.

[193] W. Maly and M. Patyra, "Design of ICs Applying Built-in Current Testing," *Journal of Electronic Testing: Theory and Applications*, Vol. 3, No. 4, pp. 397-406, Dec. 1992.

[194] W. Maly and M. Patyra, "Design of ICs Applying Built-in Current Testing," pp. 111-120. I_{DDQ} *Testing of VLSI Circuits*, Kluwer Academic Publishers, Boston, 1992. Eds: R. K. Gulati and C. F. Hawkins.

[195] K. D. Mandl, "CMOS VLSI Challenges to Test," *Proceedings of IEEE International Test Conference*, 1984, pp. 642-648.

[196] W. Mao and M. D. Ciletti, "Robustness Enhancement and Detection Threshold Reduction in ATPG for Gate Delay Faults," *Proceedings of IEEE International Test Conference*, 1992, pp. 588-597.

[197] W. Mao and R. K. Gulati, "QUIETEST: A Methodology for Selecting I_{DDQ} Test Vectors," pp. 63-71. I_{DDQ} *Testing of VLSI Circuits*, Kluwer Academic Publishers, Boston, 1992. Eds: R. K. Gulati and C. F. Hawkins.

[198] W. Mao and R. K. Gulati, "QUIETEST: A Methodology for Selecting I_{DDQ} Test Vectors," *Journal of Electronic Testing: Theory and Application*, Vol. 3, pp. 349-357, Dec. 1992.

[199] W. Mao, R. K. Gulati, D. K. Goel and M. D. Ciletti, "QUIETEST: A Quiescent Current Testing Methodology for Detecting Leakage Faults," *Proceedings of ACM/IEEE International Conference on Computer-Aided Design*, 1990, pp. 280-283.

[200] P. Maxwell, R. C. Aitken, V. Johansen and I. Chang, "The Effectiveness of Iddq, Functional and Scan Tests: How Many Fault Coverages Do We Need?" *Proceedings of IEEE International Test Conference*, 1992, pp. 168-177.

[201] P. C. Maxwell and R. C. Aitken, "I_{DDQ} Testing as a Component of a Test Suite: The Need for Several Fault Coverage Metrics," *Journal of Electronic Testing: Theory and Applications*, Vol. 3, pp. 305-316, 1992.

[202] P. C. Maxwell and R. C. Aitken, "I_{DDQ} Testing as a Component of a Test Suite: The Need for Several Fault Coverage Metrics," *I_{DDQ} Testing of VLSI Circuits*, Kluwer Academic Publishers, Boston, 1992. Eds: R. K. Gulati and C. F. Hawkins.

[203] P. Maxwell, R. Aitken, K. Kollitz and A. Brown, "I_{DDQ} and AC Scan: The War Against Unmodeled Defects," *Proceedings of IEEE International Test Conference*, 1996, pp. 250-258.

[204] S. D. McEuen, "Why Iddq?" *Proceedings of IEEE International Test Conference*, 1990, pp. 252.

[205] S. D. McEuen, "Iddq Benefits," *Proceedings of IEEE VLSI Test Symposium*, 1991, pp. 285-290.

[206] S. D. McEuen, "Reliability Benefits of I_{DDQ}," *Journal of Electronic Testing: Theory and Applications*, Vol. 3, pp. 327-335, 1992.

[207] S. D. McEuen, "Reliability Benefits of I_{DDQ}," *I_{DDQ} Testing of VLSI Circuits*, Kluwer Academic Publishers, Boston, 1992. Eds: R. K. Gulati and C. F. Hawkins.

[208] P. C. McGeer, A. Saldhana, P. R. Stephan, R. K. Brayton and A. L. Sangiovanni-Vincentelli, "Timing Analysis and Delay Fault Test Generation Using Path-Recursive Functions," *Proceedings of ACM/IEEE International Conference on Computer-Aided Design*, 1991, pp. 180-183.

[209] R. Meershoek, B. Verhelst, R. McInerney and L. Thijssen, "Functional and Iddq Testing on a Static RAM," *Proceedings of IEEE International Test Conference*, 1990, pp. 929-937.

[210] S. M. Menon, Y. K. Malaiya, A.P. Jayasumana and C. Tong, "The Effect of Built-in Current Sensors (BICS) on Operational and Test Performance," *Proceedings of Seventh International Symposium on VLSI Design*, 1994, pp. 187-190.

[211] Y. Mii, S. Wind, Y. Taur, Y. Lii, D. Klaus and J. Bucchignano, "An Ultra-Low Power 0.1 μm CMOS," *Digest of Technical Papers, IEEE Symposium of VLSI Technology*, 1994, p. 9.

[212] Y. Miura and K. Kinoshita, "Circuit Design for Built-in Current Testing," *Proceedings of IEEE International Test Conference*, 1992, pp. 873-881.

[213] S. P. Morley and R. A. Marlett, "Selectable Length Partial Scan: A Method to Reduce Vector Length," *Proceedings of IEEE International Test Conference*, 1991, pp. 385-392.

[214] E. I. Muehldorf, "A Quality Measure for LSI Components," *IEEE Journal on Solid State Circuits*, Vol. 9, No. 5, pp. 291-297, Oct. 1974.

[215] A. Mukherjee, *Introduction to nMOS and CMOS VLSI System Design.* Prentice-Hall, 1986.

[216] F. Muradali, V. K. Agrawal and B. Nadeau-Dostie, "A New Procedure for Weighted Random Built-In Self-Test," *Proceedings of IEEE International Test Conference*, 1990, pp. 660-669.

[217] S. Naik, F. Agricola and W. Maly, "Failure Analysis of High Density CMOS SRAMs Using Realistic Defect Modeling and I_{DDQ} Testing," *IEEE Design & Test of Computers*, Vol. 10, pp. 13-23, June 1993.

[218] F. Najm, R. Burch, P. Yang and I. Hajj, "CREST-A Current Estimator for CMOS Circuits," *Proceedings of ACM/IEEE International Conference on Computer-Aided Design*, 1988, pp. 204-207.

[219] F. Najm, I. Hajj and P. Yang, "Computation of Bus Current Variance for Reliability Estimation of VLSI Circuits," *Proceedings of ACM/IEEE International Conference on Computer-Aided Design*, 1989, pp. 202-205.

[220] F. Najm, I. Hajj and P. Yang, "Electronmigration Median Time-to-Failure Based on a Stochastic Current Waveform," *Proceedings of ACM/IEEE International Conference on Computer-Aided Design*, 1989, pp. 447-450.

[221] G. F. Nelson, "Parametric Tests Meet the Challenge of High-Density ICs," *Electronics*, Dec. 1975, pp. 108-111.

[222] M. Nicolaidis, F. Vargas and B. Courtois, "Design of Reliable Current Sensors for Radiation Exposed Environments," *Proceedings of the International Nuclear and Space Radiation Effects Conference*, 1993.

[223] M. Nicolaidis, "Finitely Self-checking Circuits and Their Application on Current Sensors," *Proceedings of IEEE VLSI Test Symposium*, 1993, pp. 66-69.

[224] T. M. Niermann and J. H. Patel, "HITEC : A Test Generator Package for Sequential Circuits," *Proceedings of European Design Automation Conference*, 1991, pp. 214-218.

[225] P. Nigh and W. Maly, "Test Generation for Current Testing," *Proceedings of IEEE European Test Conference*, 1989, pp. 194-200.

[226] P. Nigh and W. Maly, "Layout Driven Test Generation," *Proceedings of ACM/IEEE International Conference on Computer-Aided Design*, 1989, pp. 154-157.

[227] P. Nigh and W. Maly, "A Self Testing ALU Using Built-in Current Sensing," *Proceedings of IEEE Custom Integrated Circuits Conference*, 1989, pp. 22.1.1-22.1.4.

[228] P. Nigh and W. Maly, "Test Generation for Current Testing," *IEEE Design & Test of Computers*, Vol. 7, pp. 26-38, Feb. 1990.

[229] R. N. Noyce and M. E. Hoff Jr., "A History of Microprocessor Development at Intel," *IEEE Micro*, Vol. 11, No. 2, pp. 8 - 21, Feb. 1991.

[230] "Measuring CMOS Quiescent Power Supply Current (IDDQ) with the HP 82000," Hewlett Packard Application Note 308-3, April 1991.

[231] B. Oomman, P. Kongara and C. Mallipeddi, "Amdahl Corporation Board Delay Test System," *Proceedings of IEEE International Test Conference*, 1992, pp. 558-567.

[232] E. S. Park and M. R. Mercer, "Robust and Nonrobust Tests for Path Delay Faults in Combinational Logic Circuit," *Proceedings of IEEE International Test Conference*, 1987, pp. 1027-1034.

[233] E. S. Park and M. R. Mercer, "An Efficient Delay Test Generation System for Combinational Logic Circuits," *IEEE Transaction on Computer-Aided Design of Integrated Circuits and Systems*, Vol. 11, No. 7, pp. 926-938, July 1992.

[234] E. S. Park, M. R. Mercer and T. W. Williams, "Statistical Delay Fault Coverage and Defect Level for Delay Faults," *Proceedings of IEEE International Test Conference*, 1988, pp. 492-499.

[235] K. P. Parker, *The Boundary-Scan Handbook.* Kluwer Academic Publishers, Boston, 1992.

[236] S. Pateras and J. Rajski, "Cube Contained Random Patterns and Their Application to the Testing of Synthesized Multi-Level Circuits," *Proceedings of IEEE International Test Conference*, 1991, pp. 473-482.

[237] S. Patil and J. Savir, "Skewed-load Transition Test: Part II, Coverage," *Proceedings of IEEE International Test Conference*, 1992, pp. 714-723.

[238] R. Perry, "Iddq Testing in CMOS Digital ASICs-Putting It All Together," *Proceedings of IEEE International Test Conference*, 1992, pp. 151-157.

[239] R. Perry, "I_{DDQ} Testing in CMOS Digital ASICs," *Journal of Electronic Testing: Theory and Applications*, Vol. 3, pp 317-325, 1992.

[240] R. Perry, "I_{DDQ} Testing in CMOS Digital ASICs." *I_{DDQ} Testing of VLSI Circuits*, Kluwer Academic Publishers, Boston, 1992. Eds: R. K. Gulati and C. F. Hawkins.

[241] A. Pierzynska and S. Pilarski, "Non-Robust versus Robust," *Proceedings of IEEE International Test Conference*, 1995, pp. 123-131.

[242] I. Pomeranz and S. M. Reddy, "On the Generation of Small Dictionaries for Fault Location," *Proceedings of IEEE International Conference on Computer-Aided Design*, 1992, pp. 272-279.

[243] I. Pomeranz and S. M. Reddy, "3-Weight Pseudo Random Test Generation Based on a Deterministic Test Set for Combinational and Sequential Circuits," *IEEE Transactions on Computer-Aided Design of Integrated Circuits and Systems*, Vol. 12, No. 7, pp. 1050-1058, July 1993.

[244] D. K. Pradhan and J. Saxena, "A Design for Testability Scheme to Reduce Test Application Time," *Proceedings of IEEE VLSI Test Symposium*, 1992, pp. 55-60.

[245] A. Pramanick and S. Kundu, "Design of Scan-Based Path-Delay-Testable Sequential Circuits," *Proceedings of IEEE International Test Conference*, 1993, pp. 962-971.

[246] I. Pramanick and A. Pramanick, "Parallel Delay Fault Coverage and Test Quality Evaluation," *Proceedings of IEEE International Test Conference*, 1995, pp. 113-122.

[247] A. K. Pramanick and S. M. Reddy, "On the Detection of Delay Faults," *Proceedings of IEEE International Test Conference*, 1988, pp. 845-856.

[248] A. K. Pramanick and S. M. Reddy, "On the Design of Path Delay Fault Testable Combinational Circuits," *Proceedings of International Fault-Tolerant Computing Symposium*, 1990, pp. 374-381.

[249] A. K. Pramanick and S. M. Reddy, "On Multiple Path Propagating Tests for Path Delay Faults," *Proceedings of IEEE International Test Conference*, 1991, pp. 393-402.

[250] R. Rajsuman, *I_{DDQ} Testing of CMOS VLSI.* Artech House, 1994.

[251] R. Rajsuman, Y. K. Malaiya and A. P. Jayasumana, "Limitations of Switch Level Analysis for Bridging Faults," *IEEE Transactions on Computer-Aided Design of Integrated Circuits and Systems*, Vol. 8, No. 7, pp. 807-811, July 1989.

[252] R. Rajsuman et al., "An Analysis and Testing of Operation Induced Faults in MOS VLSI" *Proceedings of IEEE VLSI Test Symposium*, 1991, pp. 137-142.

[253] R. Rajsuman, "An Analysis of Feedback Bridging Faults in MOS VLSI," *Proceedings of IEEE VLSI Test Symposium*, 1991, pp. 53-58.

[254] R. Rajsuman, "Current Testing" Chapter 11 in *Digital Hardware Testing.* Artech House, 1992.

[255] R. Rajsuman and D. A. Penry, "Coverage of Bridging Faults by Random Testing in Iddq Test Environment," *Proceedings of the Sixth International Conference on VLSI Design*, 1993, pp. 136-139.

[256] S. M. Reddy, C. J. Lin and S. Patil, "An Automatic Test Pattern Generator for the Detection of Path Delay Faults," *Proceedings of IEEE International Conference on Computer-Aided Design*, 1987, pp. 284-287.

[257] R. S. Reddy, I. Pomeranz, S. M. Reddy and S. Kajihara, "Compact Test Generation for Bridging Faults under I_{DDQ} Testing," *Proceedings of IEEE VLSI Test Symposium*, 1995, pp. 310-315.

[258] M. K. Reddy, S. M. Reddy and P. Agrawal, "Transistor Level Test Generation for MOS Circuits at the Switch Level," *Proceedings of IEEE International Test Conference*, 1985, pp. 304-312.

[259] S. M. Reddy, M. K. Reddy and V. D. Agrawal, "Robust-Tests for Stuck-Open Faults in CMOS Combinational Circuits," *Proceedings of Fault-Tolerant Computing Symposium*, 1984, pp. 44-49.

[260] M. Renovell and G. Cambon, "Topology Dependence of Floating Gate Faults in MOS Integrated Circuits," *Electronics Letters*, Vol. 22, No. 3, pp. 150-151, 1986.

[261] M. Renovell and G. Cambon, "Electrical Analysis and Modeling of Floating-Gate Faults," *IEEE Transactions on Computer-Aided Design of Integrated Circuits and Systems*, Vol. 11, No. 11, pp. 1450-1458, 1992.

[262] M. J. Riezeman, "Wanlass's CMOS Circuit," *IEEE Spectrum*, Vol. 28, No. 5, p. 44, May 1991.

[263] M. Roca, E. Sicard and A. Rubio, "Iddq Testing of Oscillating Bridging Faults in CMOS Combinational Circuits," *IEE Proceedings*, Part G, Vol. 140, No. 1, pp. 39-44, Feb. 1993.

[264] R. Rodriquez-Montanes and J. Figueras, "Analysis of Bridging Defects in Sequential CMOS Circuits and Their Current Testability," *Proceedings of European Design and Test Conference*, 1994, pp. 356-360.

[265] R. Rodriquez-Montanes, J. Figueras and A. Rubio, "Current vs. Logic Testability of Bridges in Scan Chains," *Proceedings of IEEE European Design and Test Conference*, 1993, pp. 392-396.

[266] R. Rodrigues-Montanes, J. A. Seguera, V. H. Champac, J. Figueras and J. A. Rubio, "Current versus Logic Testing of Gate-Oxide Shorts, Floating Gate and Bridging Failures in CMOS," *Proceedings of IEEE International Test Conference*, 1991, pp. 510-519.

[267] D. Romanchik, "Iddq Testing Makes a Comeback," *Test and Measurement World*, Oct. 1993, pp. 53-58.

[268] A. Rubio, J. Figueras and J. Segura, "Quiescent Current Sensor Circuits in Digital VLSI CMOS Testing," *IEE Electronics Letters*, Vol. 26, No. 15, pp. 1204-1206, July 1990.

[269] Sandia Laboratories, Internal Memo, May 1979.

[270] E. Rudnick, W. K. Fuchs and J. H. Patel, "Diagnostic Fault Simulation of Sequential Circuits," *Proceedings of IEEE International Test Conference*, 1992, pp. 178-186.

[271] G. Ryan, W. K. Fuchs and I. Pomeranz, "Fault Dictionary Compression and Equivalence Class Computation for Sequential Circuits," *Proceedings of ACM/IEEE International Conference on Computer-Aided Design*, 1993, pp. 508-511.

[272] P. G. Ryan, S. Rawat and W. K. Fuchs, "Two-Stage Fault Location," *Proceedings of IEEE International Test Conference*, 1991, pp. 963-968.

[273] M. Sachdev, "Transforming Sequential Logic in Digital CMOS ICs for Voltage and I_{DDQ} Testing," *Proceedings of IEEE European Design and Test Conference*, 1994, pp. 361-365.

[274] M. Sachdev, "IDDQ and Voltage Testable CMOS Flip-Flop Configuration," *Proceedings of IEEE International Test Conference*, 1995, pp. 534-543.

[275] M. Sachdev, "IDDQ Test and Diagnosis in Deep Submicron," *Proccedings of IEEE Workshop on IDDQ Testing*, 1995, pp. 84-89.

[276] M. Sachdev, "Reducing the CMOS RAM Test Complexity with I_{DDQ} and Voltage Testing," *Journal of Electronic Testing: Theory and Application*, Vol. 6, pp. 191-202, 1995.

[277] J. Saxena and D. K. Pradhan, "A Method to Derive Compact Test Sets for Path Delay Faults in Combinational Circuits," *Proceedings of IEEE International Test Conference*, 1993, pp. 724-733.

[278] J. Savir, "Skewed-Load Transition Test: Part I, Calculus," *Proceedings of IEEE International Test Conference*, 1992, pp. 705-713.

[279] K. Sawada and S. Kayono, "An Evaluation of Iddq versus Conventional Testing for CMOS Sea-of-Gate ICs," *Proceedings of IEEE International Test Conference*, 1992, pp. 158-167.

[280] G. Schiesler, C. W. Spivak and S. Davidson, "Iddq Test Result on a Digital CMOS ASIC," *Proceedings of IEEE Custom Integrated Circuits Conference*, 1993.

[281] M. H. Schulz and F. Brglez, "Accelerated Transition Fault Simulation," *Proceedings of ACM/IEEE Design Automation Conference*, 1987, pp. 237-243.

[282] M. H. Schulz, K. Fuchs and F. Fink, "Advanced Automatic Test Pattern Generation Techniques for Path Delay Faults," *Proceedings of International Symposium on Fault-Tolerant Computing*, 1989, pp. 44-51.

[283] M. H. Schulz, F. Fink and K. Fuchs, "Parallel Pattern Fault Simulation of Path Delay Faults," *Proceedings of ACM/IEEE Design Automation Conference*, 1989, pp. 357-363.

[284] M. H. Schulz, E. Trischler and T. Sarfert, "SOCRATES: A Highly Efficient Automatic Test Pattern Generation System," *IEEE Transaction on Computer-Aided Design of Integrated Circuits and Systems*, Vol. 7, No. 1, pp. 126-137, 1988.

[285] C. Sechen, *VLSI Placement and Global Routing Using Simulated Annealing.* Kluwer Academic Publishers, Boston, 1988.

[286] J. A. Seguera, A. Rubio and J. Figueras, "Analysis and Modeling of MOS Devices with Gate-Oxide Shorts Failures," *Proceedings of IEEE International Symposium on Circuits and Systems*, 1991, pp. 2164-2167.

[287] J. A. Segura, V. H. Champac, R. Rodrigues-Montanes, J. Figueras and A. Rubio, "Quiescent Current Analysis and Experimentation of Defective CMOS Circuits," *Journal of Electronic Testing: Theory and Applications*, Vol. 3, No. 4, pp. 337-348, Dec. 1992.

[288] J. A. Seguera, V. H. Champac, R. Rodrigues-Montanes, J. Figueras and A. Rubio, "Quiescent Current Analysis and Experimentation of Defective CMOS Circuits," pp. 51-62. *I_{DDQ} Testing of VLSI Circuits*, Kluwer Academic Publishers, Boston, 1992. Eds: R. K. Gulati and C. F. Hawkins.

[289] Semiconductor Industry Association (SIA), *The National Technology Roadmap for Semiconductors*, 1994.

[290] S. C. Seth and V. D. Agrawal, "Forecasting Reject Rates of Tested LSI Chips," *IEEE Electron Device Letters*, EDL-2, No. 11, pp. 286-287, Nov. 1981.

[291] S. C. Seth and V. D. Agrawal, "Characterizing the LSI Yield Equation from Wafer Test data," *IEEE Transaction on Computer-Aided Design*, CAD-3, pp. 123-126, April 1984.

[292] J. J. Shedletsky and J. D. Lesser, "An Experimental Delay Test Generator for LSI Logic," *IEEE Transactions on Computers*, Vol. C-29, No. 3, pp. 235-248, 1980.

[293] T. L. Shen, J. C. Daly and J. C. Lo, "A 2-ns Detecting Time, 2μm CMOS Built-in Current Sensing Circuit," *IEEE Journal on Solid State Circuits*, Vol. 28, No. 1, pp. 72-77, Jan. 1993.

[294] A. D. Singh and J. P. Hurst, "Incorporating IDDQ Testing in BIST: Improved Coverage Through Test Diversity," *Proceedings of IEEE VLSI Test Symposium*, 1994, pp. 374-379.

[295] A. D. Singh, H. Rasheed and W. W. Weber, "IDDQ Testing for CMOS Opens: An Experimental Study," *Proceedings of IEEE International Test Conference*, 1995, pp. 479-489.

[296] G. L. Smith, "Model for Delay Faults Based upon Paths," *Proceedings of IEEE International Test Conference*, 1985, pp. 342-349.

[297] J. M. Soden and C. F. Hawkins, "Test Consideration for Gate-Oxide Shorts in CMOS ICs," *IEEE Design & Test of Computers*, Vol. 3, pp. 56-64, Aug. 1986.

[298] J. M. Soden and C. F. Hawkins, "Electrical Properties and Detection Methods for CMOS IC Defects," *Proceedings of IEEE European Test Conference*, 1989, pp. 159-167.

[299] J. M. Soden, R. K. Treece, M. R. Taylor and C. F. Hawkins, "CMOS IC Stuck-Open Fault Electrical Effects and Design Considerations," *Proceedings of IEEE International Test Conference*, 1989, pp. 423-430.

[300] J. M. Soden, C. F. Hawkins, R. K. Gulati and W. Mao, "I_{DDQ} Testing: A Review," *I_{DDQ} Testing of VLSI Circuits*, Kluwer Academic Publishers, Boston, 1992. Eds: R. K. Gulati and C. F. Hawkins.

[301] J. M. Soden, C. F. Hawkins, R. K. Gulati and W. Mao, "I_{DDQ} Testing: A Review," *Journal of Electronic Testing: Theory and Applications*, Vol. 3, pp. 291-303, 1992.

[302] C. H. Stapper, "On a Composite Model to the IC Yield Problem," *IEEE Journal of Solid State Circuits*, Vol. SC-10, pp. 537-539, Dec. 1975.

[303] T. M. Storey and W. Maly, "CMOS Bridging Fault Detection," *Proceedings of IEEE International Test Conference*, 1990, pp. 842-850.

[304] T. Storey, W. Maly, J. Andrews and M. Miske, "Stuck Fault and Current Testing Using CMOS Chip Test," *Proceedings of IEEE International Test Conference*, 1991, pp. 311-318.

[305] S. I. Syed and D. M. Wu, "Defect Analysis and Test Generation for Gate-Oxide Shorts in CMOS ICs," *Proceedings of IEEE Custom Integrated Circuits Conference*, 1990, pp. 28.6.1-28.6.4.

[306] M. Syrzycki, "Modeling of Spot Defects in MOS Transistors," *Proceedings of IEEE International Test Conference*, 1987, pp. 148-157.

[307] M. Syrzycki, "Modeling Gate-Oxide Shorts in MOS Transistors," *IEEE Transaction on Computer-Aided Design of ICs and Systems*, Vol. 8, pp. 193-202, March 1989.

[308] P. J. Thadikaran, *Evaluation, Selection and Generation of I_{DDQ} Tests.* PhD Thesis, Feb. 1997, Department of Computer Science, State University of New York, Buffalo, NY.

[309] P. J. Thadikaran and S. Chakravarty, "Fast Algorithms for Computing I_{DDQ} Tests for Combinational Circuits," *Proceedings of Ninth International Conference on VLSI Design*, 1996, pp. 103-106.

[310] P. Thadikaran, S. Chakravarty and J. H. Patel "Fault Simulation of I_{DDQ} Tests for Bridging Faults in Sequential Circuits," *Proceedings of International Symposium on Fault-Tolerant Computing*, 1995, pp. 340-349.

[311] E. Trischler, "Incomplete Scan Path with an Automatic Test Generation Methodology," *Proceedings of IEEE International Test Conference*, 1980, pp. 153-162.

[312] M. E. Turner, et al., "Testing CMOS VLSI: Tools, Concepts and Experimental Results," *Proceedings of IEEE International Test Conference*, 1985, pp. 322-328.

[313] *User Manual, CDP1802 COSMAC Microprocessor*, RCA Corporation, November 1977.

[314] B. Underwood, W. Law, S. Kang and H. Konuk, "Fastpath: A Path Delay Test Generator for Standard Scan Designs," *Proceedings of IEEE International Test Conference*, 1994, pp. 154-163.

[315] J. P. Uyemura, *Fundamentals of MOS Digital Integrated Circuits*, Addison-Wesley, Reading, MA, 1988, pp. 49-76.

[316] P. Varma, "On Path Delay Testing in a Standard Scan Design Environment," *Proceedings of IEEE International Test Conference*, 1994, pp. 164-173.

[317] J. Van Sas, U. Swerts and M. Darquennes, "Towards an Effective I_{DDQ} Test Vector Selection Strategy," *Proceedings of IEEE International Test Conference*, 1996, pp. 491-500.

[318] F. L. Vargas, M. Nicolaidis and B. Hamdi, "Quiescent Current Estimation Based Quality Requirement," *Proceedings of IEEE VLSI Test Symposium*, 1993, pp. 33-39.

[319] S. Venkataraman, J. Rajski, S. Hellebrand and S. Tarnick, "An Efficient BIST Scheme Based on Reseeding of Multiple Polynomial Linear Feedback Shift Registers," *Proceedings of the International Conference on Computer Aided Design*, 1993, pp. 572-577.

[320] S. Venkataraman, et al., "Rapid Diagnostic Fault Simulation of Stuck-at Faults in Sequential Circuits Using Compact Lists," *Proceedings of ACM/IEEE Design Automation Conference*, 1995, pp. 133-138.

[321] H. T. Vierhaus, W. Meyer and U. Glaser, "CMOS Bridges and Resistive Transistor Faults: Iddq versus Delay Effects," *Proceedings of IEEE International Test Conference*, 1993, pp. 83-91.

[322] R. L. Wadsack, "Design Verification of WE32100 CPUs," *IEEE Design & Test of Computers*, Vol. 1, No. 3, pp. 66-75, Aug 1984.

[323] R. L. Wadsack, "Fault Modeling and Logic Simulation of CMOS and MOS Integrated Circuits," *The Bell Systems Technical Journal*, Vol. 57, pp. 1449-1474, 1978.

[324] J. Waicukauski and E. Lindbloom, "Fault Detection Effectiveness of Weighted Random Patterns," *Proceedings of IEEE International Test Conference*, 1988, pp. 245-255.

[325] J. A. Waicukauski, E. Lindbloom, B. K. Rosen and V. S. Iyengar, "Transition Fault Simulation by Parallel Pattern Single Fault Propagation," *Proceedings of IEEE International Test Conference*, pp. 542-549, 1986.

[326] J. A. Waicukauski, E. Lindbloom, B. K. Rosen and V. S. Iyengar, "Transition Fault Simulation," *IEEE Design & Test of Computers*, Vol. 4, pp. 32-38, April 1987.

[327] K. M. Wallquist, *QuiC-Mon: A General Purpose Quiescent Current Monitor*, Master's Thesis, May 1994, Electrical Engineering and Computer Engineering Department, The University of New Mexico.

[328] K. M. Wallquist, "QuiC-Mon: A Simple Path to Production I_{DDQ}/I_{SSQ} Testing," *IEEE Design & Test of Computers*, Vol. 12, Fall 1995.

[329] K. M. Wallquist, A. W. Righter and C. F. Hawkins, "A General Purpose I_{DDQ} Measurement Circuit," *Proceedings of IEEE International Test Conference*, 1993, pp. 642-649.

[330] K. M. Wallquist, "On The Effect of I_{SSQ} Testing in Reducing Early Failure Rate," *Proceedings of IEEE International Test Conference*, 1995, pp. 910-915.

[331] J. Wang, T. Kuo and J. Lee, "A New Approach to Derive Robust Tests for Stuck-Open Faults in CMOS Combinational Logic Circuits," *Proceedings of ACM/IEEE Design Automation Conference*, 1989, pp. 726-729.

[332] F. M . Wanlass and C. T. Sah, "Nanowatt Logic Using field-Effect Metal-Oxide Semiconductor Triodes," *Proceedings of Solid State Circuits Conference*, 1963, pp. 32-33.

[333] P. Weiss, "Do We Need World Class Quality Standards?" *Computer Design/News Edition*, pp. 1-23, September 10, 1990.

[334] T. W. Williams and N. C. Brown, "Defect Level as a Function of Fault Coverage," *IEEE Transaction on Computers*, Vol. C-30, pp. 987-988, 1981.

[335] T. W. Williams, R. H. Dennard, R. Kapur, M. R. Mercer and W. Maly, "IDDQ Test: Sensitivity Analysis of Scaling," *Proceedings of IEEE International Test Conference*, 1996, pp. 786-792.

[336] T. W. Williams, R. Kapur, M. R. Mercer, R. H. Dennard and W. Maly, "IDDQ Testing for High Performance CMOS - The Next Ten Years," *Proceedings of IEEE European Design and Test Conference*, 1996, pp. 578-583.

[337] P. C. Wiscombe, "A Comparison of Stuck-at Fault Coverage and Iddq Testing on Defect Levels," *Proceedings of IEEE International Test Conference*, 1993, pp. 293-299.

[338] M. H. Woods, "MOS VLSI Reliability and Yield Trends," *Proceedings of IEEE*, Vol. 74, pp. 1715-1729, December 1986.

[339] H-J. Wunderlich, "Self-Test Using Unequiprobable Random Patterns," *Proceedings of International Fault-Tolerant Computing Symposium*, 1987, pp. 258-263.

[340] H-J. Wunderlich, "Multiple Distribution for Biased Random Test Patterns," *Proceedings of IEEE International Test Conference*, 1988, pp. 236-244.

[341] W. Xiaoqing, H. Tanamoto and K. Kinoshita, "Equivalence Fault Collapsing for Transistor Short Faults and its Applications to I_{DDQ} Subset Selection," *Proceedings of Fourth IEEE Asian Test Symposium*, 1995, pp. 30-34.

[342] H. Yamazaki and Y. Miura, "IDDQ Testability of Flip-Flop Structures," *Proceedings of IEEE International Workshop on IDDQ Testing*, 1996, pp. 29-35.

INDEX

C

F

L

M

N

O

P

Q

R

S

AUTHORS

Sreejit Chakravarty obtained a B.E. in Electronics and Electrical Engineering from Brila Institute of Technology and Science, Pilani, India (1980); an M. S. in Computer Science from the State University of New York, Stony Brook (1983); and a PhD in Computer Science from the State University of New York, Albany (1980). During 1980-81 he was a Development Engineer at Tata Electric Companies Research and Development Laboratories, Bombay, India. From 1986-92 he was Assistant Professor of Computer Science, State University of New York, Buffalo where he is currently an Associate Professor.

His research interests include: VLSI testing, CAD for VLSI, fault tolerant architectures and parallel CAD algorithms. He has authored over 50 papers in these areas and bulk of his work has been supported by the National Science Foundation. He was the General Chair for the Fifth ACM/IEEE Great Lakes Symp. on VLSI, 1995 and Vice Program Chair for IEEE VLSI Test Symp., 1997. He is a member of several Conference Program and Steering Committees including IEEE VLSI Test Symposium and International Symposium on VLSI Design.

Paul Thadikaran graduated with a B.E. in Electronics and Electrical Engineering from Regional Engineering College, Tiruchirapalli, India (1987); an M. S. in Computer Science from the State University of New York, Buffalo (1994); and a PhD in Computer Science at the State University of New York, Buffalo (1997). During 1987-91 he was a Development Engineer at Tata Electric Companbies Research and Development Laboratories, Bombay, India. He is currently with Intel Corporation, Santa Clara, CA. His research interests include: CAD for VLSI and Parallel CAD Algorithms.

MIX
Papier aus verantwortungsvollen Quellen
Paper from responsible sources
FSC® C105338

If you have any concerns about our products,
you can contact us on
ProductSafety@springernature.com

In case Publisher is established outside the EU,
the EU authorized representative is:
Springer Nature Customer Service Center GmbH
Europaplatz 3, 69115 Heidelberg, Germany

Printed by Libri Plureos GmbH
in Hamburg, Germany